Human Values, Ethics, and Dignity in the Age of Artificial Intelligence

Swati Chakraborty
Concordia University, Canada

IGI Global
Scientific Publishing
Publishing Tomorrow's Research Today

Vice President of Editorial — Melissa Wagner
Managing Editor of Acquisitions — Mikaela Felty
Managing Editor of Book Development — Jocelynn Hessler
Production Manager — Mike Brehm
Cover Design — Phillip Shickler

Published in the United States of America by
IGI Global Scientific Publishing
701 East Chocolate Avenue
Hershey, PA, 17033, USA
Tel: 717-533-8845
Fax: 717-533-8661
Website: https://www.igi-global.com E-mail: cust@igi-global.com

Library of Congress Cataloging-in-Publication Data

Names: Chakraborty, Swati, 1985- editor.
Title: Human values, ethics, and dignity in the age of artificial intelligence / edited by Swati Chakraborty.
Description: Hershey, PA : IGI Global Scientific Publishing, [2025] | Includes bibliographical references and index. | Summary: "This book delves into the ethical dilemmas posed by AI, the potential for both harm and good, and the philosophical questions that arise as machines become increasingly autonomous"-- Provided by publisher.
Identifiers: LCCN 2024054467 (print) | LCCN 2024054468 (ebook) | ISBN 9798369390351 (hardcover) | ISBN 9798369390368 (paperback) | ISBN 9798369390375 (ebook)
Subjects: LCSH: Artificial intelligence--Moral and ethical aspects.
Classification: LCC Q334.7 .H87 2025 (print) | LCC Q334.7 (ebook) | DDC 174/.900663--dc23/eng/20250225
LC record available at https://lccn.loc.gov/2024054467
LC ebook record available at https://lccn.loc.gov/2024054468

British Cataloguing in Publication Data
A Cataloguing in Publication record for this book is available from the British Library.

Table of Contents

Detailed Table of Contents

The rapid development of Artificial Intelligence (AI) has sparked a global conversation about its ethical implications, particularly its potential to impact human rights. This paper critically examines the intersection of AI, ethics, and human rights by exploring how AI technologies challenge the foundations of human dignity, autonomy, and equality. It discusses both the risks and opportunities posed by AI in relation to human rights, with a focus on key areas such as privacy, freedom of expression, equality, and access to resources. The paper also offers recommendations for ethical AI governance that upholds human rights. This paper structure allows for a comprehensive examination of the ethical and human rights implications of AI, providing both theoretical analysis and practical solutions to ensure AI aligns with human dignity and international human rights norms.

With the accentuating development of technology, new potentials as well as new challenges crop up that raise questions of balancing scientific temper with human rights. The proliferation of Artificial Intelligence (AI) technologies poses questions on issues of stereotyping, practice of bias and the legalities of the use and development of such technologies. It is no longer a thing of future in a utopian or dystopia, but a reality glaring us in the face of its existence. The aim of this paper is to explore the various violations of human rights that can be perpetrated by the use of AI and the complex question as to who the perpetrator is: the AI, Developer or user. The study reveals the complexities and grey areas that if not addressed, act as a peril to human rights.

The rapid integration of artificial intelligence (AI) into critical domains such as healthcare, governance, and education has sparked significant concerns about its implications for human rights. While AI holds the potential to enhance welfare, improve service accessibility, and monitor compliance with human rights standards, its unregulated use poses risks. Privacy invasion, algorithmic bias, and erosion of individual autonomy threaten to undermine fundamental human rights, raising questions about the ethical and legal frameworks governing AI technologies. Addressing this dual potential of AI has become a critical challenge for policymakers and stakeholders worldwide. This paper explores the ethical and legal mechanisms to align AI development and deployment with human rights principles. Consulting more global standards such as the International Covenant on Civil and Political Rights (ICCPR) and the Convention on the Rights of the Child (CRC) with such proposed ethical guidelines as inclusive design, regular system checks, and strict data protection.

Social justice and social equity are critical frameworks for ensuring fairness and inclusivity in societal systems. This study employs Support Vector Machines (SVM) to classify and analyze social equity across diverse domains, including access to education, healthcare, and resources. By leveraging a comprehensive dataset, features representing equity metrics were extracted and preprocessed using statistical and natural language processing techniques. The SVM model, optimized with kernel functions to handle non-linear data patterns, effectively categorized data into equity-related classes, enabling the identification of disparities. The findings highlight the potential of machine learning in promoting evidence-based decision-making to address inequities and advance social justice initiatives. This study underscores the importance of integrating computational tools with social equity research to create more inclusive and equitable societies.

AI is increasingly shaping critical sectors such as healthcare, education, commerce, and governance, offering advancements in efficiency, predictive accuracy, and personalised solutions. However, its rapid adoption raises ethical concerns, particularly regarding biases embedded in AI systems. These biases stem from data used to train machine learning models, reflecting implicit ontological, psychological, cognitive, and cultural assumptions. As such, AI biases perpetuate structural inequalities and reinforce historical discrimination. This chapter explores the urgent need for global frameworks to mitigate AI bias, highlighting policy initiatives to address these challenges. It then shifts to an interdisciplinary perspective, proposing empathy as a tool for tackling AI bias. Drawing on Baron-Cohen's Empathizing-Systemizing theory, the chapter argues that fostering empathy in AI development strengthens accountability and trust, essential for ethical AI deployment. Gender-balanced teams, with greater empathy emphasis, can offer solutions to reduce bias, supporting equitable AI systems.

This aims to establish a comprehensive framework for integrating ethics into AI research education. The mission of the chapter is to guide the development of responsible AI researchers by embedding ethical considerations at the core of their training. It explores key ideas such as the necessity of fostering a deep understanding of AI's societal impacts, the importance of promoting transparency, fairness, and accountability, and the challenges posed by biases and unintended consequences. The chapter will outline innovative methods for teaching ethics in AI, including case studies, interdisciplinary collaboration, and the integration of ethical AI design principles into research practices. Additionally, it will discuss recent advancements in AI ethics, changes in academic curricula, and the need for continuous dialogue between researchers, ethicists, and policymakers. Ultimately, this chapter seeks to prepare the next generation of AI researchers to navigate the complex ethical landscape with integrity and responsibility.

This chapter posits a combined conceptual framework for discussing the ethics of the use of artificial intelligence (AI) in K-12 schools. By combining John Dewey's pragmatism with Paulo Freire's critical pedagogy, the development of ethics guidelines for AI use in schools may be more holistic and human-centered. The potential benefits and risks associated with AI use by teachers, students, and administrators are discussed while emphasizing the need for ethical guidelines that protect the interests of all interested groups involved. Recommendations are presented based on this combined framework for ethical use of generative and non-generative AI that may be of use for teacher educators, K-12 school teachers, and administrators.

As artificial intelligence (AI) permeates various sectors, the need for robust ethical guidelines becomes increasingly urgent. This study explores the role of interdisciplinary education in fostering ethical awareness among future technology professionals. By embedding ethics within AI and advanced technology curricula, educational institutions can cultivate AI practitioners who combine technical expertise with a strong sense of social responsibility. The findings emphasize the importance of an ethics-driven framework in AI education, where integrating technology with ethics, philosophy, and social sciences can effectively address AI's ethical challenges. Using case studies and theoretical analysis, this research proposes a model for integrating ethical training at every stage of technological education. Such a model equips AI professionals to tackle issues of bias, justice, accountability, and transparency. Building a strong ethical foundation is essential for creating AI solutions that prioritize human welfare, paving the way toward a more just and responsible future in AI.

This part investigates the convergence of AI and children's rights in corporate contexts, stressing the necessity for ethical AI in corporate social responsibility.

It assesses AI's potential to bolster corporate accountability, specifically in the eradication of child labor within supply chains, and to enhance children's access to education and healthcare. The discussion encompasses the ethical and legal dilemmas associated with AI, including algorithmic biases and data privacy issues, while proposing ethical marketing strategies that protect children's digital privacy. It emphasizes the significance of collaborative AI governance, advocating for multi-stakeholder partnerships to weave children's rights into corporate frameworks. The study concludes with strategic guidance for corporations, policymakers, and civil society to embrace ethical AI frameworks, advocating for a proactive, child-centered methodology. Ultimately, it urges corporations to responsibly leverage AI, ensuring that technological progress supports and promotes children's rights in a changing environment.

This paper explores the ethical implications and challenges of using Artificial Intelligence (AI) in Refugee Status Determination (RSD), examining its potential to improve efficiency while emphasizing human dignity, fairness, and justice. The study analyzes current ethical frameworks, referencing the G20 AI Principles, SDG 16, OECD AI Principles and UNESCO Recommendation on the Ethics of AI. Through comparative insights and case studies based on interviews, core ethical issues—such as algorithmic bias, transparency, and accountability are assessed in AI's RSD applications. Human rights concerns, including privacy and equitable access, are critically discussed. Recommendations emphasize data inclusivity, transparency, human oversight, and stakeholder engagement to ensure ethical AI in RSD. Finally, future considerations address ongoing advancements, urging continued ethical scrutiny and collaborative dialogue to balance technological progress with the preservation of refugee rights.

The intersection of artificial intelligence (AI) and environmental ethics represents a critical and increasingly relevant area of inquiry as technology advances and environmental concerns escalate. AI, with its transformative potential, offers substantial benefits for environmental management, including enhanced data analysis, predictive modeling, and optimization of resource use. For example, AI-driven tools can enhance climate modeling, predict natural disasters, and optimize energy consumption. In agriculture, AI can be used to monitor crop health, manage water usage, and reduce waste through precision farming techniques. Additionally, AI applications in environmental monitoring can detect pollution levels, track wildlife populations, and support conservation efforts. These capabilities offer the promise of more effective and data-driven approaches to addressing environmental challenges and achieving sustainability goals. The benefits of AI in environmental management may not be evenly distributed across different regions and communities.

Preface

In the 21st century, few forces are reshaping human society as profoundly as Artificial Intelligence (AI). From revolutionizing industries to influencing daily life, AI is both a marvel and a challenge, raising urgent questions about its impact on our ethical frameworks, societal values, and individual dignity. As AI technologies advance at an unprecedented pace, we find ourselves at the crossroads of innovation and responsibility, where our understanding of what it means to be human is being redefined. This book explores these complex issues, offering critical insights into the ethical, philosophical, and social ramifications of AI, with the aim of helping navigate this rapidly evolving landscape.

The development of AI has brought with it unparalleled opportunities for progress. From transforming healthcare to enhancing productivity, AI has the potential to address some of the world's most pressing challenges. However, it also poses significant risks: questions of privacy, autonomy, fairness, and justice loom large. How do we ensure that AI systems respect human dignity? What ethical principles should govern the development and use of AI technologies? How do we protect vulnerable populations from the negative consequences of automation, bias, and surveillance? These questions are not just theoretical; they are urgent and essential to the future of our society.

The aim of this book is to provide a comprehensive examination of the intersection of AI and human values. We have brought together a diverse group of scholars, ethicists, technologists, and policymakers, each offering unique perspectives on the ethical dilemmas that arise from AI's integration into our lives. Throughout these chapters, we delve into the profound challenges AI presents to the moral fabric of society, from the evolving nature of work to the erosion of privacy and the potential for deepened social inequalities. We seek not only to raise awareness of these issues but also to propose pathways forward—ways to ensure that AI development is aligned with core human values and that its benefits are shared equitably across all segments of society.

This book addresses some of the most fundamental questions of our time. How does AI challenge our traditional ethical frameworks? Can machines respect and uphold human dignity, or are they fundamentally incapable of understanding what it means to be human? In exploring these questions, we confront the very essence of what it means to be moral, to be free, and to live with dignity in an age where machines are increasingly autonomous. The ethical dilemmas AI presents are not just theoretical; they touch every aspect of our lives, from healthcare to education, from law enforcement to global governance.

Our purpose in compiling this volume is twofold. First, we aim to offer a nuanced and thorough examination of the ethical challenges posed by AI, helping readers understand how these technologies might reshape human dignity and societal values. Second, we aim to contribute to an ongoing conversation—one that is not limited to academics or technologists but that also involves policymakers, industry leaders, and the public. AI is a collective challenge that demands collective solutions, and the conversation must involve a broad spectrum of voices, perspectives, and expertise. By fostering interdisciplinary dialogue, this book aspires to make a meaningful contribution to both the academic and practical discussions surrounding AI ethics.

The topics explored in this volume cover a broad range of ethical concerns. We begin with foundational philosophical questions about the role of ethics in AI, exploring how traditional theories like utilitarianism, deontology, and virtue ethics apply—or fail to apply—in the context of autonomous machines. We then move into more specific issues, such as the impact of AI on privacy, fairness, and social justice, and the implications of AI for work and economic equality. We also consider the ethical responsibilities of designers and developers of AI systems, exploring principles for ensuring that AI is built with respect for human dignity and rights.

In *Human Values, Ethics, and Dignity in the Age of Artificial Intelligence*, we embark on a journey to explore the evolving role of AI in shaping society. The book offers an interdisciplinary lens on the ethical implications of AI, while advocating for the preservation of human dignity in the face of rapid technological advancements. Below, we present an overview of each chapter, written in the voice of the editors.

CHAPTER OVERVIEW

Chapter 1: AI, Ethics, and Human Rights: A Critical Examination

Swati Chakraborty examines the critical intersection of AI, ethics, and human rights, focusing on how AI technologies challenge concepts of human dignity, privacy, and equality. This chapter provides a nuanced discussion of AI's risks and benefits, offering a call for ethical AI governance that ensures alignment with international human rights standards.

Chapter 2: AI, Stereotyping, Bias, and Law Beyond Utopia and Dystopia: Safeguarding Human Rights

Anwesha Ghosh and her co-authors examine the legal and human rights implications of AI, focusing on stereotyping, bias, and discrimination in AI systems. This chapter explores the complexities of accountability and the challenges posed by AI in violating human rights, raising critical questions about the ethics of AI development and deployment.

Chapter 3: AI and the Ethical Crossroads- Navigating Privacy, Personhood, and Autonomy in the Digital Age

Muhammad Usman Tariq's chapter addresses the ethical dilemmas arising from the integration of AI into everyday life, particularly concerning privacy, personhood, and autonomy. As AI systems gather and process vast amounts of personal data, concerns over privacy violations have become increasingly urgent. Tariq argues for the need to develop robust ethical frameworks to safeguard privacy and dignity in an age of pervasive data collection. The chapter also delves into the shifting understanding of personhood in the context of intelligent machines, exploring the implications of AI's decision-making capabilities and its impact on individual autonomy. Through this exploration, the author calls for a careful reexamination of the societal norms and ethical standards that govern our relationship with AI technologies.

Chapter 4: Evolution of Human Rights through Literature and Law: Role of AI in its Exploration in Modern Ages

Shrabana Chattopadhyay, Ria Roy Choudhury, and Prithwijit Chakraborty explore how artificial intelligence (AI) intersects with human rights, particularly in the realms of literature and law. They emphasize the need for flexible legal frameworks to address the challenges AI poses to human rights. While AI introduces significant potential for protecting rights and enhancing societal well-being, it also presents risks to individual freedoms, including privacy violations and discrimination. The authors argue for the integration of human rights considerations into AI research and development, urging governments and stakeholders to take proactive steps in protecting rights. Additionally, they highlight the influence of literature in shaping public perceptions and ethical guidelines surrounding AI.

Chapter 5: AI, Ethics, and Human Rights: Navigating the Intersection of Technology and Human Dignity

Sakshi and Shivani Dhand's chapter examines the complex intersection of artificial intelligence (AI), ethics, and human rights. As AI becomes integral to sectors such as healthcare, governance, and education, concerns arise about its potential to infringe on fundamental human rights, including privacy, autonomy, and equality. The authors explore the risks associated with unregulated AI, including algorithmic bias and invasions of privacy, and discuss global frameworks such as the International Covenant on Civil and Political Rights (ICCPR) and the Convention on the Rights of the Child (CRC) as potential guiding principles. They propose ethical guidelines for AI development that focus on inclusive design, rigorous system monitoring, and stringent data protection, emphasizing the need to align AI advancements with human dignity and rights.

Chapter 6: A Machine Learning Approach to Social Equity Analysis

Bhimavarapu Usharani's chapter focuses the social justice and social equity, which are critical frameworks for ensuring fairness and inclusiveness in societal systems. This study employs Support Vector Machines (SVM) to classify and analyze social equity across diverse domains, including access to education, healthcare, and resources. By leveraging a comprehensive dataset, features representing equity metrics were extracted and preprocessed using statistical and natural language processing techniques. The SVM model, optimized with kernel functions to handle non-linear data patterns, effectively categorized data into equity-related classes, enabling the identification of disparities. The findings highlight the potential of machine learning in promoting evidence-based decision-making to address inequities and advance social justice initiatives. This study underscores the importance of integrating computational tools with social equity research to create more inclusive and equitable societies.

Chapter 7: Addressing the Critical Challenge of AI Bias through Empathy: The Role of Empathizing-Systemizing Theory in AI Teams

Rita Tegon's chapter focuses on one of the most pressing ethical concerns in AI development: the presence of bias. AI systems, which are increasingly used in healthcare, education, and governance, can perpetuate existing inequalities by reflecting biases in the data used to train them. Tegon argues for the importance of establishing global frameworks to mitigate AI bias, proposing empathy as a vital tool in this process. By applying Baron-Cohen's Empathizing-Systemizing theory, the chapter suggests that fostering empathy within AI development teams—especially those that are gender-balanced—can reduce bias and foster more equitable outcomes. The discussion is centered on the ethical responsibility of developers to create AI systems that are not only efficient but also fair and just.

Chapter 8: Building Ethical Foundation: Educating Researchers for Responsible Use of AI

This chapter, authored by Debajit Karmakar and colleagues, outlines a framework for embedding ethics into AI research education. It calls for interdisciplinary collaboration and the integration of ethical principles such as transparency, fairness, and accountability in AI curricula to foster a generation of responsible AI researchers.

Chapter 9: Ethics Implications of Artificial Intelligence for K-12 Schools: Combining Dewey's Pragmatism with Freire's Critical Lens

Adam Attwood's chapter proposes a comprehensive ethical framework for integrating AI in K-12 schools, blending the pedagogical philosophies of John Dewey and Paulo Freire. The chapter emphasizes the importance of human-centered AI guidelines in education, focusing on the ethical challenges and opportunities AI presents for teachers, students, and administrators, with an emphasis on safeguarding all stakeholders' interests.

Chapter 10: Fostering Ethical Awareness in AI Practitioners: The Role of Interdisciplinary Education in Shaping Future Technologists

Hemlata Parmar and Utsav Murari explore how interdisciplinary education can promote ethical awareness among future AI practitioners. By integrating ethics into technical curricula, they argue that AI professionals can be better equipped to address challenges related to bias, justice, and accountability in AI systems.

Chapter 11: Navigating AI and Children's Right in Corporate Practices: Ethical AI Integration in Corporate Responsibility

This chapter investigates the convergence of AI and children's rights in corporate contexts, stressing the necessity for ethical AI in corporate social responsibility. It assesses AI's potential to bolster corporate accountability, specifically in the eradication of child labor within supply chains, and to enhance children's access to education and healthcare. The discussion encompasses the ethical and legal dilemmas associated with AI, including algorithmic biases and data privacy issues, while proposing ethical marketing strategies that protect children's digital privacy. It emphasizes the significance of collaborative AI governance, advocating for multi-stakeholder partnerships to weave children's rights into corporate frameworks. The study concludes with strategic guidance for corporations, policymakers, and civil society to embrace ethical AI frameworks, advocating for a proactive, child-centered methodology. Ultimately, it urges corporations to responsibly leverage AI, ensuring that technological progress supports and promotes children's rights in a changing environment.

Chapter 12: AI, Refugee Status Determination, and Human Dignity: Ethical Challenges in Refugee Migration Management with Comparative Insights

Abhinash Mohanta and his co-authors explore the ethical challenges of using AI in refugee status determination processes. The chapter examines how AI can enhance efficiency while upholding human dignity, fairness, and justice, with a focus on human rights concerns and recommendations for ethical oversight in refugee management.

Chapter 13: AI and Environmental Ethics the Role of AI in Addressing Climate Change Ethical Considerations in AI-Driven Environmental Solutions the Environmental

The intersection of artificial intelligence (AI) and environmental ethics represents a critical and increasingly relevant area of inquiry as technology advances and environmental concerns escalate. AI, with its transformative potential, offers substantial benefits for environmental management, including enhanced data analysis, predictive modeling, and optimization of resource use. For example, AI-driven tools can enhance climate modeling, predict natural disasters, and optimize energy consumption. In agriculture, AI can be used to monitor crop health, manage water usage, and reduce waste through precision farming techniques. Additionally, AI applications in environmental monitoring can detect pollution levels, track wildlife populations, and support conservation efforts. These capabilities offer the promise of more effective and data-driven approaches to addressing environmental challenges and achieving sustainability goals. The benefits of AI in environmental management may not be evenly distributed across different regions and communities.

While the challenges are significant, so too are the opportunities for positive change. By applying ethical frameworks to the design, deployment, and regulation of AI systems, we can ensure that these technologies serve to enhance, rather than diminish, human dignity. We can also ensure that AI contributes to a more just, equitable, and humane society, where the benefits of technological advancement are shared broadly and its risks mitigated. The chapters in this book are not just explorations of problems, but also proposals for solutions—ways to ensure that AI remains a force for good, aligned with the best values of humanity.

As AI continues to evolve, so must our ethical and philosophical understanding of it. This book offers a foundation for that evolving conversation, encouraging readers to reflect on their own role in shaping the future of AI and to engage with the ethical questions it raises. Whether you are a scholar, a policymaker, a developer, or simply a concerned citizen, the insights shared here will provide valuable tools for understanding and addressing the moral challenges of AI. It is our hope that, by exploring these issues, we can contribute to a future where technology enhances human dignity and advances the common good.

As we close this collection of thought-provoking chapters, we are reminded of the profound and far-reaching impact that artificial intelligence (AI) is having on the world. From reshaping economic landscapes and challenging traditional legal concepts to rethinking human rights, privacy, and ethics, AI's presence is a defining feature of the modern age. The contributions in this volume collectively highlight both the incredible potential and the critical challenges AI presents across a variety of sectors.

Through these insightful perspectives, we have explored how AI is transforming industries and economies, yet also unearthed the ethical, legal, and societal complexities that must be navigated as we integrate this technology into everyday life. The chapters urge a careful balancing act—harnessing AI's capabilities for growth and innovation while mitigating its risks, particularly in terms of fairness, privacy, and human dignity.

As this book attests, the future of AI is not one of inevitability but of careful shaping. It is clear that robust, forward-thinking strategies are needed to ensure that AI development remains aligned with the values of equity, justice, and human rights. The road ahead will require collaboration across disciplines—law, ethics, technology, and policy—to address the ever-evolving challenges and possibilities that AI brings.

We hope that this volume serves as a valuable resource for scholars, policymakers, and practitioners seeking to deepen their understanding of AI's transformative effects and to contribute to the ongoing dialogue about its responsible development and deployment. In this fast-moving field, it is only through such multidisciplinary engagement that we can ensure AI serves the common good and supports a sustainable, inclusive future for all.

Swati Chakraborty
Concordia University, Canada

Chapter 1
AI, Ethics, and Human Rights: A Critical Examination

Swati Chakraborty
https://orcid.org/0000-0003-0799-1954
Concordia University, Canada

ABSTRACT

The rapid development of Artificial Intelligence (AI) has sparked a global conversation about its ethical implications, particularly its potential to impact human rights. This paper critically examines the intersection of AI, ethics, and human rights by exploring how AI technologies challenge the foundations of human dignity, autonomy, and equality. It discusses both the risks and opportunities posed by AI in relation to human rights, with a focus on key areas such as privacy, freedom of expression, equality, and access to resources. The paper also offers recommendations for ethical AI governance that upholds human rights. This paper structure allows for a comprehensive examination of the ethical and human rights implications of AI, providing both theoretical analysis and practical solutions to ensure AI aligns with human dignity and international human rights norms.

1. INTRODUCTION

The rapid growth of AI technologies has brought about both opportunities and challenges for human rights. AI can potentially support human rights goals, such as advancing access to information, improving healthcare, and enhancing public safety. However, AI systems can also pose significant risks if not designed and implemented ethically. This paper aims to highlight the critical aspects of AI that

DOI: 10.4018/979-8-3693-9035-1.ch001

intersect with human rights, outline the ethical dilemmas involved, and propose methods for aligning AI with global human rights principles (Binns, R., 2018).

Artificial Intelligence (AI) has rapidly transformed various aspects of society, influencing everything from healthcare and education to finance and entertainment. As these technologies become increasingly integrated into daily life, the ethical implications surrounding their use have garnered significant attention. Ethics, in this context, refers to the moral principles that govern a person's or group's behavior, while human rights pertain to the basic rights and freedoms to which all individuals are entitled. The intersection of AI, ethics, and human rights is crucial to ensuring that technological advancements do not infringe upon individual liberties or exacerbate social inequalities (Cath, C., 2018).

Examining these themes is of utmost importance in contemporary society, where the pace of innovation often outstrips the establishment of regulatory and ethical frameworks. As AI systems can influence decisions regarding employment, surveillance, and access to services, it is imperative to consider how these technologies may uphold or violate human rights. For instance, biases embedded within AI algorithms can lead to discriminatory outcomes, affecting marginalized communities disproportionately. Thus, a thorough exploration of AI ethics is essential in safeguarding human rights and promoting social justice.

The objectives of this research paper are to analyze the ethical considerations in the development and deployment of AI technologies, investigate the implications for human rights, and propose recommendations for ethical AI practices. The structure of the paper will unfold as follows: first, we will review the current landscape of AI technologies and their applications. Next, we will delve into specific ethical dilemmas and human rights concerns associated with AI. Finally, we will discuss potential frameworks and guidelines that can foster responsible AI development, ensuring alignment with ethical principles and human rights standards (Calo, R., 2017).

History and Background of AI and its Development

The roots of artificial intelligence (AI) can be traced back to ancient civilizations, where myths and stories depicted the creation of intelligent beings. However, the formal foundation of AI as a discipline emerged in the mid-20th century. In 1956, the Dartmouth Conference, organized by John McCarthy, Marvin Minsky, Nathaniel Rochester, and Claude Shannon, marked the official birth of AI. This event gathered prominent researchers who aimed to explore the possibilities of creating machines that could simulate human intelligence.

Throughout the 1960s and 1970s, AI research progressed with the development of early algorithms and programs capable of solving mathematical problems and playing games like chess. One significant milestone was Allen Newell and Herbert

A. Simon's work on the Logic Theorist in 1955, which solved problems in symbolic logic. However, the field faced challenges, leading to periods of reduced funding and interest, commonly referred to as "AI winters."

The resurgence of AI in the 1980s was prompted by advancements in machine learning and the introduction of expert systems, which utilized knowledge-based approaches to solve complex problems. The 1997 victory of IBM's Deep Blue over chess champion Garry Kasparov symbolized a pivotal moment in AI's evolution, showcasing the potential of machine learning and computational power (Eubanks, V., 2018).

In recent years, AI has experienced exponential growth driven by advancements in neural networks, data availability, and computational resources. Breakthroughs in natural language processing and computer vision, exemplified by models like OpenAI's GPT and Google's DeepMind, have propelled AI into mainstream applications, including virtual assistants, image recognition, and autonomous vehicles.

Despite its advancements, AI has also sparked controversies and ethical dilemmas. Concerns about privacy, surveillance, and algorithmic bias have raised questions about the implications of AI on human rights. Debates surrounding the ethical use of AI technologies emphasize the need for responsible development and the establishment of regulatory frameworks to ensure fairness, accountability, and transparency in AI systems.

2. THE INTERSECTION OF AI AND HUMAN RIGHTS

AI and human rights intersect in several core areas:

Privacy: AI's ability to collect, process, and analyze massive amounts of data can infringe upon individual privacy rights. Surveillance technologies, for instance, may lead to unauthorized data collection, profiling, and invasion of privacy.

Equality and Fairness: AI algorithms can inadvertently perpetuate or exacerbate biases, leading to discrimination in areas such as employment, finance, and criminal justice.

Freedom of Expression: Content moderation systems powered by AI often lack the nuance to distinguish between harmful and non-harmful content, risking undue censorship and limitations on free speech.

The integration of artificial intelligence (AI) into modern society has profound implications for human rights. As AI technologies evolve, they increasingly affect fundamental rights such as privacy, freedom of expression, non-discrimination, and the right to work. Each of these rights faces unique challenges in the context of AI deployment, leading to potential violations that must be addressed.

One significant concern is privacy. AI systems often rely on vast amounts of personal data to function effectively. For instance, facial recognition technologies, used by law enforcement and corporations alike, can infringe upon individuals' rights to privacy. Reports of surveillance systems employing AI to track citizens without consent highlight the potential for abuse. In countries like China, extensive use of AI surveillance has raised alarms over the erosion of privacy and civil liberties, with citizens being constantly monitored.

Freedom of expression is also at risk due to AI. Content moderation algorithms on social media platforms can inadvertently suppress diverse voices by flagging or removing posts deemed inappropriate. In some cases, these algorithms may disproportionately target content from marginalized groups, hindering their ability to express themselves freely. The case of Facebook's content moderation, where certain political speech was censored, illustrates how AI can inadvertently stifle free expression (Whittaker, M., et al. 2018).

Moreover, the right to non-discrimination is jeopardized by biased AI algorithms. Discriminatory practices can arise when AI systems are trained on datasets that reflect existing societal biases. For example, predictive policing algorithms have been criticized for perpetuating racial profiling and targeting specific communities based on flawed data. Such practices not only violate the principle of non-discrimination but also deepen existing social inequalities.

Finally, the right to work faces challenges from AI automation. As machines increasingly take over tasks traditionally performed by humans, there are fears of widespread job displacement. Industries such as manufacturing and customer service have already seen significant changes due to AI integration, leading to concerns about economic security for workers. The transition to an AI-driven economy necessitates a reevaluation of labor rights and protections to ensure that all individuals can benefit from technological advancements.

Addressing these implications requires a comprehensive approach that prioritizes human rights in AI development and implementation (Kroll, et. el, 2017).

3. ETHICAL FRAMEWORKS FOR AI

To bridge AI's benefits with human rights protection, several ethical frameworks have been proposed.

Transparency and Accountability: AI developers and organizations must ensure their systems are transparent, enabling users and impacted individuals to understand how decisions are made.
Fairness and Non-Discrimination: AI systems must be designed to treat all individuals equitably, requiring careful attention to data sources, training methodologies, and testing for biases.
Inclusivity and Human-Centered Design: Developing AI that accounts for diverse perspectives, especially from marginalized groups, is essential for equitable AI systems that respect human rights.

The rapid advancement of artificial intelligence (AI) necessitates a careful examination of the ethical frameworks that guide its development and deployment. Three prominent ethical theories—utilitarianism, deontological ethics, and virtue ethics—offer diverse perspectives on how AI should be approached in terms of fairness, accountability, and transparency.

Utilitarianism, rooted in the principle of maximizing overall happiness, evaluates actions based on their consequences. When applied to AI, this framework encourages developers to design systems that produce the greatest good for the greatest number. For instance, utilitarian principles may lead to prioritizing AI applications that enhance public welfare, such as healthcare innovations that improve patient outcomes. However, this approach may also risk justifying harmful practices if they benefit a majority, raising concerns about marginalized groups who might be adversely affected (Russell, S., & Norvig, P., 2020).

In contrast, deontological ethics focuses on the morality of actions themselves rather than their outcomes. This framework emphasizes adherence to rules and duties, which translates into a strong commitment to fairness and accountability in AI systems. For example, a deontologist would argue against the use of biased algorithms, regardless of potential benefits. This approach underscores the importance of establishing ethical guidelines and regulations that ensure AI respects individual rights and promotes justice, thereby highlighting the need for transparency in decision-making processes.

Virtue ethics, on the other hand, emphasizes the character and intentions of the developers behind AI systems. This framework promotes the cultivation of virtues such as honesty, integrity, and empathy in technology design. By fostering a culture of ethical responsibility among AI practitioners, virtue ethics advocates transparency and accountability as intrinsic values guiding AI development. This perspective encourages ongoing reflection on the moral implications of AI technologies and their societal impacts.

Each of these frameworks provides valuable insights into the ethical challenges posed by AI, emphasizing the importance of balancing technological innovation with moral responsibility. By integrating these ethical theories, developers and policymakers can work towards creating AI systems that not only advance human capabilities but also uphold fundamental ethical principles (O'Neil, C., 2016).

4. CHALLENGES IN BALANCING AI ADVANCEMENTS WITH HUMAN RIGHTS

Several obstacles complicate efforts to align AI with human rights principles:

Data Privacy Concerns: Data collection for AI often occurs at scale, raising concerns about consent, security, and user control.

Bias and Discrimination: AI systems can reproduce and magnify historical biases present in training data, leading to unfair treatment of certain groups.

Lack of Accountability: The complexity of AI models often results in a "black-box" effect, where neither users nor developers fully understand how decisions are reached.

Bias and Discrimination in AI Systems

The issue of bias in artificial intelligence (AI) systems has emerged as a significant concern, particularly as these technologies become more prevalent in decision-making processes across various domains. Bias in AI can originate from several sources, including the data used to train algorithms, the design of the algorithms themselves, and the societal contexts in which they are developed. When AI systems are trained on historical data that reflects existing prejudices—whether related to race, gender, or socio-economic status—these biases can perpetuate discriminatory outcomes.

For instance, a notable case is that of a hiring algorithm used by a major tech company, which showed a preference for male candidates over female candidates. The algorithm was trained on resumes submitted to the company over a ten-year period, a dataset that inherently reflected the tech industry's gender disparity. As a result, the AI system inadvertently learned to favor resumes that exhibited traits associated with male candidates, leading to systemic discrimination in the hiring process.

Another example involves facial recognition technology, which has been criticized for its disproportionate error rates among people of color, particularly Black individuals. Studies have demonstrated that commercial facial recognition systems misidentified Black faces at a significantly higher rate than white faces. This discrepancy stems from training datasets that lack sufficient diversity, resulting in systems that reinforce existing societal biases. Such misidentifications can have dire consequences, including wrongful arrests and heightened surveillance of marginalized communities.

The impact of biased AI systems extends beyond individual cases. When such technologies are widely adopted in law enforcement, hiring practices, or loan approvals, they can exacerbate existing inequalities and perpetuate discrimination on a systemic level. Therefore, addressing bias in AI is critical not only for ensuring fairness in automated decision-making but also for upholding human rights and fostering an equitable society.

As these examples illustrate, the challenge of bias in AI systems requires a multifaceted approach, involving diverse data representation, algorithmic transparency, and ongoing evaluation to mitigate discriminatory practices and their far-reaching effects.

Accountability and Responsibility in AI

As artificial intelligence (AI) becomes increasingly embedded in various sectors, the question of accountability and responsibility arises with greater urgency. The actions and decisions made by AI systems can have significant consequences, raising the issue of who should be held accountable when these systems produce harmful outcomes. This complexity is compounded by the interplay between developers, users, and organizations involved in the AI lifecycle.

Developers of AI technologies play a crucial role in establishing the ethical guidelines and functionality of these systems. They are responsible for creating algorithms that are fair, transparent, and free from bias. However, the challenge lies in the difficulty of predicting how AI will behave in real-world scenarios, which can lead to unforeseen consequences. For example, an AI system trained on biased data may perpetuate discriminatory practices, yet the developers may argue that they cannot foresee every outcome. This raises the question of whether accountability should rest solely on the developers or if it should also extend to the organizations deploying these technologies (Whittaker, M., Crawford, K., Dobbe, R., et al., (2018).

Users of AI systems also hold a degree of responsibility, particularly when it comes to understanding the limitations and potential biases of these technologies. For instance, if a user relies on an AI-driven hiring tool without critically assessing its outputs, they may inadvertently contribute to discriminatory practices. This highlights the need for users to be informed and cautious, as they play a pivotal role in the decision-making processes that involve AI.

Organizations that implement AI systems bear significant responsibility as well. They must ensure that they are deploying AI in a manner that aligns with ethical standards and human rights. This includes establishing oversight mechanisms and accountability frameworks to monitor AI performance and address any issues that arise. However, the decentralized nature of AI development—often involving multiple stakeholders—complicates the establishment of clear lines of accountability.

The challenges of accountability in AI are further exacerbated by legal and regulatory frameworks that have not yet fully caught up with the rapid advancements in technology. As AI systems evolve, it becomes increasingly imperative for policymakers to develop robust guidelines that delineate accountability across all stakeholders involved in the AI ecosystem.

Regulatory Approaches and Policies

The regulation of artificial intelligence (AI) is an evolving landscape that reflects the urgency of establishing ethical standards and protecting human rights in the face of rapid technological advancements (Eubanks, V., (2018). Governments and international organizations are increasingly recognizing the need for comprehensive frameworks that govern the development, deployment, and use of AI technologies. These regulatory efforts aim to address the ethical concerns associated with AI, particularly those related to privacy, discrimination, and accountability.

In recent years, several countries have proposed or enacted legislation aimed at AI governance. For example, the European Union (EU) introduced the Artificial Intelligence Act, which seeks to create a legal framework for AI that categorizes applications based on risk levels. This legislation emphasizes the importance of transparency, requiring AI systems to disclose their capabilities and limitations, particularly in high-risk applications such as healthcare and law enforcement. The EU's approach reflects a commitment to ensuring that AI technologies operate within ethical boundaries, promoting human rights and user safety.

Internationally, organizations such as the United Nations (UN) and the Organization for Economic Cooperation and Development (OECD) have launched initiatives to establish global AI principles. The UN's "AI for Good" initiative aims to harness AI's potential to achieve sustainable development goals while safeguarding human rights. The OECD's Principles on Artificial Intelligence advocate for inclusive growth, sustainability, and well-being, emphasizing the need for responsible AI practices that respect human rights.

In addition to these regulatory frameworks, ongoing discussions around AI ethics are prompting national governments to adopt more nuanced policies. Countries like Canada and the United Kingdom are exploring the establishment of regulatory bodies dedicated to AI oversight, which would monitor AI systems for compliance with ethical standards and human rights protections.

As the dialogue surrounding AI regulation continues to evolve, it is essential for policymakers to engage with stakeholders, including technologists, ethicists, and civil society—to ensure that regulations reflect diverse perspectives and uphold fundamental human rights. The challenge lies in balancing innovation with the safeguards necessary to protect individuals and communities from the potential harms of AI technologies.

5. CASE STUDIES: POSITIVE AND NEGATIVE IMPACTS OF AI

Artificial Intelligence (AI) has the potential to significantly impact society, offering both benefits and challenges. Examining specific case studies helps illuminate these dual aspects, highlighting instances where AI has enhanced human rights and well-being, as well as scenarios where it has led to abuses or infringements.

One positive case study is the use of AI in healthcare. A notable example is the development of AI algorithms that assist in diagnosing diseases more accurately and swiftly. For example, Google's DeepMind created an AI system capable of analyzing retinal scans to detect diabetic retinopathy, a condition that can lead to blindness. By enabling earlier detection, this technology enhances patient outcomes and supports the right to health. Hospitals utilizing such AI-driven tools have reported reductions in misdiagnosis and improved treatment plans, showcasing how AI can uphold human rights by facilitating access to quality healthcare.

Conversely, the negative impact of AI is evident in the realm of surveillance and privacy violations. In countries with authoritarian regimes, AI technologies have been employed for mass surveillance, leading to severe infringements on civil liberties. For instance, in China, the deployment of facial recognition systems allows the government to monitor citizens continuously, suppress dissent, and target specific ethnic groups, such as the Uyghurs. This case illustrates how AI can be weaponized to infringe upon human rights, creating an environment of fear and oppression.

Another example of negative impact is found in the criminal justice system, where predictive policing algorithms have been criticized for perpetuating bias. These systems often rely on historical crime data that reflect existing societal prejudices, disproportionately targeting marginalized communities. A study in the United States revealed that such algorithms led to increased police presence in neighborhoods with higher minority populations, exacerbating tension and distrust between law enforcement and communities. This case highlights the urgent need for ethical considerations in AI development to prevent discriminatory practices.

Through these case studies, it becomes evident that while AI possesses the power to enhance human rights and well-being, it also poses significant risks that must be addressed to prevent abuses and ensure equitable outcomes for all individuals (Zuboff, S., 2019).

6. TECHNOLOGICAL SOLUTIONS TO ETHICAL CHALLENGES

The rapid advancement of artificial intelligence (AI) brings forth a myriad of ethical challenges, necessitating the development of technological solutions aimed at mitigating these issues. As AI systems increasingly influence decision-making processes across various sectors, the need for accountability, transparency, and fairness has never been more critical. Several tools and frameworks have been developed to address these ethical challenges, including algorithmic audits, bias detection mechanisms, and responsible AI design frameworks.

Algorithmic audits serve as a vital tool in ensuring the ethical deployment of AI systems. These audits involve systematic evaluations of algorithms to assess their performance and impact, particularly concerning fairness and bias. By scrutinizing the underlying data and decision-making processes of AI systems, organizations can identify and rectify potential biases that may lead to discriminatory outcomes. For example, auditing a hiring algorithm can reveal whether it disproportionately favors certain demographics over others, thereby enabling stakeholders to make necessary adjustments to promote equity.

Bias detection mechanisms also play a crucial role in addressing ethical challenges posed by AI. These technologies focus on identifying and mitigating biases within AI models, ensuring that they reflect diverse perspectives and do not perpetuate existing societal inequalities. Techniques such as fairness-aware machine learning can help developers create models that prioritize fairness while maintaining accuracy. By integrating bias detection into the AI development process, organizations can proactively address potential ethical concerns before these systems are deployed.

Responsible AI design frameworks provide comprehensive guidelines for the ethical development and implementation of AI technologies. Such frameworks emphasize the importance of incorporating ethical considerations into every stage of the AI lifecycle—from conception to deployment. They promote principles such as transparency, accountability, and inclusivity, ensuring that AI systems are designed to respect human rights and uphold ethical standards. Initiatives like the IEEE's Ethically Aligned Design and the EU's Guidelines for Trustworthy AI exemplify the growing recognition of the need for responsible AI design practices.

These technological solutions are instrumental in fostering an ethical approach to AI, ultimately contributing to a more equitable and just society. By prioritizing transparency, accountability, and fairness, stakeholders can work towards harnessing the potential of AI while mitigating its ethical risks.

7. FUTURE DIRECTIONS AND CONSIDERATIONS

As artificial intelligence (AI) continues to evolve, its future developments are likely to shape not only technological landscapes but also ethical and human rights considerations that require urgent attention. Emerging trends in AI, such as increased automation, enhanced natural language processing, and more sophisticated machine learning algorithms, promise to revolutionize various sectors. However, these advancements also raise significant ethical dilemmas that must be navigated thoughtfully.

One prominent future direction is the rise of autonomous systems, particularly in areas like transportation and healthcare. While self-driving vehicles have the potential to reduce traffic accidents and improve efficiency, they also introduce questions about liability and accountability in the event of an accident. Similarly, AI-driven healthcare diagnostics can enhance patient outcomes, yet the reliance on these systems must be scrutinized to avoid over-dependence and ensure that human oversight remains central to patient care.

Furthermore, the increasing use of AI in personalized content delivery—such as news feeds and advertisements—may inadvertently create echo chambers, limiting exposure to diverse perspectives. This raises concerns about the erosion of freedom of expression and the potential manipulation of public opinion, particularly in the context of political discourse. As AI algorithms curate information, the need for transparency regarding how these algorithms operate and make decisions becomes paramount to uphold democratic values.

Ethical implications extend to privacy rights as well. The growing capability of AI to analyze vast datasets poses significant risks to individual privacy, particularly as surveillance technologies become more pervasive. Emerging frameworks must prioritize the protection of personal data, ensuring that consent remains a cornerstone of AI deployment.

In light of these developments, ongoing dialogue among stakeholders, including technologists, ethicists, policymakers, and civil society—is crucial. Adaptation in policy and ethical considerations must be a continuous process, allowing responses to technological advancements to evolve alongside them. Establishing adaptive regulatory frameworks that reflect diverse perspectives and prioritize human rights will be essential in navigating the complex landscape of future AI developments.

8. CONCLUSION

As AI continues to evolve, its impact on human rights will deepen, making it critical for developers, policymakers, and society to establish ethical standards. Balancing the technological potential of AI with the need for human rights protections is essential. An ethical approach to AI that prioritizes transparency, fairness, and accountability can help foster a future where technology enhances human dignity and freedom, rather than undermining it.

This paper has explored the intricate relationship between artificial intelligence (AI), ethics, and human rights, underscoring the urgent need to address these critical issues as AI technologies continue to proliferate. Key findings indicate that while AI holds tremendous potential to enhance various aspects of society—from healthcare to finance—its integration also raises significant ethical dilemmas and human rights concerns. Notably, issues of bias, discrimination, privacy violations, and accountability have emerged as major challenges that threaten to exacerbate existing societal inequalities.

The importance of integrating ethical considerations into AI development cannot be overstated. It is imperative that stakeholders recognize the potential ramifications of AI technologies on individual rights and freedoms. Therefore, a robust ethical framework is essential to guide AI practices, ensuring that human dignity and rights are upheld. This involves not only understanding the ethical implications of AI systems but also actively working to mitigate risks associated with their deployment.

To move forward effectively, several recommendations can be proposed for stakeholders involved in AI development and implementation. First, organizations must prioritize transparency in their AI systems, providing clear information on how algorithms operate, the data used, and the decision-making processes involved. Second, rigorous algorithmic audits should be established to identify and rectify biases, ensuring that AI applications do not perpetuate discrimination. Third, regulatory bodies should be created to oversee AI development, ensuring compliance with ethical standards and human rights protections.

Finally, fostering collaborative dialogues among technologists, ethicists, policymakers, and civil society is essential to create adaptive regulatory frameworks that reflect diverse perspectives. By taking these steps, stakeholders can work towards harnessing the benefits of AI while safeguarding ethical principles and human rights in an increasingly automated world.

REFERENCES

Binns, R. (2018). Fairness in Machine Learning: Lessons from Political Philosophy. In *Proceedings of the 2018 Conference on Fairness, Accountability, and Transparency* (pp. 149-159). ACM.

Calo, R. (2017). Artificial Intelligence Policy: A Primer and Roadmap. *UCLA Law Review. University of California, Los Angeles. School of Law*, *51*, 199–217.

Cath, C. (2018). Governing Artificial Intelligence: Ethical, Legal, and Technical Opportunities and Challenges. *Philosophy & Technology*, *31*(4), 655–680. PMID: 30873343

Eubanks, V. (2018). *Automating Inequality: How High-Tech Tools Profile, Police, and Punish the Poor*. St. Martin's Press.

European Commission. (2021). *Proposal for a Regulation of the European Parliament and of the Council Laying Down Harmonised Rules on Artificial Intelligence (Artificial Intelligence Act)*. European Commission.

Kroll, J. A., Huey, J., Barocas, S., & Felten, E. W. (2017). Accountable Algorithms. *University of Pennsylvania Law Review*, *165*(3), 633–705.

O'Neil, C. (2016). *Weapons of Math Destruction: How Big Data Increases Inequality and Threatens Democracy*. Crown Publishing Group.

Organization for Economic Cooperation and Development. (2019). *OECD Principles on Artificial Intelligence*. OECD Publishing.

Russell, S., & Norvig, P. (2020). *Artificial Intelligence: A Modern Approach* (4th ed.). Pearson.

UNESCO. (2021). *Recommendation on the Ethics of Artificial Intelligence*. United Nations Educational, Scientific and Cultural Organization.

Whittaker, M., Crawford, K., & Dobbe, R.. (2018). *AI Now 2018 Report*. AI Now Institute.

Zuboff, S. (2019). *The Age of Surveillance Capitalism: The Fight for a Human Future at the New Frontier of Power*. PublicAffairs.

Chapter 2
AI, Stereotyping, Bias, and Law Beyond Utopia and Dystopia:
Safeguarding Human Rights

Anwesha Ghosh
https://orcid.org/0000-0003-4167-2121
Amity University, India

Narem Usha Amulya
GITAM University, India

Sachin Singh Parihar
https://orcid.org/0009-0007-5965-012X
Central University of Punjab, India

ABSTRACT

With the accentuating development of technology, new potentials as well as new challenges crop up that raise questions of balancing scientific temper with human rights. The proliferation of Artificial Intelligence (AI) technologies poses questions on issues of stereotyping, practice of bias and the legalities of the use and development of such technologies. It is no longer a thing of future in a utopian or dystopia, but a reality glaring us in the face of its existence. The aim of this paper is to explore the various violations of human rights that can be perpetrated by the use of AI and the complex question as to who the perpetrator is: the AI, Developer or user. The study reveals the complexities and grey areas that if not addressed, act as a peril to human rights.

DOI: 10.4018/979-8-3693-9035-1.ch002

INTRODUCTION

In an age where the tools of artificial intelligence have become regularly used by a layman in everyday life, new perspectives (Yu-Cheng Wang et al., 2023) and issues (Scherer, 2015) arise that were seldom envisioned before by the experts of various fields. From healthcare (Adam Bohr & Kaveh Memarzadeh, 2020), marketing (Haleem et al., 2022), banking (OECD, 2020) to investment (Mou, 2019), and research and innovation to market deployment (Soni et al., 2020), cherry picking fruits (Kollewe, 2022) and real world applications, (Sarker, 2021) artificial intelligence is constantly operating around us, for us, and sometimes, against us. Even if one does not deliberately experiment with tools of artificial intelligence (AI), any person who is in social media will inevitably have been part of the AI revolution. (Vemuri, 2014) The tailormade search and product suggestions that one regularly finds in one's social media accounts are AI crafted if not handcrafted by the social media giants. Whether one is aware or unaware, AI has become part and parcel of our lives and impacting our life decisions without explicit realisation of it. In as much as one would find relief in product suggestions in your social media that you were searching in your search engine, for it makes your life easier and help you in indulge in its personalised experience (Bronzin et al., 2021) (Lin, Huang, and Lu, 2023), it raises pertinent questions on how the data you searched (Lu, 2022) will create a web of surveillance that could be used by state and private players in perpetuating rampant human rights violations and widespread stereotypes.

The Emergence of Bias, and Stereotyping as Critical Ethical Concerns

Bias and stereotyping have become prominent issues in the field of AI, influencing the direction of technical progress in important ways that are ethical. While AI has been praised for its ability to transform businesses and enhance decision-making, it has also raised concerns about the ethical framework supporting its development and use by highlighting the possibility that it could reinforce and magnify social biases (Gaur and Sahoo, 2022).

When systemic prejudices that are common in the datasets that these systems are trained on are unintentionally included and reinforced, it is referred to as bias within AI systems (Stanford and Wilson, 2022). AI relies significantly on algorithms and machine learning processes that are based on historical data, which unfortunately often contain hidden biases reflecting existing disparities (Cardie, 1999). As a result, AI systems pick up on and reinforce these biases, which affect decision-making in a variety of fields, including economics, employment, and criminal justice. Another major ethical issue with AI is stereotyping, which is assuming or generalizing about people or groups based on insufficient or outdated data, often resulting in biased results.

The lack of diversity in AI development teams is an important factor in the rise of bias and stereotyping in the field (Wong, 2023). There are blind spots and societal prejudices that go unnoticed because different voices and experiences are underrepresented in the design and management of AI systems (Wolfram and Kienesberger, 2023). In the context of AI development, addressing these ethical issues calls for an integrated approach that includes more knowledge, accountability, and preventative measures. It is therefore essential for AI teams to promote diversity in order to promote inclusivity and prevent prejudice (Gunn et al., 2020). To ensure that AI systems are being used ethically and in compliance with anti-discrimination laws and societal standards, periodic evaluation processes to identify and address biases within the systems are essential (McDowall, 2015).

DATA BIAS, AI BIAS, AND DEPLOYMENT BIAS

Historical Causes of Bias in AI

The formation of AI biases relies heavily on historical data. Machine learning models are frequently built using prejudiced historical data as the basis (Van Giffen, Sternberg, Meijer, and Batchelor, 2022). This information might represent historical injustices and discriminatory practises, giving AI systems an incorrect view of reality (Zhai, 2022). For instance, figure 1(Leslie et al., 2021) below illustrates how careless data selection techniques can have a significant downstream impact on an entire system (in this case, the health industry). The creation and application of machine learning (ML) models have a significant potential for revealing biases in the absence of a sufficiently representative dataset and selection techniques that take diversity into consideration. Applying ML to some of the most crucial aspects of our lives, like healthcare, has the potential to foster or reinforce existing inequalities within those systems. The repercussions of bias in machine learning can range from higher project overhead to more significant societal consequences and systematic discrimination.

Figure 1. An infographic from the British Medical Journal (Leslie et al., 2021) illustrating the long-term effects of bias entering data in one area, the health sector

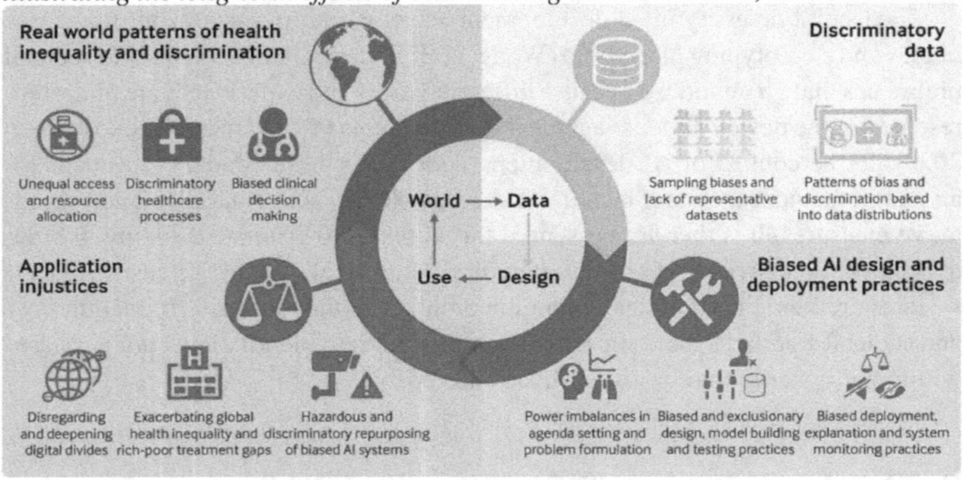

Another example is the "gorilla" scandal (figure 2) involving Google Photos from 2015 (Zhang, 2015). This significant and well-publicized incident brought attention to the problem of racial bias in AI systems. In this sad event, a user uploaded a selfie to Google Photos that included both himself and his friend both of whom were black. Google Photos' image labelling technology mistakenly identified the image as containing "Gorillas." In addition to being a technical error, this incident was a highly offensive and distressing illustration of AI bias. In addition to being embarrassing for Google, the erroneous labelling of the photograph as "Gorillas" also caused significant damage to those directly affected by it and offended a large number of people. Critical issues like data bias, a lack of diversity in AI development, and ethical and societal concerns became apparent by the event.

Figure 2. Google Photos "gorilla" scandal: In 2015, Google Photos' photo-tagging system incorrectly categorised a user and his friend's selfie—both of whom were black—as "Gorillas" (Zhang, 2015)

The "gorilla" incident involving Google Photos serves as an unsettling reminder of the difficulties and commitments involved in the creation and application of AI technologies. It emphasizes the significance of ongoing initiatives to recognize and minimise prejudice in AI systems, as well as the necessity of diversity and ethical concerns in the development process to guarantee that AI technologies are just, inclusive, and respectful to all users.

AI Bias: Types and Causes

Artificial intelligence bias is a common and serious problem that is receiving more and more attention. It refers to the possibility of unfair and biased outcomes or predictions generated by AI systems, which may have serious repercussions. The following are a few of the biases mentioned below:

Firstly, data bias is among the most pervasive forms of AI bias. When the training data used to create AI, models is unrepresentative or distorted in any manner, data bias develops. Such biases may have crept into the data over time through discriminatory acts or societal prejudices. For instance, data bias emerges when an information set is inaccurate, skewed, or incomplete, failing to faithfully represent the diversity and characteristics of the entire population it seeks to depict as is illustrated in Figure 3 (WallStreetMojo, 2023). Secondly, on the other side, algorithmic bias results from the decision-making and design processes that underlie AI algorithms (Raghavan, 2023). This kind of bias might be brought on by biased feature selection, incorrect model layouts, or optimization criteria that unintentionally include biases seen in training data. As a result, algorithms may produce inaccurate forecasts or suggestions, such as showing prejudice towards certain racial or gender groups while making hiring choices. Thirdly, a type of bias known as interaction bias develops over time as AI systems interact with humans (Shin, 2022). AI systems can learn and later reinforce users' biases if they engage in prejudiced behaviour or preferences. For instance, if social media algorithms prioritise user involvement, they may unintentionally magnify extreme viewpoints, thus fostering polarisation and spreading false information. Fourthly, labelling bias happens when training data are annotated or labelled (Labelvisor, 2023). When labelling data, human annotators can unintentionally introduce their own prejudices, which would instruct and reinforce these biases in AI models. Labelling bias frequently requires subjective judgements, making its detection and mitigation extremely difficult. Fifthly, evaluation bias is concerned with the biases that could appear while evaluating the effectiveness of AI models. Evaluation measures and benchmarks might not fully take into account all forms of bias, which would leave the fairness of an AI system unclear. This may result in the use of prejudiced systems that behave fairly in testing but not in real-world situations. Sixthly, deployment bias occurs when AI systems perform differently in real-world situations than they do in testing environments. Bias can be introduced during deployment due to a number of variables, such as modifications in user behaviour, changing data quality, or the complex interactions between AI and human users. Even if the model performed well during its development and testing phases, this can still happen.

Figure 3. Data bias: A distorted reflection of the population (McDowall, 2015)

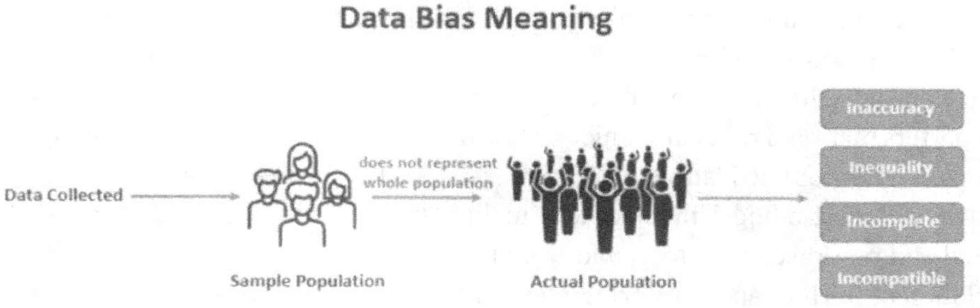

Bias in AI has many root causes, many of which are related. As a result of historical biases in data gathering techniques or data sources that reflect societal preconceptions, distorted training datasets might result from bias in the data collection process. Particularly in industries like medical, underrepresentation of some groups in training data might result in AI systems that are inaccurate and unfair to particular populations.

THE MAGNIFICATION OF DISPARITIES AND PREJUDICES

The interaction between gender and technology in the field of AI is a complex and important subject matter that has significant repercussions on society. Numerous applications of AI systems have shown gender biases, which raises concerns about the systems' potential to reinforce and exacerbate already-existing societal imbalances. Because AI depends on data, algorithms, and human inputs, it has shown biases that both reflect and occasionally aggravate gender inequality. For example, AI-driven technologies have shown biases favouring male candidates during the recruitment process, which has perpetuated gender inequities in the workforce.

AI-powered virtual assistants and natural language processing (NLP) systems have the potential to unintentionally propagate gender stereotypes. There have been cases of biases in speech recognition, translation, or language generation that have reinforced social prejudices about the roles and skills of women. Although these biases may appear subtly, they can have a significant overall effect.

HUMAN RIGHTS IN THE DIGITAL AGE

Human rights are inherent in every human being. They constitute alienable rights which apply to all of us irrespective of nationality, ethnicity, color, class, gender, language or status. In furtherance of this, various international and regional conventions or charters, national rules in municipal law have been enacted. Various principles of Human rights (United Nations, 2003) are envisaged by the international conventions, which includes, being Universal and inalienable, Interdependent and indivisible, Equal and non-discriminatory, and Both Rights and Obligations. Historically, as in violation of the human rights principles, human rights and its violations extended to discrimination on the basis of nationality (Yazici, 2018), ethnicity (Sharma, Yang, and Wu, 2017), racism (Bradley, 2019), gender (Krivenko, 2020), language (Paz, 2014) (Skutnabb-Kangas and Phillipson, 2023), and other forms of discrimination as can be seen with the evolution (Luminita, 2013) of world order.

In the digital era that we are living in, whilst digital platforms provide for various opportunities for advancement and growth in the society, it comes with challenges of its own. Seldom does it provide for equal representation or freedom from discrimination. On one side of the dice, stands the delicate balance of individual rights vis-a-vis government privilege in monitoring and surveillance for protection of public interest. On one of the many other faces of the dice, is the question of human rights violations in the digital environment (Monshipouri, 2016). The society is not far from becoming the one in the Dystopian novel of George Orwell where people live in surveillance by the government. It would not be long before certain groups or individuals or classes are targeted for non-conformity with state. In a digital era, new technologies bring in new rights as well as corresponding duties (Razmetaeva, Barabash, and Lukianov, 2022). At this juncture, the concept of digital human rights ought to be incorporated in the generation of human rights. In fact, most countries consider Digital rights as fourth generation human rights (Risse, 2021). This is especially the case, as this digital transformation can become a tool for the state and the corporations/companies to extend the welfare or economic growth of the society, in the event of being used improperly. Moreover, the scope of human rights have expanded with the inclusion of digital rights, including concerns raised for mass surveillance (Nyst and Falchetta, 2017) (Kosta,2020) and stereotyping. In AI Global Surveillance Index (AIGS) (Feldstein, 2019), one of the first of its kind as conducted by Carnegie Endowment for International Peace, over 75 countries out of the 176 countries, in which the empirical studies is conducted, use Artificial intelligence for surveillance purposes. This surveillance is conducted through smart/safe city platforms, facial recognition systems, and smart policing. Within this arena, of contentious issue is the geopolitical conflicts and origin of manufacturers that could further exemplify the problem. For instance, China (Roberts et al., 2020)

(Li, Tong, and Xiao, 2021) and the (US Hine and Floridi, 2022), are considered to be the world leaders in the provision of artificial intelligence. China has emerged as one of the global leaders (Karpa, Müller, and Ernst, 2022) along with the US in surveillance technologies with the use of artificial intelligence. It is dubious as to what extent the digitalisation of social security has met its intended objectives or if the extent of intervention of AI in social security programmes can lead to predictive policy making which result in exemplification of biases. This raises further questions on the notion of democracy if it leads to voter manipulation and the grey areas of targeted election campaigns (OECD, 2022) (Human Rights Group, 2020). Human rights in the digital age come with complex issues that seldom find mention in the existing framework of human rights. Thus, it becomes imperative that these contemporary issues will have to be looked upon with much caution for what it holds for the future of enforcing human rights and the potential lapses.

Implications of AI on Human Rights: From Dystopian Fiction to Dystopian Reality

The world of Artificial intelligence is not in the future, we are already living in it. The world that we entered seems to be straight out of a science fiction, except only that it isn't. Numerous science fictions and series have already cautioned us to tread with caution. For instance, the science fiction classic Neuromancer (Greer, 2010) by William Gibson released in 1984, is credited with popularizing the cyberpunk subgenre. The dystopian future in which the story is set explores issues like artificial intelligence, virtual reality, and the fuzziness of technological and human boundaries. In this world, where big companies and technology hold power, there are mercenaries and hackers that want to help or hurt humanity. It explores the murky, frequently unethical aspects of a future that is shaped by rapid technological advancements and the pervasiveness of corporate power. As much as Deep Blue (Campbell, Hoane Jr., and Hsu, 2002) won the world chess championship, AlphaGo (Holcomb et al., 2018) established its victory in the game Go or even recently, when ChatGPT has entered the world through prompt engineering and machine learning. With these algorithms, only time will tell before AI can develop a will of its own against its creator and it presents new opportunities and challenges that come with it (Gurkaynak, Yilmaz, and Haksever, 2016). Skynet from Terminator (Palumbo, 2008), the Colossus system from The Forbin Project (Berman and Dalvi, 2011), Control from Star Trek: Discovery (Patterson et al., 2017), the dominance of the machines in the Matrix trilogy led by Deus Ex Machina (Arnheim, 1992) are all examples of this (Hermann, 2021). The world is transitioning from an evolutionary process of homo sapiens to robo sapiens (Raveh and Tamir, 2018). These narratives about AI present a "Frankenstein Complex" (McCauley, 2007) and stories of strong and omnipotent AI systems governing humans and mankind perfectly mirror the individual's fear of impotence and helplessness in the presence of superior structures. They frequently utilize historical references to colonialism and totalitarian administrations as a springboard for discussions about the dangers of dictature, anti-democratic society, and the repression of freedom of choice.

AI, Marginalized Communities and Decision Making: To Include or Not to Include?

AI has a potential of resulting in compounding of existing inequalities and targeting marginalised communities. Predictive policing systems that are used by law enforcement agencies, can contribute to bias and lead to rampant human rights violations. Although artificial intelligence relies on machine learning model and demonstrates the output based on the data that is inputted, seldom is it free from errors. For instance, in cases where there is more focus on a certain community in a country with police patrolling due to lack of trust in a certain demographic constituency in a state, the data that is fed into AI could be considered as biased. When artificial intelligence feeds on this data, it is likely that it would lead to inaccurate predictive policing that could result in discrimination and violation of human rights.

One such real life case study that shows this potential peril in action with racism (Li, 2021) being perpetrated. In Chicago, Unannounced, a Police Department commander made a visit to Robert McDaniel, a black man residing on Chicago's South Side, to warn him not to commit any further crimes. McDaniel was caught aback by the unexpected visit. He had never committed a crime, had no violent criminal past, and had had no recent run-ins with the authorities. Based on their geographic location and arrest histories, these individuals were all projected to be involved in violent crime. The bias is worsened by a lack of transparency from both police departments and private firms about how predictive policing (McDaniel and Pease, 2021) models are built, how departments utilize their data, and if the algorithms unfairly target certain demographics (Lum and Isaac, 2016). The idea that people with certain facial characteristics are more prone to commit crimes and that the program can identify them has been promoted using AI-powered face analysis software. For coloured people, who are already seen as essentially criminal, this opens up perilous chances. The development of AI may be more equitable if under-represented groups were represented more frequently (Johnson et al., 2022). Given that the algorithms that drive the artificial intelligence are a human construction which could be subjective in nature, seldom can the artificial intelligence display objectivity. This shows that artificial intelligence could result in racial discrimination, thereby violating the basic tenets of human rights. Nevertheless, this is merely one of the various other facets of discrimination and the consequence of making use of artificial intelligence by state agencies.

Exacerbation of Existing Inequalities by Artificial Intelligence

The outcomes produced because of AI could end up exacerbating existing inequalities if there is a bias in data that serves as an input to artificial intelligence. This could lead to perpetuation of inequalities or even magnification of these inequalities (Alston, 2019). This makes it necessary to understand the social dynamics of these inequalities and label the problem areas to find a suitable measure to combat such counterproductive output by AI systems. Although there is lack of any longitudinal research as to how AI perpetuates or diminishes existing inequalities, this can be attributed to the fact that such intervention of Artificial intelligence in daily lives of people is a phenomenon of the present era. With passing time, the impact of AI with longitudinal research can be identified in a profound way. This is not to say that such research on the impact of artificial intelligence and its algorithms is completely absent in the framework. Several studies, Artificial Intelligence Index Report 2021 (Zhang et al., 2021) and 2022 (Zhang et al., 2022) by Stanford University, on surveillance through Artificial intelligence (Zhai, 2022) (Feldstein, 2019) by Carnegie Endowment for International Peace which sought to examine the perpetuation of inequalities through biased inputs that result in AI bias, gender (UNESCO, 2020), Technology And Innovation Report 2021: Catching technological waves innovation with equity by UNCTAD (The *Technology and Innovation Report 2021*,2021). These reports sought to understand the impact of AI and its inherent bias owing to biased data input as indicated in figure 4. Moreover, it must be observed that issues and the concept of matrix, with trust and distrust come not merely with a single challenge or vulnerability but a set of challenges and vulnerabilities, sometimes overlapping yet varying at times depending on the varying nature and concerns of stakeholders (Lockey et al., 2021).

Figure 4. UNCTAD (The Technology and Innovation Report 2021,2021)

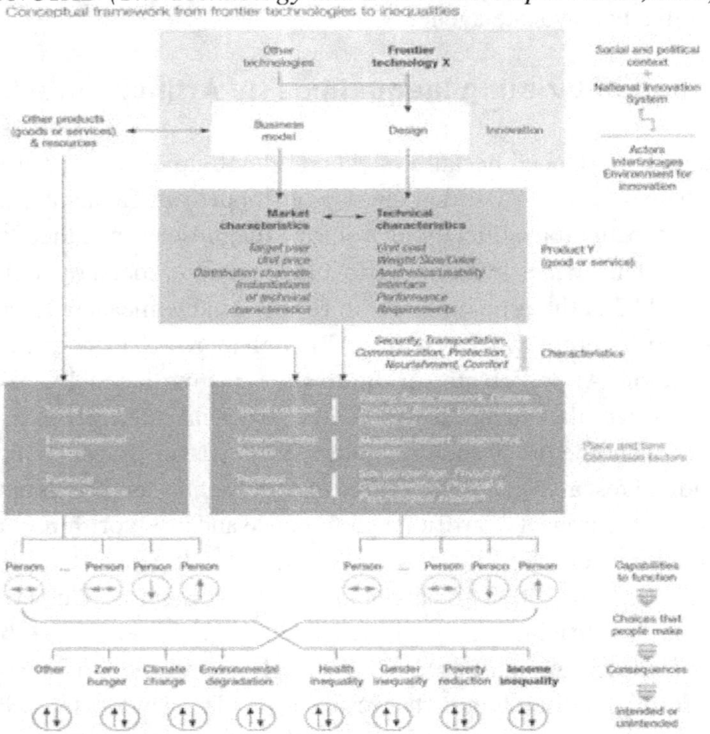

In case of delivery of public services (Ingrams et al., 2021), the states using artificial intelligence for delivery of welfare schemes can end up overrepresented or underrepresented a section of the population, which could undermine the pertinence of sensitive issues due to systematic bias. Current social and economic inequalities have an impact on how technology is used, and contemporary technologies may both maintain and exacerbate existing imbalances (Schwartz et al., 2022). Therefore, it is doubtful that technology applications alone will lead to extensive societal reforms toward more equal outcomes (*The Role of Data in AI*, 2020). Defects in data representativeness (Kuhlman, Jackson, and Chunara, 2020) (for example, undersampling of vulnerable populations) and biases emerge throughout the development and implementation lifecycle (for example, failure to include relevant demographic variables in the model results in disparate performance for vulnerable populations) are all examples of how discriminatory structures become ingrained in datasets used to train systems (Leslie et al., 2021).

Figure 5. AI inclusion life cycle (The World Economic Forum, 2022)

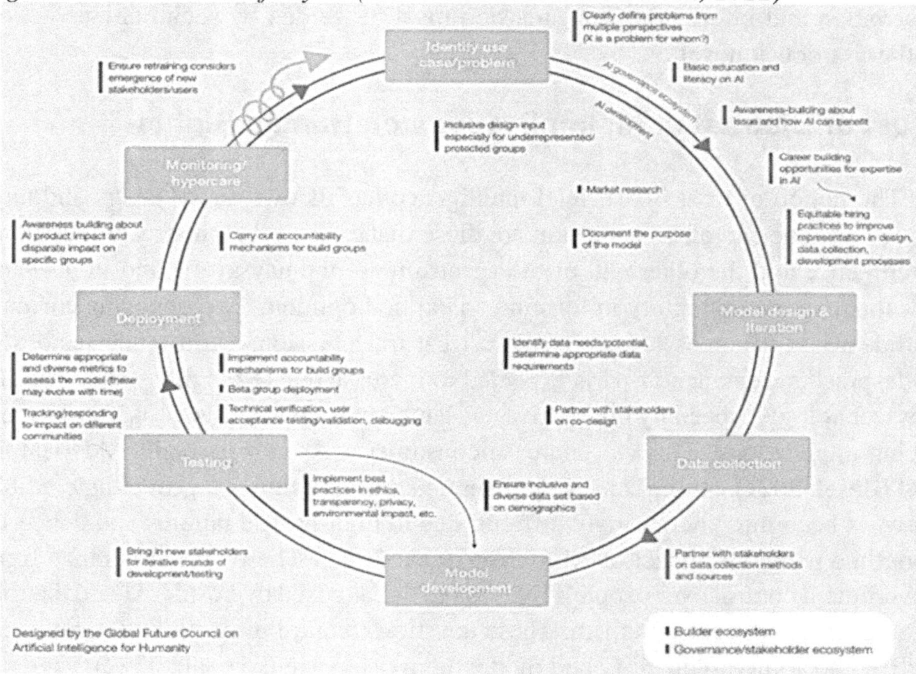

The AI inclusion Life cycle as depicted in figure 5 provides a framework for diminishing, if not elimination of discrimination and inclusivity of underrepresented groups as well as providing for a diverse data set, tested through its impact on different communities to diminish exacerbation of inequalities by artificial intelligence. It will be argued that the UN Special Rapporteur on Extreme Poverty and Human Rights is correct in stating that the primary goal should be to reduce the possibility of a digital welfare dystopia (Yazici, 2018) (Alston, 2019), even though there are numerous evaluative criteria for channeling and shaping AI deployment in welfare. When constructed, monitored, and enhanced in accordance with best practice standards (The *Oxford Commission on AI & Good Governance*, 2021), AI in administration may undoubtedly increase consumer accessibility, give more accurate and responsive options, and contribute to greater efficiency. AI may also be used to ensure citizen entitlement, such as detecting and contacting those who might otherwise be refused a benefit. Poorly built AI systems, on the other hand, jeopardize ideals like procedural fairness, the rule of law (Huq, 2021), and government accountability (Zajko, 2022). This underscores the importance of conducting more study on topics such as how to define fairness in the context of AI systems and how to detect and counteract discrimination in AI administration (Carney, 2020).

In order to improve social inclusion in this context (Haddad and Chedrawi, 2019), innovation and information transmission must be guided by social citizen action and grassroots innovation.

Trust or Distrust in AI: Implications on Human Rights

The notion of trust in artificial intelligence has its own set of takers and naysayers. On one end of the spectrum are those that express their distrust of artificial intelligence and the other end of the spectrum are the naysayers, and in between lies that express ambiguity in forming an explicit opinion. Two opposing currents collide when it comes to macro-micro trust transmission. On the one hand, the widespread acceptance of AI is expected to increase due to the many successes of this technology, especially in sensitive and large-scale sectors as well as in customer-facing applications. This can increase consumer trust in a particular AI (Hamm and Klesel, 2021). At the same time, skepticism is growing and providing credible views is becoming increasingly difficult due to high-profile failures and concerns about the potential impact of AI (Cheng et al., 2022). These concerns range from government control over people's lives to the threat posed by hostile AI and the role of AI in surveillance capitalism. These questions require in-depth investigation as well as the ability to quantify and model them (Lukyanenko et al., 2022).

A global study (Gillespie et al., 2023) conducted by KPMG shows that some countries trust Artificial intelligence more than others. However, the report concludes that most of the population have considerable trust in artificial intelligence although they opinionize that a robust regulatory mechanism is necessary for building trust and maintaining it. However, on a critical analysis of this report, it is startling that the notion of trust as a dependent variable is based on the independent variable of age. For instance, countries such as India where there is a large extent of young adults, show more trust in artificial intelligence as compared to Japan which has a growing aged population.

Figure 6. Trajectories of trust for robotic, virtual and embedded AI as reflected by most of the reviewed studies (Glikson and Woolley, 2020)

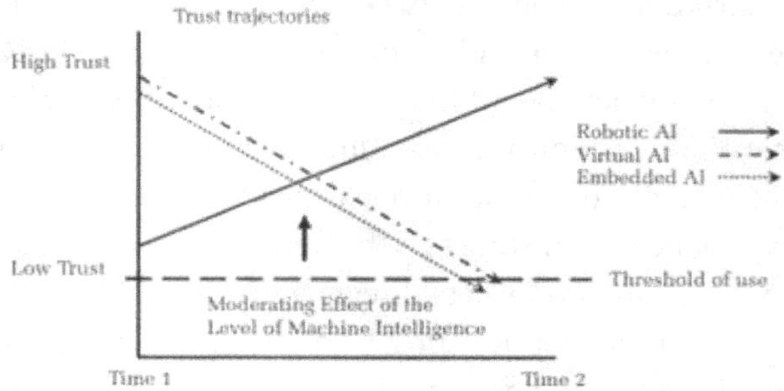

Existing study on the evolution of people's confidence in robotic AI shows that it is initially low and gradually increases as reflected in figure 6*(Glikson and Woolley, 2020)*. Existing study on the evolution of people's confidence in robotic AI shows that it is initially low and gradually increases (Zerilli et al., 2022). The decision to acquire trust lies at the heart of digital trust. People's expectations that enterprises supplying digital products and services would adhere to social standards and protect the interests of all stakeholders are referred to as digital trust.The elements of digital trust, which include cybersecurity, safety, transparency (Zerilli et al., 2022), interoperability, auditability, redressability, fairness, and privacy, can help to meet the objectives of the digital trust framework (The World Economic Forum, 2022).

However, instances where contact tracing, face, and voice recognition allow for the mass collection of personal data, using AI inevitably creates privacy issues. The conflict between AI's advantages and disadvantages is a major social problem (Ryan, 2020). When technology behemoths create and consume enormous volumes of user data between a service user and a supplier, this problem becomes crucial in the sharing economy. For instance, privacy compliance issues arise when artificial intelligence is used for tracking, surveillance, and facial recognition. A negative perception of AI technology may raise doubts among the public (Wong et al., 2023). This shows that corporate responsibility in dealing with data plays a crucial factor in case of private enterprises and robust government data infrastructure is necessary in case of public corporations. There is an accentuating necessity for socially responsible Artificial intelligence to protect human rights. At this juncture, it has been theoretically proposed by several experts that the pyramid of corporate social responsibility as proposed in Caroll's pyramid ought to be extended to Artificial Intelligence social responsibility (Cheng et al., 2021). This includes extending the scope of Caroll's CSR pyramid consisting of functional, legal, ethical and philanthropic responsibilities to Artificial intelligence.

The Right to Privacy

In a quest for striking a balance between human rights and innovation through artificial intelligence, is a most widely debated argument of privacy. Seldom are we free from loss of privacy even though we do not deliberately choose artificial intelligence, for we are implicitly consenting when we use certain applications in our daily life. One instance of common yet surprising concern clothed in delivering a personalized experience to consumers is the frequency of advertisements appearing on applications concerning products and services we searched for topics we entered in the search engines in other applications. Furthermore, the suggestions that we find in feed based on what we see, hear and do, is stifling the right to privacy (Dwork & Minow, 2022). A few of the privacy concerns that may arise as a result of AI abuses are intrusion of privacy or isolation, interference into private matters, public publication of humiliating sensitive data, erroneous false exposure, or misappropriation of name or likeness. Other ideas of privacy exist in addition to the common law roots of privacy and individuality (Manheim & Kaplan, 2019). These include physical privacy, behavioral privacy, decisional privacy, and informational privacy. Our contacts, intimate connections, and activities, political inclinations, governmental records, hereditary, biometric, and health information (from the moment of conception to death), as well as educational and employment histories, messages, calling, and electronic mail, social media platforms preferences, close companions, and likes and dislikes browsing activity, surroundings, and movements, consumers choices, finance, insurance, and other financial data (Marsch, 2019).

This not only impacts the decisions, whether public or private, that we make, it can also lead to discrimination of certain nationalities, ethnic or religious groups, various communities or groups that could be targeted by various state or non-state actors, based on the data collected through artificial intelligence by those who commission them. AI in the information domain enables non-state actors to produce credible fake information on a large scale, which may then be utilized to alter the opinions of a demographic group for political purposes. AI may be used to automate some operations, such as sniping or drone attacks (United Nations Office of Counter-Terrorism & United Nations Interregional Crime and Justice Research Institute, 2021), in order to target high-value targets for assassination or minority groups inside a country. Because state actors control information under uncertain regimes, this is especially dangerous in the context of the global system for monitoring and reporting on human rights. If a government changes, there is a risk that information and data gathered with the intention of advancing rights will end up in the wrong hands. It is critical to address this risk as part of a risk assessment when adopting AI technology (Dulka, 2023). In case of non-state actors, use of AI happens to be even more dangerous. AI may be used to help and magnify the impact of cyberattacks by concentrating on those who are financially or psychologically weak and utilizing machine learning on massive datasets. This results in weaponization of AI by state or non-state actors (Brundage et al., 2018) for nefarious activities or to gain leverage of hegemony amongst others. AI may be used to create artificially created news or deep fakes in order to influence public opinion (Kreps, 2021). The fact that it is not a dystopia but occurring at the moment even as we exist right now is alarming (Yamin et al., 2021). Nevertheless, Right to privacy can be protected through progressive regulatory framework which provides for a delicate balance of individual rights without stifling innovation. Hence, the onus is on the states to protect not merely the right to privacy, but human rights for the preservation of eras of struggle to uphold principles of human rights against all forms of violations of human rights.

LEGAL AND REGULATORY FRAMEWORK

Hohfeld's Classification and AI

Hohfeld formalised the conceptualisation of rights and duties (Hohfeld, 1917). This can be extended to understand the rights and duties of or towards Artificial intelligence (Markovich, 2019) (Villata et al., 2022). Theorists of law and jurists are certainly interested in the clear formulation of the intricate Hohfeldian links in this context, but correctly detecting these jural admixtures has immense practical significance for the developing digital economy employing legal toolkits (Sichelman, 2017). At this juncture, it becomes pertinent to bring out that in this interdisciplinary analysis of law and sciences (Izzidien, 2023), a language was invented to extend the application of Hohfeld's classification which improves legal reasoning (Krogh & Herrestad, 1999) and shows how research on artificial intelligence issues in computer science continuously feeds back to law (Allen & Saxon, 1995) (Boella et al., 2001).

Figure 7. Hohfeldian classification of rights (Lima et al., 2021)

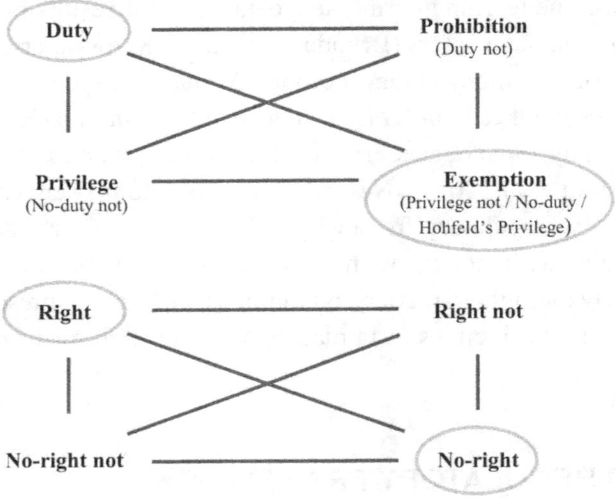

Understanding application to Hohfeld's classification of rights and duties as has been given in figure 7 can be perceived from two different context. One, rights and its jural postulates and correlatives of human beings from/towards artificial intelligence and two, rights and its jural postulates and correlatives of artificial intelligence from/towards human beings. The first question concerns the ambit of human beings, the second questions pull us towards personhood of AI (Jaynes, 2019). In this study, the authors are concerned with the first question, for the paper is with respect to human rights, law and ethics. Although the second questions is certainly interesting to research upon, it will not find mention in the scope of this study.

Addressing, rights of human beings, it can be extended to consider several forms of human rights that has been discussed in this paper. At this juncture, only the jural opposites and jural correlatives of "right" are concerned in this paper. Firstly, the jural correlative of right is duty. Therefore, this is a question as to what duties human beings have towards AI, in as much as human beings seek to protect human rights against violation by artificial intelligence. These duties could be taken to mean the duty to provide data inputs and algorithms that are free from bias and discrimination. Similarly, the jural opposite of right is no-right. one instance of looking at it, is similar to how human beings have the right to protect human rights, they have no right to infringe upon the rights of Artificial intelligence. This will inevitably take one to the understanding of legal personhood of Artificial intelligence and whether AI has rights (Pusca, 2020). Given that this scope of this study is limited to human rights and not the rights of AI, the concept of rights of AI requires a separate analysis (Novelli et al., 2021) (Mocanu, 2022). Furthermore, an interesting question to consider in the realm of rights, is rights that are subject to judicial review or review by courts. In the judicial jurisprudentia, Right to non-discrimination, Privacy and equal protection of laws is usually guaranteed through municipal courts in country specific scenarios, and specific international courts according to classification of the kind of right that has been violated. However, in the backdrop of the age of artificial intelligence, the question of rights assumes complex set-up. whilst it is not complex, as to determination of whose or what rights are violated, complex questions emerge as to affixing the liability to determine who is the "perpetrator?" This comes at a context that AI has power beyond the ordinary human intelligence, capable of complex tasks and therefore, answers regarding, "liability", its jural correlative becomes pertinent.

Before considering whether an AI, or autonomous technology for that matter, might be held legally culpable, it is important to consider who has legal responsibility for the negligent or illegal actions of an AI, the AI, the algorithm programmer, or its owner. Can one bring a lawsuit against AI? The flipside of this question is whether the AI is eligible to file a lawsuit? It deals with giving AI legal rights as well as laying legal duties on it. In circumstances when the AI owner has minimal awareness of and control over the AI's actions, placing all legal responsibility on the AI owner might hinder legal accountability and make it challenging to establish a causal link (Brown, 2020). What makes Alan Turing most famous today is the test he devised to determine if a machine could think. Since then, the Turing test has taken center stage AI discussions. A computer is deemed clever if it can persuade the person asking the inquiry that it is human halfway as frequently as a person can (Bublitz, 2022). It has yet to be demonstrated that a computer or piece of program can pass the Turing test (Carney, 2020) (Brown, 2020). Nevertheless, it is pertinent to understand who the parties are involved before delving into more perplexing questions.

Figure 8. Potential perpetrators in AI-driven actions

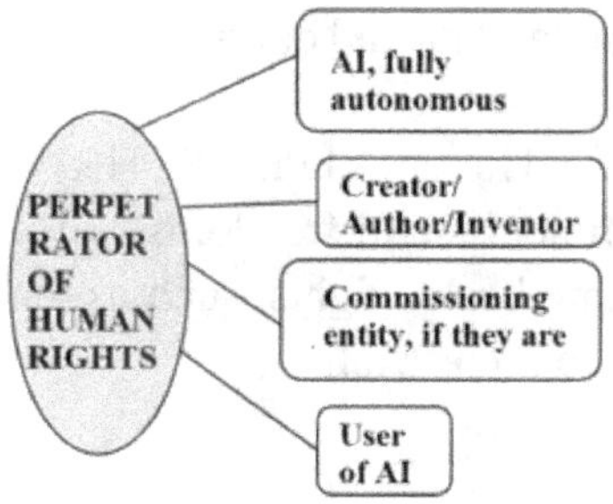

As can be seen from the above figure 8, proposed by the authors of this study, these complex questions as to liability involves determination of "who is the perpetrator"? This is one of the issues that could be baffling the judicial system in the near future in the justice administration. This is important especially in case of violation of human rights. For instance, Human right violations are subject to judicial review in most of the countries (Katyal, 2022). However, in case of violation of human rights by artificial intelligence, intricate questions such as who ought to be the party to suit.

This requires a re-look in the framework of intellectual property law, to determine, who is the "creator" as well as the level of autonomy given to the AI, whether it operates as assistive technology or a fully autonomous machine. Nevertheless, it can be concluded that a decade old Hohfeld's classification of rights is as relevant today in application towards artificial intelligence, as it was back then. It is necessary that this formalisation of classification is necessary to define the corresponding rights and duties for a robust regulatory framework as well as policy.

Current Legal Frameworks, Human Rights Conventions and Their Adequacy to AI

The relationship between legal frameworks and AI as well as the incorporation of human rights norms, presents a dynamic and intricate environment. AI technology is developing at a rapid pace, which raises a number of ethical and legal questions, particularly in relation to human rights conventions and current legal frameworks. It is crucial to consider whether the legal frameworks in place are sufficient to control and accommodate artificial intelligence's wide-ranging applications and possible effects on human rights.

The legal and human rights frameworks mentioned below influence the regulatory environment that surrounds artificial intelligence and its ethical implications.

General Data Protection Regulation (GDPR): The GDPR, which is enforced throughout the European Union, focuses on protecting individuals' privacy and data within the EU and the European Economic Area. It applies to companies and organizations that handle personal data and controls how that data is processed and transferred.

The European Convention on Human Rights (ECHR), which was created by the Council of Europe, protects fundamental freedoms and human rights throughout the continent. It addresses a number of rights, such as the freedom of speech, the right to privacy, and the right to a fair trial.

The Convention for the Protection of Individuals with regard to Automatic Processing of Personal Data (Convention 108): This convention, which is enforced by the Council of Europe, is concerned with data protection when it comes to automatically processing personal data.

Fair Credit Reporting Act (FCRA): The FCRA governs consumer information collection, dissemination, and usage in the United States, including credit reports and background checks. Its objectives are to protect consumers' rights to privacy, fairness, and accuracy.

The Algorithmic Accountability Act is a measure that was introduced in the United States Congress with the intention of regulating automated decision-making systems and mandating that companies evaluate their AI systems for discriminatory and biased practises.

The Convention on Cybercrime, often known as the Budapest Convention, is an international agreement governed by the Council of Europe that attempts to combat cybercrime by promoting global cooperation, streamlining national legal systems, and enhancing investigation techniques.

These are a few of the well-known norms, conventions, and frameworks that have an impact on artificial intelligence (AI) and the ethical concerns raised by its advancement and use. As AI technology evolves and poses new problems to ethical and legal governance, these laws and conventions must be continuously adapted and interpreted in the context of AI.

Promise or the Compromise of Ethics in AI: Navigating Moral Frontiers

Whilst ethics is often times looked at as addressing philosophical questions, it seldom is a standalone discipline since its branches reach to sciences, social sciences and much beyond (Stahl et al., 2023). The concept of AI ethics has emerged in response to growing concern about the possible impacts of artificial intelligence. It is a subset of the wider field of digital ethics, which addresses worries posed by the creation and use of new digital technologies such as AI, analysis of big data, and the technology of blockchain (von Struensee, 2021).

This requires a deeper development and embedding of ethics informed choices as compared to merely technically informed choices (Koniakou, 2023). This ought not to be seen as a choice of protection vs performance as this is a question of ethical vs scientific temper, but scientific temper with ethics for a reliable and trustworthy AI (Brusseau, 2023). Furthermore, ethics requires a macro and micro outlook of policy considerations seeping into government regulations as well as the protection of individual rights (Trotta, Ziosi, & Lomonaco, 2023). Given, that AI has now a technology that is part and parcel of our daily lives, it requires a new relook at applicability of ethics and redefining the boundaries of legislative framework in domestic and international law.

CONCLUSION

Addressing AI bias and stereotyping is an urgent and critical imperative in the development of artificial intelligence systems. As AI technologies become increasingly integrated into our lives, their potential to reinforce and perpetuate societal biases has become a growing concern. The consequences of AI bias and stereotyping are far-reaching, affecting decision-making processes across diverse domains such as employment, criminal justice, healthcare, and finance.

The lack of diversity in AI development teams is a contributing factor to these biases. Underrepresentation from diverse backgrounds in the creation and omission of AI systems leads to lapses and blind spots concerning societal biases. The absence of varied perspectives inhibits a comprehensive understanding of societal nuances and specific needs, contributing to the perpetuation of biases within AI systems. To address these urgent concerns, concerted efforts are needed across multiple fronts. Ethical AI development requires rigorous testing, continuous evaluation, and the implementation of frameworks that promote fairness, transparency, and accountability. Diversity in AI development teams is crucial to incorporate a spectrum of perspectives, ensuring more inclusive and unbiased AI systems. Additionally, enhanced awareness, education, and regulatory measures are imperative to mitigate the impact of bias and stereotyping in AI technologies.

While significant strides have been made in formulating policies and regulations for AI, several critical legal gaps persist. These gaps present substantial challenges in effectively addressing the ethical, social, and practical implications of AI technology. It is imperative to acknowledge and focus on these gaps to ensure that AI policies evolve to meet the demands of an increasingly AI-driven world. One of the fundamental legal gaps in AI policy revolves around the lack of standardized international regulations. The global nature of AI and its applications calls for cohesive and harmonized international standards. Currently, the absence of universally accepted and consistent regulations allows for disparities and varying levels of accountability across different jurisdictions. This discrepancy can lead to challenges in enforcement, potentially creating loopholes that could be exploited. Furthermore, existing data protection laws may not comprehensively cover the unique challenges posed by AI technology. AI systems depend heavily on vast datasets, raising concerns about the collection, usage, and privacy of personal data. The evolving nature of data processing and AI necessitates continuous adaptation of data protection laws to ensure robust safeguards and prevent misuse of sensitive information.

REFERENCES

Allen, L. E., & Saxon, C. S. (1995). Better language, better thought, better communication: The A-Hohfeld language for legal analysis. In *Proceedings of the Fifth International Conference on Artificial Intelligence and Law* (p. 219). ACM Press. DOI: 10.1145/222092.222245

Alston, P. (2019). Extreme poverty and human rights: Seventy-fourth session Agenda item 70(b) Promotion and protection of human rights: Human rights questions, including alternative approaches for improving the effective enjoyment of human rights and fundamental freedoms (A/74/493). United Nations General Assembly. https://undocs.org/en/A/74/493

Arnheim, R. (1992). Deus ex machina. *British Journal of Aesthetics*, *32*(3), 221–226. DOI: 10.1093/bjaesthetics/32.3.221

Berman, M., & Dalvi, R. (2011). *Heroes, monsters and values: Science fiction films of the 1970s*. Cambridge Scholars Publishing.

Boella, G., van der Torre, L., & Verhagen, H. (2001). An action-based ontology of legal relations. In *Proceedings of the 8th International Conference on Artificial Intelligence and Law* (pp. 227–228). ACM Press.

Bohr, A., & Memarzadeh, K. (Eds.). (2020). *Artificial Intelligence in healthcare [1 ed.] 0128184388, 9780128184387*. dokumen.pub. https://dokumen.pub/artificial-intelligence-in-healthcare-1nbsped-0128184388-9780128184387.html

Bradley, A. S. (2019). Human rights racism. *Harvard Human Rights Journal*, *32*, 1–45.

Bronzin, T., Milinovic, M., & Maricic, D. (2021). Artificial Intelligence (AI) brings enhanced personalized user experience. *2021 44th International Convention on Information, Communication and Electronic Technology (MIPRO)*. https://doi.org/ DOI: 10.23919/MIPRO52101.2021.9370446

Brown, R. D. (2020). Property ownership and the legal personhood of artificial intelligence. *Information & Communications Technology Law*, *30*(2), 208–234. DOI: 10.1080/13600834.2020.1861714

Brown, R. D. (2020). Property ownership and the legal personhood of artificial intelligence. *Information & Communications Technology Law*, *30*(3), 208–234.

Brundage, M., Avin, S., Clark, J., Toner, H., Eckersley, P., Garfinkel, B., Dafoe, A., Scharre, P., Zeitzoff, T., Filar, B., & Anderson, H. (2018). The malicious use of artificial intelligence: Forecasting, prevention, and mitigation. *arXiv preprint*, arXiv:1802.07228.

Brusseau, J. (2023). From the ground truth up: Doing AI ethics from practice to principles. *AI & Society*, *38*(4), 1651–1657. DOI: 10.1007/s00146-021-01336-4

Bublitz, J. C. (2022). Might artificial intelligence become part of the person, and what are the key ethical and legal implications? *AI & Society*. Advance online publication. DOI: 10.1007/s00146-022-01432-7

Campbell, M., Hoane, A. J.Jr, & Hsu, F. (2002). Deep Blue. *Artificial Intelligence*, *134*(1–2), 57–83. DOI: 10.1016/S0004-3702(01)00129-1

Cardie, C. (1999). Integrating case-based learning and cognitive biases for machine learning of natural language. *Journal of Experimental & Theoretical Artificial Intelligence*, *11*(3), 297–337. DOI: 10.1080/095281399146445

Carney, T. (2020). Artificial intelligence in welfare: Striking the vulnerability balance? *Monash University*, 25-26. DOI: 10.26180/13370369.v2

Cheng, L., Varshney, K. R., & Liu, H. (2021). Socially responsible AI algorithms: Issues, purposes, and challenges. *Journal of Artificial Intelligence Research*, *1139-1140*, 1137–1181. Advance online publication. DOI: 10.1613/jair.1.12814

Cheng, X., Lin, X., Shen, X.-L., Zarifis, A., & Mou, J. (2022). The dark sides of AI. *Electronic Markets*, *32*(1), 11–15. DOI: 10.1007/s12525-022-00531-5 PMID: 35600917

Chowdhury, M. H., Mahmud, M., Islam, M. T., & Rahman, M. M. (2022). Unfair AI: It isn't just biased data. In *2022 IEEE International Conference on Data Mining (ICDM)*. DOI: 10.1109/ICDM54844.2022.00058

Dulka, A. (2023). The use of artificial intelligence in international human rights law. *Stanford Technology Law Review : STLR : An Online High-Technology Law Journal from Stanford Law School*, *26*(2), 357.

Dwork, C., & Minow, M. (2022). Distrust of artificial intelligence: Sources & responses from computer science & law. *Daedalus*, *151*(2), 309–321. DOI: 10.1162/daed_a_01918

Feldstein, S. (2019). *The global expansion of AI surveillance* (Working Paper). Carnegie Endowment for International Peace. https://carnegieendowment.org/files/WP-Feldstein-AISurveillance_final1.pdf

Feldstein, S. (2019). The global expansion of AI surveillance. *Carnegie Endowment for International Peace* (Working paper). https://carnegieendowment.org/files/WP-Feldstein-AISurveillance_final1.pdf

Fourneret, E., & Yvert, B. (2020). Digital normativity: A challenge for human subjectivation. *Frontiers in Artificial Intelligence*, *3*, 27. Advance online publication. DOI: 10.3389/frai.2020.00027 PMID: 33733146

Gaur, L., & Sahoo, B. M. (2022). Explainable AI in ITS: Ethical concerns. In *Explainable Artificial Intelligence for Intelligent Transportation Systems* (pp. 79–90). Springer. https://doi.org/DOI: 10.1007/978-3-030-95923-5_5

Gillespie, N., Lockey, S., Curtis, C., Pool, J., & Akbari, A. (2023). *Trust in artificial intelligence: A global study*. The University of Queensland and KPMG Australia., DOI: 10.14264/00d3c94

Glikson, E., & Woolley, A. W. (2020). Human trust in artificial intelligence: Review of empirical research. *The Academy of Management Annals*, *14*(2), 627–660. DOI: 10.5465/annals.2018.0057

Greer, C. (2010). *Crime and media: A reader* (1st ed.). Routledge.

Gunn, F., Haegele, M., Langellotti, A., & Zanolini, F. (2020). *An exploration of how artists use immersive technologies to promote inclusivity, diversity, and deep public engagement in ethical ways*. Electronic Workshops in Computing., DOI: 10.14236/ewic/EVA2020.36

Gurkaynak, G., Yilmaz, I., & Haksever, G. (2016). Stifling artificial intelligence: Human perils. *Computer Law & Security Report*, *32*(5), 749–758. DOI: 10.1016/j.clsr.2016.05.003

Haddad, G., & Chedrawi, C. (2019). Artificial intelligence and the inclusion-exclusion paradox: A sociomaterial perspective. *ICTO Conference*.

Haleem, A., Javaid, M., Asim Qadri, M., Pratap Singh, R., & Suman, R. (2022). Artificial Intelligence (AI) applications for marketing: A literature-based study. *International Journal of Intelligent Networks*, *3*, 119–132. DOI: 10.1016/j.ijin.2022.08.005

Hamm, P., & Klesel, M. (2021). Success factors for the adoption of artificial intelligence in organizations: A literature review. *AMCIS 2021 Proceedings*, 10. https://aisel.aisnet.org/amcis2021/art_intel_sem_tech_intelligent_systems/art_intel_sem_tech_intelligent_systems/10

Hermann, I. (2021). Artificial intelligence in fiction: Between narratives and metaphors. *AI & Society*, *38*(2), 319–329. DOI: 10.1007/s00146-020-01119-y

Hine, E., & Floridi, L. (2022). Artificial intelligence with American values and Chinese characteristics: A comparative analysis of American and Chinese governmental AI policies. *AI & Society*. Advance online publication. DOI: 10.1007/s00146-022-01425-9

Hohfeld, W. N. (1917). Fundamental legal conceptions as applied in judicial reasoning. *The Yale Law Journal*, *26*(8), 710–770. DOI: 10.2307/786270

Holcomb, S. D., Porter, J., & Ballard, A. (2018). Overview on DeepMind and its AlphaGo Zero AI. In *Proceedings of the 2018 International Conference on Big Data and Education* (pp. 7–11). DOI: 10.1145/3206157.3206174

Human Rights Group. (2020). *Human rights in the digital age: Making digital technology work for human rights*. Glion Human Rights Dialogue., DOI: 10.1787/7ec52ca6-

Huq, A. Z. (2021). Artificial intelligence and the rule of law. SSRN *Electronic Journal*. DOI: 10.2139/ssrn.3794777

Ingrams, A., Manoharan, A., & Kang, D. (2021). In AI we trust? Citizen perceptions of AI in government decision making. *Policy and Internet*, *14*(3), 390–409. DOI: 10.1002/poi3.249

Izzidien, A. (2023). Using the interest theory of rights and Hohfeldian taxonomy to address a gap in machine learning methods for legal document analysis. *Humanities & Social Sciences Communications*, *10*(1), 251. Advance online publication. DOI: 10.1057/s41599-023-01693-z

Jaynes, T. L. (2019). Legal personhood for artificial intelligence: Citizenship as the exception to the rule. *AI & Society*, *35*(2), 343–354. DOI: 10.1007/s00146-019-00897-9

Johnson, T. L., Labbé, S., & Dunn, M. A. (2022). Facial recognition systems in policing and racial disparities in arrests. *Government Information Quarterly*, *39*(4), 101753. DOI: 10.1016/j.giq.2022.101753

Karpa, D., Müller, M., & Ernst, E. (2022). Artificial intelligence, surveillance, and big data. In *Advanced Studies in Diginomics and Digitalization* (pp. 145–172). Springer., DOI: 10.1007/978-3-030-97759-8_7

Katyal, S. K. (2022). Democracy & distrust in an era of artificial intelligence. *Daedalus*, *151*(2), 322–334. DOI: 10.1162/daed_a_01919

Kollewe, J. (2022, June 1). *World's first raspberry picking robot cracks the toughest nut: Soft fruit*. The Guardian. https://www.theguardian.com/business/2022/jun/01/uk-raspberry-picking-robot-soft-fruit

Koniakou, V. (2023). From the "rush to ethics" to the "race for governance" in artificial intelligence. *Information Systems Frontiers*, *25*(1), 71–102. DOI: 10.1007/s10796-022-10300-6

Kosta, E. (2020). Algorithmic state surveillance: Challenging the notion of agency in human rights. *Regulation & Governance*, *16*(2), 212–224. DOI: 10.1111/rego.12250

Kreps, S. (2021, November). *Democratizing harm: Artificial intelligence in the hands of nonstate actors*. Foreign Policy at Brookings. https://www.brookings.edu/wp-content/uploads/2021/11/FP_20211122_ai_nonstate_actors_kreps.pdf

Krivenko, E. Y. (2020). Gender and human rights: Expanding concepts. In *Gender and Human Rights* (pp. 21–43). Edward Elgar Publishing.

Krogh, C., & Herrestad, H. (1999). Artificial intelligence and legal reasoning: A survey of the field. *Artificial Intelligence and Law*, *7*(2), 81–96. DOI: 10.1023/A:1008367514393

Kuhlman, C., Jackson, L., & Chunara, R. (2020). No computation without representation: Avoiding data and algorithm biases through diversity. In *Proceedings of the 26th ACM SIGKDD International Conference on Knowledge Discovery & Data Mining* (pp. 1257–1265). DOI: 10.1145/3394486.3411074

Labelvisor. (2023). Ensuring label fairness and bias reduction in data labeling. *Labelvisor*. https://www.labelvisor.com/ensuring-label-fairness-and-bias-reduction-in-data-labeling/#:~:text=Label%20bias%20can%20result%20from,unfair%20outcomes%20for%20certain%20groups

Leslie, D., Mazumder, A., Peppin, A., Wolters, M. K., & Hagerty, A. (2021). Does "AI" stand for augmenting inequality in the era of COVID-19 healthcare? *BMJ (Clinical Research Ed.)*, *372*(304), n304. Advance online publication. DOI: 10.1136/bmj.n304 PMID: 33722847

Leslie, D., Mazumder, A., Peppin, A., Wolters, M. K., & Hagerty, A. (2021). Does "AI" stand for augmenting inequality in the era of COVID-19 healthcare? *BMJ (Clinical Research Ed.)*, *372*(304), n304. Advance online publication. DOI: 10.1136/bmj.n304 PMID: 33722847

Li, D., Tong, T. W., & Xiao, Y. (2021). Is China emerging as the global leader in AI? *Harvard Business Review*. https://hbr.org/2021/02/is-china-emerging-as-the-global-leader-in-ai

Li, X. (2021). Analysis of racial discrimination in artificial intelligence from the perspective of social media, search engines, and future crime prediction systems. *Advances in Social Science, Education and Humanities Research, 575*, 172–177. DOI: 10.2991/assehr.k.210902.029

Lima, J. A. de O., de Oliveira, L. C., & Nascimento, F. L. (2021). Casting the light of the theory of opposition onto Hohfeld's fundamental legal concepts. *Legal Theory, 27*(1), 2–35. DOI: 10.1017/S1352325221000070

Lin, C.-C., Huang, A. Y., & Lu, O. H. (2023). Artificial intelligence in intelligent tutoring systems toward sustainable education: A systematic review. *Smart Learning Environments, 10*(1), 41. Advance online publication. DOI: 10.1186/s40561-023-00260-y

Lockey, S., Gillespie, N., Holm, D., & Someh, I. A. (2021). A review of trust in artificial intelligence: Challenges, vulnerabilities and future directions. In *Proceedings of the 54th Hawaii International Conference on System Sciences* (pp. 5468–5469). DOI: 10.24251/HICSS.2021.664

Lu, S. (2022). Data privacy, human rights, and algorithmic opacity. *California Law Review, 110*(4), 2081–2103. DOI: 10.15779/Z38804XM07

Lukyanenko, R., Maass, W., & Storey, V. C. (2022). Trust in artificial intelligence: From a foundational trust framework to emerging research opportunities. *Electronic Markets, 32*(4), 1993–2020. DOI: 10.1007/s12525-022-00605-4

Lum, K., & Isaac, W. (2016). To predict and serve? *Significance, 13*(5), 14–19. DOI: 10.1111/j.1740-9713.2016.00960.x

Luminita, D. (2013). Evolution of the human rights issue. *International Journal of Academic Research in Economics and Management Sciences, 2*(1), 45–58. DOI: 10.6007/IJAREMS/v2-i1/738

Manheim, K., & Kaplan, L. (2019). Artificial intelligence: Risks to privacy and democracy. *The Yale Journal of Law & Technology, 21*, 118–119.

Markovich, R. (2019). Understanding Hohfeld and formalizing legal rights: The Hohfeldian conceptions and their conditional consequences. *Studia Logica, 108*(1), 129–158. DOI: 10.1007/s11225-019-09870-5

Marsch, N. (2019). Artificial intelligence and the fundamental right to data protection: Opening the door for technological innovation and innovative protection. In *Regulating Artificial Intelligence* (pp. 33–52).

McCauley, L. (2007). The Frankenstein complex and Asimov's three laws. *AI & Society*. Advance online publication. DOI: 10.1007/s00146-006-0047-2

McDaniel, J., & Pease, K. (2021). *Predictive policing and artificial intelligence* (1st ed.). Routledge., DOI: 10.4324/9780429265365

McDowall, R. (2015). Periodic evaluation: Independent review to ensure continued validation of computerized systems. In *EU Annex 11 Guide to Computer Validation Compliance for the Worldwide Health Agency GMP* (pp. 91–110). Springer., DOI: 10.1201/b18315-15

Mocanu, D. M. (2022). Gradient legal personhood for AI systems—Painting continental legal shapes made to fit analytical molds. *Frontiers in Robotics and AI*, *8*, 740500. Advance online publication. DOI: 10.3389/frobt.2021.788179 PMID: 35127836

Monshipouri, M. (2016). Human rights in the digital age: Opportunities and constraints. *Public Integrity*, *19*(2), 123–135. DOI: 10.1080/10999922.2016.1230690

Mou, X. (2019). Artificial-intelligence-investment-trends-and-selected- ... https://documents1.worldbank.org/curated/en/617511573040599056/pdf/Artificial-Intelligence-Investment-Trends-and-Selected-Industry-Uses.pdf

Novelli, C., Di Marco, A., & Mazzoccoli, M. (2021). A conceptual framework for legal personality and its application to AI. *Jurisprudence*, *13*(2), 194–219. DOI: 10.1080/20403313.2021.2010936

Nyst, C., & Falchetta, T. (2017). The right to privacy in the digital age. *Journal of Human Rights Practice*, *9*(1), 104–118. DOI: 10.1093/jhuman/huw026

OECD. (2020). Artificial-intelligence-machine-learning-big-data-in-finance. ... https://www.oecd.org/finance/financial-markets/Artificial-intelligence-machine-learning-big-data-in-finance.pdf

OECD. (2022). Rights in the digital age: Challenges and ways forward. *OECD Digital Economy Papers*, (347), 12–13. DOI: 10.1787/7ec52ca6-en

Oxford Commission on AI & Good Governance. (2021, December). *AI in the public service: From principles to practice* (p. 13). https://oxcaigg.oii.ox.ac.uk/wp-content/uploads/sites/11/2021/12/AI-in-the-Public-Service-Final.pdf

Palumbo, D. (2008). The monomyth in James Cameron's The Terminator: Sarah as monomythic heroine. *Journal of Popular Culture*, *41*(3), 413–427. DOI: 10.1111/j.1540-5931.2008.00528.x

Patterson, D., Laing, P., & Johnson, T. (2017, September 27). How today's smart tech inspired *Star Trek: Discovery. TechRepublic*. https://www.techrepublic.com/article/review-how-todays-smart-tech-inspired-star-trek-discovery/

Paz, M. (2014). The tower of Babel: Human rights and the paradox of language. *European Journal of International Law*, *25*(2), 473–496. DOI: 10.1093/ejil/chu037

Pusca, C. A. (2020). Should we share rights and obligations with artificial intelligence robots? In *Lecture Notes of the Institute for Computer Sciences, Social Informatics and Telecommunications Engineering* (pp. 412–419). Springer. DOI: 10.1007/978-3-030-51005-3_33

Raghavan, M. (2023). Mitigating bias in algorithmic decision-making: Evaluating claims and practices. In *The Societal Impacts of Algorithmic Decision-Making*. Springer., DOI: 10.1145/3603195.3603212

Raveh, A. R., & Tamir, B. (2018). From homo sapiens to Robo sapiens: The evolution of intelligence. *Information (Basel)*, *10*(2), 2. DOI: 10.3390/info10010002

Razmetaeva, Y., Barabash, Y., & Lukianov, D. (2022). The concept of human rights in the digital era: Changes and consequences for judicial practice. *Access to Justice in Eastern Europe*, *5*(3), 41–56. DOI: 10.33327/AJEE-18-5.3-a000327

Risse, M. (2021). The fourth generation of human rights: Epistemic rights in digital lifeworlds. *Moral Philosophy and Politics*, *8*(3), 351–378. DOI: 10.1515/mopp-2020-0039

Roberts, H., Cowls, J., Morley, J., Taddeo, M., Wang, V., & Floridi, L. (2020). The Chinese approach to artificial intelligence: An analysis of policy, ethics, and regulation. *AI & Society*, *36*(1), 59–77. DOI: 10.1007/s00146-020-00992-2

Ryan, M. (2020). In AI we trust: Ethics, artificial intelligence, and reliability. *Science and Engineering Ethics*, *26*(5), 2749–2767. DOI: 10.1007/s11948-020-00228-y PMID: 32524425

Sarker, I. H. (2021). Machine learning: Algorithms, real-world applications and Research Directions. *SN Computer Science*, *2*(3), 160. Advance online publication. DOI: 10.1007/s42979-021-00592-x PMID: 33778771

Scherer, M. U. (2015). Regulating artificial intelligence systems: Risks, Challenges, competencies, and strategies. SSRN *Electronic Journal*. DOI: 10.2139/ssrn.2609777

Schwartz, R., Vassilev, A., White, C., Grady, N., & Johnson, J. (2022). *Towards a standard for identifying and managing bias in artificial intelligence* (NIST Special Publication 1270). National Institute of Standards and Technology. DOI: 10.6028/NIST.SP.1270

Sharma, K., Yang, Y., & Wu, Q. (2017). A complex network analysis of ethnic conflicts and human rights violations. *Scientific Reports*, *7*(1), 1–12. DOI: 10.1038/s41598-017-09101-8 PMID: 28811628

Shin, D. D. (2022). Algorithmic bias. In *Algorithms, Humans, and Interactions* (pp. 103–125). Springer. DOI: 10.1201/b23083-6

Sichelman, T. M. (2017). Very tight 'bundles of sticks': Hohfeld's complex jural relations. In S. Balganesh, T. Sichelman, & H. Smith (Eds.), *The legacy of Wesley Hohfeld: Edited major works, select personal papers, and original commentaries* (Forthcoming). Cambridge University Press. *San Diego Legal Studies Paper No. 17*. https://doi.org/DOI: 10.2139/ssrn.2950912

Skutnabb-Kangas, T., & Phillipson, R. (2023). *Handbook of linguistic human rights*. John Wiley & Sons.

Stahl, B. C., Brooks, L., Hatzakis, T., Santiago, N., & Wright, D. (2023). Exploring ethics and human rights in artificial intelligence: A Delphi study. *Technological Forecasting and Social Change*, *191*, 123439. DOI: 10.1016/j.techfore.2023.122502

Stanford, S., & Wilson, C. (2022). Bias in schools. In *Systems of Thought and Behaviors within Bias and Inequity*. Springer., DOI: 10.4135/9781071900611

Technology and innovation report 2021: Catching technological waves innovation with equity. (2021). United Nations Publications. https://unctad.org/webflyer/technology-and-innovation-report-2021

The Role of Data in AI. (2020). *Report for the Data Governance Working Group of the Global Partnership on AI*. Digital Curation Centre, Trilateral Research, School of Informatics, The University of Edinburgh.

Trotta, A., Ziosi, M., & Lomonaco, V. (2023). The future of ethics in AI: Challenges and opportunities. *AI & Society*, *38*(2), 439–441. DOI: 10.1007/s00146-023-01644-x

UNESCO. (2020). *Artificial intelligence and gender equality: Key findings of UNESCO's Global Dialogue*. (GEN/2020/AI/2 REV). https://unesdoc.unesco.org/ark:/48223/pf0000374174

United Nations. (2003). *Human rights-based approach*. United Nations Sustainable Development Group. https://unsdg.un.org/2030-agenda/universal-values/human-rights-based-approach

United Nations Office of Counter-Terrorism (UNOCT) & United Nations Interregional Crime and Justice Research Institute (UNICRI). (2021). *Algorithms and terrorism: The malicious use of artificial intelligence for terrorist purposes: A joint report.*

Van Giffen, B., Sternberg, Z., Meijer, S., & Batchelor, D. (2022). Overcoming the pitfalls and perils of algorithms: A classification of machine learning biases and mitigation methods. *Journal of Business Research*, *144*, 93–106. DOI: 10.1016/j.jbusres.2022.01.076

Vemuri, V. K. (2014). *the second machine age: Work, progress, and prosperity in a time of brilliant technologies*, by Erik Brynjolfsson and Andrew McAfee. *Journal of Information Technology Case and Application Research*, *16*(2), 112–115. DOI: 10.1080/15228053.2014.943094

Villata, S., Araszkiewicz, M., Ashley, K., Bench-Capon, T., Branting, L. K., Conrad, J. G., & Wyner, A. (2022). Thirty years of artificial intelligence and law: The third decade. *Artificial Intelligence and Law*, *30*(4), 561–591. DOI: 10.1007/s10506-022-09327-6

von Struensee, S. (2021). *Artificial intelligence and corporate social responsibility*. Global Research Initiative.

Wolfram, M., & Kienesberger, M. (2023). Gender in sustainability transition studies: Concepts, blind spots, and future orientations. *Environmental Innovation and Societal Transitions*, *46*, 100686. DOI: 10.1016/j.eist.2022.100686

Wong, C. (2023). AI 'fairness' research held back by lack of diversity. *Nature*. Advance online publication. DOI: 10.1038/d41586-023-00935-z PMID: 36997714

Wong, L.-W., Cheng, T. C. E., & Tong, Y. (2023). The role of institutional and self in the formation of trust in artificial intelligence technologies. *Internet Research*. Advance online publication. DOI: 10.1108/INTR-09-2022-0455

World Economic Forum. (2022). *A blueprint for equity and inclusion in artificial intelligence* (White Paper). https://www.weforum.org/reports/a-blueprint-for-equity-and-inclusion-in-artificial-intelligence

World Economic Forum. (2022). *Earning digital trust: Decision-making for trustworthy technologies.* In collaboration with Accenture, KPMG, and PwC. https://www.weforum.org/reports/earning-digital-trust-decision-making-for-trustworthy-technologies

Yamin, M. M., Ahmed, F., Shaikh, M. A., & Majeed, A. (2021). Weaponized AI for cyber attacks. *Journal of Information Security and Applications*, *57*, 102722. DOI: 10.1016/j.jisa.2020.102722

Yazici, E. (2018). Nationalism and human rights. *Political Research Quarterly*, *72*(1), 147–161. DOI: 10.1177/1065912918781187

Zajko, M. (2022). Artificial intelligence, algorithms, and social inequality: Sociological contributions to contemporary debates. *Sociology Compass*, *16*(2), e12966. DOI: 10.1111/soc4.12962

Zerilli, J., Knott, A., Maclaurin, J., & Gavaghan, C. (2022). How transparency modulates trust in artificial intelligence. *Patterns (New York, N.Y.)*, *3*(7), 100455. DOI: 10.1016/j.patter.2022.100455 PMID: 35465233

Zhai, X. (2022). Artificial AI bias. In *American Educational Research Association (AERA)2022 Annual Meeting*. https://doi.org/DOI: 10.3102/9781234567890

Zhang, D., Maslej, N., Brynjolfsson, E., Etchemendy, J., Lyons, T., Manyika, J., Ngo, H., Niebles, J. C., Sellitto, M., Sakhaee, E., Shoham, Y., Clark, J., & Perrault, R. (2022). *The AI index 2022 annual report*. AI Index Steering Committee, Stanford Institute for Human-Centered AI, Stanford University. https://hai.stanford.edu/ai-index-2022

Zhang, D., Mishra, S., Brynjolfsson, E., Etchemendy, J., Ganguli, D., Grosz, B., Lyons, T., Manyika, J., Niebles, J. C., Sellitto, M., Shoham, Y., Clark, J., & Perrault, R. (2021). *The AI index 2021 annual report*. AI Index Steering Committee, Stanford Human-Centered AI Institute, Stanford University. https://hai.stanford.edu/ai-index-2021

Zhang, M. (2015, July 1). Google Photos tags two African-Americans as gorillas through facial recognition software. *Forbes*. https://www.forbes.com/sites/mzhang/2015/07/01/google-photos-tags-two-african-americans-as-gorillas-through-facial-recognition-software/?sh=40c3811c713d

Chapter 3
AI and the Ethical Crossroads:
Navigating Privacy, Personhood, and Autonomy in the Digital Age

Muhammad Usman Tariq
https://orcid.org/0000-0002-7605-3040
Abu Dhabi University, UAE & University College Cork, Ireland

ABSTRACT

This chapter explores the difficult moral issues that artificial intelligence (AI) is bringing about as it changes society. It examines the points where AIs expanding influence raises new ethical and legal conundrums: privacy autonomy and personhood. As the chapter looks at how AI systems gather and use personal data it raises questions regarding data ownership and control and changes privacy standards. It also discusses the idea of AI personhood raising issues such as whether sophisticated AI systems ought to have moral or legal standing and the ramifications for accountability rights and responsibility. The chapter also examines human and AI autonomy examining how AIs decision-making powers affect human freedom and control.

INTRODUCTION

In the current society, Artificial Intelligence AI has been adopted into almost every task performed in a day be it in health care, education, transport, and even entertainment. with the development of new technologies at an ever-increasing rate it calls into question a range of ethical issues concerning the dignity of the human person, privacy and liberty. The rise in smart systems education causes significant

DOI: 10.4018/979-8-3693-9035-1.ch003

concerns of how these technologies engage with ethical concerns and societal values. In this chapter, the author tries to examine the ethical potentialities on one hand and AI ethics and human dignity on another hand, and also focuses on the great need for effective ethical trustworthy theories about the use and function of AI in this digital life (Arciniegas Gome, 2023). The principle that underpins much of the ethical debate about AI is that of human dignity. Human dignity is included as the intrinsic and absolute value of human beings that should be accorded respect and. Being predisposed to assume decisions, to analyse data and to explore interaction with individuals, the AI systems are developed in structures that can potentially erode human dignity. For example, the machines can be over-biased depending on the sample obtained for training because discriminatory patterns are also found in these samples. An example of this can be witnessed in use of facial recognition where proofs of biometric algorithms disproportionately fail to identify certain people from the marginalized groups. When #AI fails, it results in a violation of that dignity, which raises major questions to do with equity in artificial intelligence (Nersessian & Mancha, 2020).

Another human dignity right that is slippery in the current generation pervaded by artificial intelligence is privacy. In a world where information gathering and monitoring tools of all sorts are becoming increasingly available and widespread, people simply can no longer understand how much of their data is being collected, analysed and used. Think about the latest vendor's applications through smart, wearable, and portable devices with monitoring and profiling features to advertise related products, services, or events by using relevant social networks and the Web (Harasimiuk & Braun, 2021). While these technologies increase the level of user experiences, they at the same time make consent and privacy to be more an ambiguous affair. Several of the users gave informed consent based on limited awareness of the consequences of exposing their data, and hence they could be surrendering their autonomy as well. The ethical issue of concern here can be summed up in the fact of how to and to what extent information and technology can be used to enhance innovation while at the same not infringing on the rights of individuals to privacy and self-determination (Arciniegas Gome, 2023). Personhood is also applicable in ethical consideration on Artificial Intelligence. The more advanced the AI systems become, the closer a machine's ability will start resembling the ability of a human or human brain. The questions that arise, in this case, are whether an AI entity can be given rights or obligations comparable to a human being. For example, when thinking about the design of the new generation of companion or care robots (Harasimiuk & Braun, 2021).Though these robots have social patterning capability and able to interact with human being like a human being they do not have mind, feelings just like human being have. There are moral implications of the utilization of diverse AI entities and perception of whether such entities deserve some regard that is standardized for conscious beings. As for example, establishment of rapport between peoples and the AI comes with the question of how people with emotional disorders may be manipulated by the AI they grow to love (Nersessian & Mancha, 2020).

This, we have seen, is the case as the field of AI changes and as the importance of proper ethical frameworks becomes more apparent. Ethical models are concepts that may be used as reference for solving multifaceted ethical issues connected with the use of AI systems. These frameworks can help sort out questions of bias, privacy, issues of accountability and transparency to ensure that artificial intelligence systems are implemented properly and according to our social norms. For example, there are initiatives such as the Partnership on AI who come together to ensure proper AI practices persist. Their work focuses on integrated designs in their machine with solutions embracing fairness and transparency as well as embracing diversity in matters concerning AI ethicality, this means that it requires input from various fields (Tsakalogianni, 2023). However, policymakers play the significant part of developing ethical directions for AI's use. Political leaders around the globe are slowly acknowledging institutions that need to create guidelines to moderate the application of artificial intelligence applications; it is paramount to appreciate the impacts of ethical questions in policy formulation. For example, for AI regulation, the European Union has put its own law forward that can primarily address the matter of high risk AI applications and their ethical points. This approach will be on designing AI with intention to meet human rights and ethical benchmarks. If the rules are clarified, politicians create the conditions in which AI solutions can be developed for the sake of the common good while minimizing the violation of people's rights (Arciniegas Gome, 2023). Other than government and policymakers, therefore, technologists, ethicists, and wider society must be part of the conversation regarding AI. Finally, cooperation enables organization to find potential ethical dilemmas and develop strategies that respect a human being dignity and his or her right to self-determination. For instance, engaging stakeholders in its conceptualization and the execution of the technologies means that everyone stands an equal chance of benefiting. When it comes to AI, Ensmenger and Niele use different examples of how, by including other stakeholders, organizations can avoid an unintentional alignment with pre-existing disparities and prejudices (Nersessian & Mancha, 2020).

In addition, it will be seen that the ethical issues involved are not simply questions of rights but are organizational and systemic in nature and scope. When AI technologies redefine industries and employment opportunities there are always questions concerning employment, economic divide, and social integration. This brings in to question the consequence management responsibility of the AI implementers and creators. For instance, if corporations embrace AI technologies, they should also take measures to ensure that employees that stand to be impacted will be trained to adapt to other things proving that the company is being ethical (Cheong, 2020).

It is also important to give ethic consideration to AI impact on decision making. AI solutions are used in various fields to help with decision-making; these fields comprise health care and criminal justice. Although such technologies can in theory enhance the quality and productivity of the investigative process, they also raise the issue of responsibility and openness of the situation. For instance, if an AI system made a medical diagnosis or decided on the legal consequences of a case, who is to blame? It is necessary to develop the methods of addressing injustice of artificial intelligence products so that people may appeal against unjust decisions. This underlines the need for complexity, which exists to explain the procedures and allow people to have faith in those algorithms that govern different aspects of their lives (Milic-Frayling, 2023). Ethical issues are also related to human rights in the continuum of the entire society. When the AI system is deployed in all facets of life it becomes essential to guarantee that the systems do not violate human rights. The Universal Declaration of Human Rights states what rights belong to the members of the human family and again, AI technologies should only be developed in a way that protects these rights. For example, application of artificial intelligence in surveillance to monitor the general public is industries the right to privacy and freedom of speech. Accountability frameworks must allow for the generation of positive AI systems that will not only serve the existing human rights but will also enhance the revelation of positive innovations (Arciniegas Gome, 2023).

While finding our way through these ethical concerns of AI, there needs to be a continued conversation and similar self-reflection. Due to the high rate of technological innovation, debates concerning the role and impact of AI on society and particular person's rights should occur continuously. It is for this reason that it is essential to have discourses on ethic in AI not only with stakeholders but indeed with the public. Informed citizens to their rights when it comes to AI and get them involved in creating the society that will be served by the advanced technology (Horváth, 2022). Consequently, the relation between AI, ethics, and human dignity constitutes emerging and challenging territory. This might be true but as we further advance our society to make use of AI technologies, we have to advance the ethical principles that safeguards the density human beings, our privacy and our freedom. The major issues that need to be confronted include the need for sound ethical solutions, conclusive stakeholders, and this calls for proactive polices in the facet of AI (Lyreskog et al., 2023). Therefore, by promoting an environment that offers consideration for ethical consideration and human rights we are able to bring optimization of AI and its correlation to the society within a strengthened societal culture. Ethical AI development is a continuous process, and it is everyone's common endeavour to contribute to a process that will make the use of technology to make human a better entity and enable individuals to benefit from the creations of a digitized world (Milic-Frayling, 2023).

BACKGROUND

This chapter shall focus on two- In the current world where technology changes almost daily, privacy is changing in equal measures, particularly with the introduction of Artificial Intelligence (AI). Personal privacy previously being the stewardship of personal information, is now fundamentally intertwined with data gathering, analysis, and application, in regard to AI systems. Accompanying features of AI mirror on penalties on personal privacy and social standards and can be not only analysed as a problem of privacy in a digital society but also as an issue which redefines the very nature of the term 'private'. This new definition of privacy makes one to consider the ethical dimensions in the issues of data gathering, monitoring and permission besides searching for the best ways of protecting privacy in this new dispensation (Wong, 2021). AI has affected personal privacy most especially in ways that many people might not have the ability to fully appreciate. For instance, the core functionality of social media sites can monitor and process user activity, likes, shares, and messages, amongst others to generate a comprehensive profile that defines the posts we see. This not only poses questions about personalized advertising but also ab out the level of privacy control the users have. Previously, people were in charge of determining when, how and by whom their data will be used. In today's world where all sorts of decisions are made through various calculated AI, the users too often find themselves in a situation where their privacy is being invaded unconsciously (Rodrigues, 2024).

Because of AI, the expectations of privacy in society are gradually changing as well. Often, users knowingly or unknowingly reveal their personal data to online platforms in situations, perhaps, for convenience' sake or in some entertainment, but not knowing the implications of their actions. For example, when a user uploads a photo on a social networking account, they may think they are sharing it with friends and family. However, AI algorithms can very fast process, classify and disseminate that content to other recipients other than the intended ones, which can result to violation of user's personal information. Consequently, the privacy levels of cultures are relaxed with people willingly surrendering their right to privacy in favour of convenience. This transition makes people wonder whether the privacy relating to artificial intelligence can be protected or not and whether the existing manners of defining this issue meet the point (Hansen, 2022). Dilemmas concerning data collection and surveillance form the central controversy in privacy disputes in an Artificial Intelligence context. The means of obtaining the information are also often opaque, meaning that people cannot understand how their data is collected and used within firms. For instance, there is currently the common requirement, in many applications or devices, of accepting lengthy privacy policies before utilization. However, most of these documents are extremely long, written in legal language, and are designed to make the user consent when they may hardly understand the consequences. This lack of informed consent highlights an ethical dilemma: users can consequently accept certain conditions that enable organizations to generate revenues from their data; this is contrary to fairness as well as self-governance (Wong, 2021).

In addition, there is a problem with surveillance technologies that are based on the AI systems' usage in dangering people's privacy. Autonomous AI surveillance has become routine through facial recognition and data analytics in governmental and organizational campaigning in public domains. As advocates will claim, technologies proposed for implementation are capable of increasing security and effectiveness, yet at the same time negatively impact the privacy of individuals (Chen et al., 2024). For example, take face recognition systems that are being put in practice in different public spaces including airports and streets. That said, such systems are useful, if only to flag possible perils; nevertheless, they engender an omnipresent climate of surveillance that discourages public liberties. Some people may be discouraged from freely enjoying their rights and pumping their opinions in the public domain since they know they are being recorded. This generates rather essential ethical issues such as security and privacies as well as whether such Surveillance ought to be made (Askin, 2024). Consent, in the era of artificial intelligence, responds to some difficulties, especially as concerns data gathering processes. Inappropriate about such concerns as the exploitation of personal data as a source of income, traditional concepts of consent fail to suffice. For example, many web-based services provided are tie-in, meaning that users have to explicitly unsubscribe in order to decline sharing of personal data. This model presupposes that the consumers comprehend what they are agreeing to, while in point of fact, many of them shift through the OSBA options with little knowledge of what they are doing. This throws into moral ambiguous territory the possibility of consent where people are not given further details, regarding how their data will be utilized, arguing for more enhanced and effective ways of getting consent (Hansen, 2022).

For realigning the ethical principle of privacy within the AI context and tackling these ethical concerns, one need strategies to privacy one's own private affairs. Four strategies can be employed, which include asserting to increase the transparency of the procedures used to gather data. An organization must ensure that potential users of data are clear on the collection, storage, and use of their data. Companies can ensure that user control is affected through the provision of easily understandable privacy policies hence helping the users to have control over their data. Moreover, the organizations could incorporate privacy-favouring practices as the fundamental concepts of their AI systems from the ground up. This means that privacy is not an addition of technology but rather part and parcel of most artificial intelligence products (Wong, 2021). The second useful approach in the area of privacy preservation is a focus on increasing user's data control. Allowing people to control their data, for instance, enter, modify, or erase inputs within a database entails motivational factors of ownership. For instance, business organizations could provide interfaces that show people what data is gathered and its application. When organizations offer their users ways through which they can control their information, then they are placing their commitment in the right perspective as an ethical organization (Estrada, 2020).

In this case, regulatory measures also have vital role to determine new dimension of privacy with aid of artificial intelligence. Federal governments across the globe are now putting in place data protection laws to protect individuals' rights in cyberspace. The European Union's General Data Protection Regulation (GDPR) is one of such laws, that focuses on explicit consent, data subject's right to data portability and the right to erasure. Comparable rules may be utilized to set guidelines to data gathering, handling, and sharing while demanding liability from such organizations. To ensure that abuses of these newfound freedoms are meted out in equal measure, the policy makers need to enforce these regulations to best provide a template that preferences privacy (Krysa & Impett, 2022). In addition, efforts by organization and governments to increase the awareness of the people will go a long way in creating a culture of privacy. People need to be made aware of the dangers of putting out their details on the social media blog and the likely consequences of the AI enabled technologies. Awareness creating campaigns related to user's data privacy rights or how one can protect their data from misuse can enable a person and help him regain control over his/her data. This, in return promotes change and encourages the citizen to fight for his/her rights as well as calling the organizations or the policymakers to order (Estrada, 2020).

It is also important to involve stakeholders for a better approach to hand the ethical issues of AI and privacy. The development and implementation of a set of ethical rules for using artificial intelligence should be based on cooperation between IT corporations and startups, governmental bodies and NGOs. This approach needs to be cooperative so that all those affected can be protected by their rights and their needs can be met. For example, projects that engage the tech firms and the NGOs can promote dialogue on how data can be collected responsibly, how consent maybe achieved, less surveillance methods and the like that will yield better and more ethical results (Horváth, 2022). Due to the progressive advancements in the field of AI technologies, constant discussions and research is need in order to gain further knowledge on their impact to patients' privacy as well as other social taboos. To expand our knowledge of what is going to be ahead we can study in an interdisciplinary way possible applications of AI, the ethical aspect of using AI, and the question of privacy. Scholars can examine the effect of AI technology on personal privacy, evaluate the existing body of privacy laws, and advance ideas to strengthen the privacy regime in the new technological environment (Krysa & Impett, 2022).

Therefore, it is hard and challenging to describe and redefine privacy in an AI-driven world. As with PGR, AI impacts individual and community privacy hence, there arises the concern in applying and discussing the ethical issues that arise with data aggregation, monitoring and consent. Thus, measures can be ensured as follows to maintain privacy in such a developing environment: Increased transparency, better control by user's side, regulatory control, awareness among users, and cooperation among the user and other stakeholders. Through elimination of these problems, one can paint the picture of the future in which the decentralization of power will start from privacy from the Fourth Amendment, rights of citizens will be defended, and the spirit of ethical use of AI will be maintained. By doing so, we are able to steer clear of the negative impact that technology can have on human dignity while, at the same time, make full use of artificial intelligence as part of solution-finding processes (Horváth, 2022).

Personhood and Autonomy: The Human-AI Relationship

AI has an ever increasingly position within numerous spheres of people's live and this seminal raises several important and imperative issues concerning human dignity and personhood and especially autonomy. This paper examines social interactions between the human and AI in extending the manner in which intelligent machines enter the lives of individuals and influence its choices. It is thus important to navigate these dimensions as a way of making a worth understanding of the ethical issue related to AI technology and its relation to society (Nourbakhsh & Keating, 2020). Human dignity is one of the fundamental principles which advocates for worth of a human being. It defines creating a worldwide society for the respect, recognition, protection of individual human rights and value. Speaking of AI, Human dignity requires that no element of human processing be supplanted by an inanimate one. For example, in the health sector, AI can help the physicians to diagnose ailments more efficiently and effectively, but it cannot help as a physician does when alleviating a patient's suffering. In the design and development of such systems, certain measures should be put in place so as to respect the dignity of users. That is to say technology should help enrich people's experiences instead of detracting from them (Lovell, 2023). AI capability to participate in a process of decision making presents a major problem to the concept of autonomy. Self-governance means an ability to decide and manage a person and his or her affairs. Given that it enables people to express their values, interests, and opinions, it is a crucial component of human dignity. But as artificial intelligence (AI) systems proliferate in decision-making procedures, concerns concerning their capacity to uphold and respect human autonomy grow

(Nourbakhsh & Keating, 2020). Think about AI's application in online platforms that provide content recommendations based on user preferences. Although these suggestions can improve the user experience, they may also restrict the range of information that people can come across, which may lead people to adopt particular opinions and compromise their capacity for free will (Elliott, 2021).

Another illustration is the use of AI technology in autonomous cars, which allows it to make crucial driving decisions. These technologies can decrease human error, which increases safety, but they also provide moral conundrums about decision-making in cases where tragedies cannot be prevented. Who establishes the moral standard by which an autonomous car must decide whether to damage onlookers or its occupants? The intricate relationship that exists between AI decision-making and human values emphasizes how crucial it is to incorporate ethical principles into AI systems in order to guarantee that they uphold individual autonomy and human dignity (Hansen, 2022). The development of intelligent machines has a big impact on how we think about people. Personhood has always been connected to humans, but as AI becomes more sophisticated, discussions have arisen concerning whether some highly developed AI systems qualify as "people" in and of themselves. Take conversational AI as an example, which can communicate with humans like a person and show emotional reactions. These systems can mimic human speech, but they are devoid of true awareness, feelings, and comprehension. This begs the question of whether AI can have human-like personality or rights (Lovell, 2023).

It is critical to distinguish between the simulation of human-like behaviours and the true characteristics that determine personality as society wrestles with these issues. Even while AI is capable of mimicking empathy and understanding, its imitations are based on preprogrammed reactions rather than real emotional experiences. It is critical to understand this difference in order to protect human dignity and ensure that AI enhances human experiences rather than takes the place of interpersonal relationships (Nourbakhsh & Keating, 2020). Furthermore, society ramifications as well as individual relationships are affected by AI's influence on personhood. There is a chance that human input will be undervalued when AI technologies are incorporated more into decision-making procedures. For example, there is a risk of prejudice and discrimination in the workplace when artificial intelligence (AI) is used for recruiting or performance reviews if the systems are not properly developed and supervised. The importance of human intuition and judgment might be diminished if decisions are made only by robots, which would diminish human dignity (Elliott, 2021).

It is essential to create ethical frameworks that direct the creation and application of AI technology in order to solve these issues. These frameworks ought to put human autonomy and dignity first by making sure AI systems are made to support human decision-making, not to replace it. AI need to be utilized, for instance, to supplement human talents by offering suggestions and insights, while letting people make the ultimate judgments. By working together, individuals may take advantage of AI's benefits while still having influence over their decisions (Herzfeld, 2023). The interaction between humans and AI is also significantly shaped by public discourse and education. People need to be aware of their rights and obligations when engaging with AI systems as these technologies advance. A better educated society may be achieved via raising knowledge of the ethical implications of AI and encouraging conversations on how it affects autonomy and dignity. With this knowledge, people will be able to interact with AI in a way that maintains their autonomy and sense of dignity (Lovell, 2023).

Thus, the connection between humans and artificial intelligence is a dynamic and complicated one that brings up significant issues with personhood, autonomy, and dignity. Prioritizing human dignity is crucial as AI technologies permeate more aspects of daily life. This may be achieved by making sure that technology enhances rather than detracts from unique experiences. As AI systems should assist human decisions while being guided by ethical concerns, respecting autonomy in decision-making is essential. The changing concept of personhood calls into question long-held beliefs and need close consideration in order to honour and cherish each person's distinctive characteristics. Through the development of ethical frameworks, awareness-raising campaigns, and meaningful interactions with AI, society may successfully traverse this new environment while respecting each person's individuality and dignity (Herzfeld, 2023).

Algorithmic Fairness and Social Justice

Algorithmic justice has become a critical field of study as artificial intelligence (AI) technology permeate more facets of our daily lives. The idea of "algorithmic fairness" states that AI systems need to function impartially and treat every person equally. However, as biases can originate from a variety of sources, comprehending and reducing prejudice in AI systems is a challenging task. These prejudices may be ingrained in the data that AI models are trained on, in the algorithms themselves, or in the social norms that shape the creation and application of these technologies (Chesterman, 2020). The data that AI systems are educated on is one important source of bias in those systems. Should the training data mirror preexisting social prejudices, the artificial intelligence system might reinforce or even magnify such biases. A recruiting algorithm trained on past employment data from a business with a history of discriminatory hiring practices, for instance, may be the subject of this discussion. In the event that this data contains fewer applications from particular demographic groups, the algorithm may be trained to give preference to such applicants, hence maintaining current disparities in hiring procedures. This case emphasizes how critical it is to train AI systems on representative and varied datasets in order to prevent the reinforcement of past biases and the production of unjust results (Gaviria, 2022).

It's also necessary to regularly audit and test AI systems for fairness in order to mitigate bias in these systems. Frequent reviews can assist in locating potential bias hotspots and enable developers to make the required corrections. Tech businesses, for example, may use fairness measures to evaluate how well their algorithms work for various demographic groups. Developers can look into the underlying reasons and make changes to an AI model used in lending choices if it rejects applications from particular racial or ethnic groups disproportionately. This will ensure that all applicants are treated fairly. Artificial intelligence engineers can strive for more egalitarian results by proactively tackling prejudice (Stahl, 2021). Accountability and transparency are crucial elements of moral AI design. Transparency is even more necessary when AI systems have a greater impact on important decisions in fields like healthcare, criminal justice, and education. People who are impacted by these choices have a right to know how AI systems operate and why the results they generate make sense. An applicant should be provided with a detailed explanation of the reasons behind the AI system's decision, for example, if it rejects their loan application. When appropriate, this transparency can assist people in contesting unjust results and pursuing reparations (Gaviria, 2022). Making sure that organizations and developers have processes in place to hold them accountable for the effects of their technology is a key component of accountability in AI systems. There ought to be easy ways for people to report problems and seek redress when an AI system discriminates or causes harm. Regulations may be put in place, for instance, requiring businesses to reveal the workings of their algorithms and to submit periodic evaluations on their fairness. This accountability makes sure that companies prioritize justice and fairness in their designs and cultivates a culture of ethical responsibility in AI research (Chesterman, 2020).

In order to ensure justice and equality via ethical AI design, proactive measures that consider the wider social ramifications of AI technology must be taken. Involving a variety of stakeholders in the development process, such as members of underrepresented groups that could be disproportionately impacted by AI systems, is one successful tactic. Developers may recognize possible biases and address them early in the design phase by considering various viewpoints. Including community members in the development of an AI system for predictive policing, for instance, can help guarantee that the algorithm does not perpetuate preexisting prejudices in law enforcement procedures (Stahl, 2021). The idea of "fairness by design" is another crucial component of moral AI design. This method highlights that fairness should not be an afterthought, but rather considered early on in the AI development process. Developers may guarantee that AI systems function fairly across various demographic groups by putting strategies like algorithmic adjustments into practice. In the event that an AI model utilized for healthcare recommendations exhibits bias towards specific patient demographics, developers may employ strategies such as adjusting the decision thresholds or re-weighting the training data to get more equitable results. Organizations may develop AI systems that are consistent with societal ideals of justice and equity by including fairness into the design process (Gaviria, 2022). Furthermore, constant observation and assessment are necessary for the continual development of AI technology. Our conception of justice and fairness in the context of artificial intelligence must change along with society standards. Organizations must remain up to date on new ethical issues and be prepared to modify their processes accordingly. For example, some firms have started to adopt "bias bounties," which incentivize academics to find and fix prejudice in AI systems, as awareness of algorithmic bias has risen. This cooperative strategy encourages continuous advancements in algorithmic fairness by fostering a culture of innovation and responsibility (Marino et al., 2024).

Thus, as AI technologies develop, it is critical to concentrate on the nexus between algorithmic fairness and social justice. It is essential to comprehend and reduce bias in AI systems in order to provide fair results for every person. Organizations may cultivate a culture of ethical responsibility in AI development and establish trust with the communities they serve by placing a high priority on accountability and transparency. Adopting moral design guidelines that consider many viewpoints, and continual assessment is another way to guarantee that AI innovations advance equality and justice in society. It is critical that we don't let up in our attempts to develop systems that respect the ideals of justice and fairness for all while we negotiate the difficulties of artificial intelligence (Gaviria, 2022).

THE FUTURE OF WORK: BALANCING AUTOMATION AND HUMAN DIGNITY

The current nature of work has continued to change at great pace due to the progress in AI and automation trends. It would be impossible to not see that there is an effect between AI and employment and the job markets now present in industries like manufacturing and health care. In that sense AI is a complex of opportunities regarding innovations, improvement of efficiency and productivity it introduces great concerns about job deletion, as well as the ethical issue of implementing labour automation. With increasing versatility of its application, it is pertinent to come to terms with this shift thus safeguarding human dignity as the world goes increasingly Agilities (Breczko et al., 2021). The subject of employment and job markets as influenced by AI forms one of the main points of discussion on the future of work. AI has already started influencing industries through automating most of the tasks that were initially handled by people. Robotic systems have been developed in ways that they can perform repetitive tasks in production lines than human operators. These changes have raised worry regarding employment loss, primarily to those employees in low skill or routine occupations that are most sensitive to automation. In retail, for example, smart checkouts and rationally controlled trolleys that constantly videos shelves for misplaced products are manifestations of AI that take the place of human beings and when such innovations become widespread, pose a threat to jobless citizens (Marino et al., 2024).

Atomisation through the advancement on AI driven tools can make workers redundant, but at the same time it also opens new forms of employment. The implementation of AI-system may lead to the fact that workers will be occupied with more interesting, challenging and requiring more emotional activity such as decision-making work. For instance, in the healthcare industry numerous administrative problems are solved with the help of AI leaving more time to doctors and other medical personnel. In customer service, simple questions are answered by intelligent and efficient chatbots while agents deal with what the bots cannot. This shift gives a meaning of a change in the kind of jobs instead of eliminating everyone's job as it is common believe, new jobs are likely to be created in industries revolving around the development, maintenance and regulation of artificial intelligence (Brusseau, 2023). However, the confrontation with the AI workforce increases the roles and contradictions of ethical issues. One of the major issues for concern is the possible increase in income disparity this will result since this is a move that will lock out those in lower paid or less skilled employment positions. Holders of high-skill employment – especially in the knowledge-intensive areas including technology, data, and programming – stand to gain in this new reality while holders of low skills, particularly in conventional nondigital fields, will struggle. For example, employees in the transport sector such as truck drivers and factory employees and the sales employees such as retail workers can find it tough to adapt to new positions without formal training and education. To overcome such impacts there is the need for governments, educational institutions and organizations to employ training solutions for existing workers to introduce them into the new market situation (Breczko et al., 2021).

A second ethical concern is the impact on human dignity due to overuse of automation. When tasks are delegated to machines, then the value that people give to work is at risk of being minimized to only mechanical value. Many employees, for example, in the logistics industry may spend most of their time responding to machines rather than dealing with people, and thus lose the meaning of their job. In order to prevent this, it should be important to find a reasonable middle ground to implement smart services into work environment with a due type of human interaction and participation. AI can and should be understood as the way to expand human potential, and not as the means to replace interaction with people. Prioritizing tasks that would need empathy, creativity, and critical thinking—it is possible to retain the humanity in work despite the ongoing automation policy. Reimagining of work with AI is an apparent chance to keep dignified the roles of a workforce by redesigning expectations and expectations with focusing on strengths of human. The foremost advantage of AI technology is data processing where it is capable of transforming human experience in data massaging and high calculations by handling them. This further permits the workers to dedicate their effort strictly on those creative, analytical, and thinking tasks that demand essential skills such as thinking or emotional processing skills. For example, with reference to the creative worker industries human beings are able to come up with ideas and AI can help in other issues like analysis of certain trends and also in offering first drafts of certain designs and other related work, but the final decisions and enhancements should be taken by the creative workers. Therefore, AI does not eliminate creativity from the process but pins it to work with it (Breczko et al., 2021).

Continuously, technology enhances the experience of AI in the enhancement of systems in the provision of safety and the wellbeing of workers. For example, in construction, mining, and production industries, AI robots or drones can be used for tasks that involve risks of the work accidents. This not only leads to safer work environment but also provide protection to workers by protecting them from unfavourable and dangerous situations. Thus, AI will allow the workers to carry out the riskiest parts of these occupations in order to concentrate on performance of such functions as monitoring, supervision, and decision-making, which will increase the feelings of the workers' importance and responsibilities (Brusseau, 2023). Nevertheless, apart from these benefits, the application of AI at the workplace requires a careful consideration with due attention being paid to the substantive rights of the workers or the potential end users of AI. Perfect integration can be attained through cooperative artificial intelligence which focuses on the use of integrated, joint and harmonized intelligence where man and machine are integrated. For example, in medicine AI can also be used to help doctors to look through large amount of medical information and give potential diagnosis, but the final decision will always be made by the man doctor. This approach makes it possible for workers to harness the value affordances of AI, while at the same time preserving their feelings of control, as well as their roles within their various professions. With the idea of the synthesis of human and artificial intelligence, the future of work can be redesigned and promoted in accordance with human rights and improved work quality (Breczko et al., 2021).

Another critical factor of automation is that you must call it the right thing to ensure workers are treated fairly in an economy that is dominated by artificial intelligence. But as the automation gains ground in the economies the necessary requirement should be sought to avoid leaving the workers behind. Employers and governments have a duty in creating labour market policies that give employees equal wages, steady work, and social security despite the mobile nature of work. For example, if due to automation, new fulltime traditional jobs or employment are done away with, then other forms of employment like gig work or freelance, part-time employment or contract employment may become widespread in the market. In such a context, it is important that the non-standard workers are provided adequate workplace benefits including health insurance, retirement and unemployment insurance, which are usually connected with the standard employment (Brusseau, 2023). Organisations such as workers unions will also play a crucial role determine how the future of work with AI will look like. Such organizations can have a crucial role in bargaining reasonable working conditions, demanding that employees be heard on the use of the AI technologies. For example, some unions might demand that organisations allow the people being governed by AI to know how the system is making decisions about them, such as how one was ranked on a performance or merit scale. These groups can therefore play a role in asserting fairness within the workplace by encouraging worker input in how AI implementation decisions will be made (Breczko et al., 2021).

Last but not the least; education, and particularly lifelong learning will also remain a fundamental part of remunerative human dignity with the help of which, mankind can prepare to continue to face the challenges posed by an ever-rolling advance in the domain of AI. With the increase of demands for coding abilities, data analysis, and understanding artificial intelligence, workers need to be supplied with educative tools that can make them succeed in the new world of work. Equally important, education needs to prepare learners for soft skill, which is hard for AI to mimic, and ever relevant in the labour market like communication skills, teamwork, and understanding of other people's emotions. Thus, the relationship between AI and work in the future form has become a great question with opportunities and risks. Many people claim that automation is likely to replace certain jobs, but, at the same time, it opens new opportunities for employees in terms of concentrating on those activities that are more creative, imaginative, and intricately invulnerable. To make sure that we have the requisite number of people working while protecting human dignity in this new era of job displacement, low wages, and dehumanizing work, ethical concerns need to be well handled. Thus, the society can overcome AI opportunistic effect and build a fair future of work as long as new technologies will foster collaboration between humans and AI, support fair labour practices, invest in education and lifelong learning (Brusseau, 2023).

GLOBAL GOVERNANCE AND ETHICAL AI

Another is artificial intelligence (AI) is becoming a robust force; it is changing the relations between nations hence the international relations. It impacts diplomatic as well as security policies, and even warfare not just limited to technology. Given the fact that the AI is setting up its own specialty in these fields, the world wonders how AI should be regulated. The role of AI is growing more and more important for the international relations; many countries apply AI system in such operations as decision-making, data, and cyber operations. Furthermore, mine and counter-mine detection employing AI technology in national security and military affairs has other larger and more serious ethical issues of enabling deployment of lethal autonomous weapons and pervasive surveillance technologies. The groups seeking to avoid or minimize the risk of misuse of the technology propose that there ought to be set universal ethical standards on how the Artificial intelligence systems are governed to foster fairness, stability or avert conflict in the future (Ward, 2024). This interesting development comes at a time when most governments are warming up to the power of AI in facilitating diplomatic strategies. The use of AI in this capability will pertinent a tremendous of data within the shortest time possible, thereby helping the diplomats to arrive at some informed decisions based on intelligence. For example, using political indexes, economical situations and social networks it is possible to forecast diplomatic tensions or cooperation prospects. Information systems can also be used to support negotiations since AI technologies translate messages in real-time while diplomats from different parts of the world engage in talks. One of this is the use of AI to be involved in the peace making process by passing on experience of previous negotiations and recommending effective approaches (Graef et al., 2023). But it also came with ethical issues because the use of AI in international diplomacy is on the rise. This one could be seen as that while AI is extremely effective, it does not possess the ability to make proper judgment calls concerning even political and cultural endeavours. Bargaining skills are developed considering historical relations, the emotional aspect, and the awareness of cultural difference – all of which AI can still lack. For example, an AI system might recommend the economic advantage of a particular course of action in a trade conflict even if they consider the political impact or the effects it will have on the countries involved in the conflict. This limitation of the AI models has some implications of whether the AI should be the chair in diplomatic decisions, or if the final decision rests in the hands of AI but supported with options by AI only (Ward, 2024).

It is especially more ethical concern to incorporate the use of AI in the national security, and especially in war. Modern unmanned systems leaned on artificial intelligence, which can lead to the creation of systems where hardware primarily controls and makes choices for operational missions without human input. Every one of these weapons systems is run by AI capabilities that can determine targets, initiate an attack, and change strategies depending on the environment on the contemporary theatre of war. Although this technology may minimize soldiers' losses by avoiding direct exposure of the combatants, the technology has numerous ethical concerns. The biggest issue related to AI use in warfare is whether or not machines should be able to kill humans (Graef et al., 2023). An example of AI application in a country's defence mechanisms is the creation of precision assaults 'eye in the sky' drones. Such drones are capable of functioning autonomously with little or no supervision by a human operator and relying instead on the data fed to them such as heat indication or motion. However, such emergence and usage of technology will call into question accountability of such practice and possibilities of blunders. If an unmanned system fails to distinguish a non-combatant, what happens? Who is at fault—the developers behind the applications, the military operators involved or is it the actual AI system? Such questions call for precisely formulated ethical standards and measures of responsibility in the usage of AI in warfare, already today (Ward, 2024).

The next area I will discuss were AI influences national security concerns surveillance. Autonomous surveillance technologies consist of intelligence systems implemented by governments in order to monitor areas, people and even crime rates. Though these technologies may help increase safety and security by minimizing crimes and risks resulting from crime, these technologies raise deepest concern on privacy and civil liberties. For instance, China is currently applying intelligent facial recognition to mass surveillance an issue that is violating people's rights to privacy. These issues include monitoring and potentially controlling people's movements and activity and predicting these same people's actions in accordance with certain patterns and raises the obvious questions of where national security meets the liberties and rights of the individual (Murgia, 2024). Lack of homogeneity of views on the part of countries in relation to AI as a factor of national security is one of the key obstacles to the formation of a unified global ethics for the regulation of artificial intelligence. While some may wish to see AI developed for military use, there may be others who wish to see more rigid control placed on use of the technology in the wrong way. It is a serious and commendable endeavour if there is unanimous agreement among governments, both locally and internationally, international organizations, and the business community to agree on ethical standards (Ward, 2024). Having regarded these views, it is imperative to build the global ethical frameworks in order to govern the AI around the visibility, crossing accountability together with fairness of AI. One way is to sign treaties much like the Geneva Conventions are treaties and agreements that set standards of behaviour during warfare. They could provide a prescription of how AI will be used in war, ban on creation of autonomous weapons and guarantee that the AI technology used in wars will not be destructive to the civilians. Such arrangements could be assisted by International institutions like United Nations in arrangement of as well as supervision of implementation of these pacts.

It is another vital area of innovation that require transparency of AI around the world. Everyone requires the companies and governments to be clearer on exactly how the AI systems are trained, what the data set included, it may even contain biases. For instance, when a government deploys an AI system that performs threat recognition with data inputs then the system should show how it defines threat and whether it is capable of profiling people based on their ethnicity. Such openness is necessary to avoid prejudice and to establish that artificial intelligence is used appropriately (Ward, 2024). Accountability is one more crucial aspect of ethical AI Governance responsibility. The paper further posits that there have to be observable procedures for ensuring that the culpable of the actions of AI are answerable. For instance, if an autonomous weapon ends up harming civilian, then there should be a mechanism of identifying who to blame, is it the military using the system, the people who designed it, or the politicians who allowed its use. With no oversight, AI systems could be applied irresponsibly, and people won't be able to be held liable for the resulting consequences (Qorbani, 2020). It would be also worthwhile to mention the role of the private sector in the global management of AI. Quite a number of the emerging AI technologies are the discoveries of private firms, most of which cannot be held to the conventional regulatory regimes of national security. Multinational corporations such as Google, Microsoft corporation, and Amazon are among some of the biggest investors in the field AI and their technologies are being employed for several reasons by governments: for instance, national security. Corporations applying AI must use ethical measures when coming up with the AI and must clarify concerning the usage of such technologies (Qorbani, 2020).

More of such efforts include The European Union's Artificial Intelligence Act, which seeks to regulate the use of high-risk AI applications and the OECD/IFCPAR Principles on Artificial Intelligence which seek to guide the development of AI. They can also give structures of how exactly AI should be designed and implemented and stress on the accountable and transparent approaches. But getting global cooperation on such subjects is difficult also, especially with the geopolitical tensions between countries such as the USA and China that are spending a lot of money to employ artificial intelligence for defense goals. Thus, AI's impact in the international relations, national security as well as war is ethical problematic that pursues global governance solutions (Ward, 2024). AI in diplomacy improves decision making as well as communication since these are a few of the significant executive applications of artificial intelligence to diplomacy but care has to be taken in employing artificial intelligence especially without the human component considering cultural differences. In defense, AI Applied Systems making autonomous weapons or surveillance systems are some of the hardest questions of ethics in terms of accountability, privacy as well as measurement of human value. Joint effective international cooperation with governments, intergovernmental organizations, the business community, and civil society will be needed for setting and implementing new ethic for AI governance and its utilization that will have positive impact on human and state security, and will guarantee freedom, equality and human dignity of people all around the world (Murgia, 2024).

CONCLUSION

As the utilization of AI increases in nearly every sector in the world there is no doubt that the things it brings with it in terms of ethical dilemmas do not conform to one particular field of study. AI ethics cannot be the job of technologist, nor can it be the job of policymakers; rather, it needs to be a multi-disciplinary humanitarian endeavour with people from fields including philosophy, law, sociology and computer science all working proactively to cater for the greater good. This is where interdisciplinary methods are called for since only cross-disciplinary approaches supply the diversification of viewpoints and backgrounds necessary for tractability of the ethical issues that AI presents. For instance, while using algorithms and AI models, engineers will be more concerned with its efficiency in terms of speed and many other factors, ethicists can ensure that the used AI models are fitting to the society's norms and human rights. That is why I believe that important lessons for building responsible AI system could be learned through interdisciplinary cooperation to develop comprehensive frameworks for implementation of AI which was the case of this paper.

This concept aims at the creation of the future where artificial intelligence complements human value and promotes human dignity to the maximum. To this end, we have to guarantee that an AI is not only strong and fast but ethical as well as maintains the sanctity of human life. This means developing AI technologies which remain compliant with user's privacy, free from bias, and give the freedom of choice. For instance, AI systems in healthcare could be created to improve the treatment of patients and protect people's privacy at the same time so that medical choices continue to be made by doctors and patients, not by the system itself. Likewise, AI applied in education can be developed such that it promotes student agency, even as it grants the students customized learning mechanisms. Being an assistant, but also a critical co-worker in the decision-making process, AI should be purposed and operated in such a way that deepens rather than erodes the key components of our culture (Murgia, 2024).

AI brings significant ethical concerns as well incredible opportunities, which consequently serve as significant potential to influence the shape of the future for the better. However, besides the concerns that result from privacy infringement, bias, and the use of artificial intelligence in confuse warfare there is much hope as AI can help address major world issues. For example, AI can help reverse the effects of climate change by modelling how energy should be efficiently consumed, medical health can be enhanced by providing accurate data of diseases, and international relation can be enhanced through efficient translation of language barriers. The suggestion here is that by paying attention to the ethical factors that undergird the use of this technology, one can be productive in its application while at the same reducing the potential harm of the technology in equal measure.

All in all, the question of practical means of making acquisitions of ethical AI is a continuous discussion, international collaboration, and acting in the protection of human worth. There is no fixed path set for the future of AI, but the course that we are going to take in future depends upon our decisions in present. Ensuring ethical considerations are placed at the centre of development of artificial intelligence mush create a future in which technology serves humans and the total good, centrally fairness and dignity of the person. At the same time, concern with the ethical transformation of AI is the concern that requires people to be as careful as they are creative with the ways in which these technologies are developed and deployed so that they to become the means by which humanity is empowered to bring the best within itself into the world (Murgia, 2024).

REFERENCES

Arciniegas Gome, M. (2023). *Reframing privacy in the digital age: The shift to data use control* (Doctoral dissertation, Carnegie Mellon University).

Askin, J. (2024). From data co-opting to data co-oping: Using new corporate structures, data corporate-hood, and data personhood to prioritize data privacy. *UC Law Business Journal, 20*(2), 177.

Breczko, A., Filipkowski, W., & Kraśnicka, I. (2021). Ethics and the development of artificial intelligence: Challenges and dilemmas in the context of the 2030 United Nations agenda for sustainable development. In *Crime Prevention and Justice in 2030* (pp. 355–380). The UN and the Universal Declaration of Human Rights. DOI: 10.1007/978-3-030-56227-4_17

Brusseau, J. (2023). AI human impact: Toward a model for ethical investing in AI-intensive companies. *Journal of Sustainable Finance & Investment, 13*(2), 1030–1057. DOI: 10.1080/20430795.2021.1874212

Chen, A., Hannon, O., Koegel, S., & Ciriello, R. (2024, March). Feels like empathy: How "emotional" AI challenges human essence. In *Australasian Conference on Information Systems*.

Cheong, P. H. (2020). Rethinking piety, practices, and pedagogy in the era of artificial intelligence. In *Religion in the Age of Digitalization: From New Media to Spiritual Machines* (pp. 75-89). Routledge.

Chesterman, S. (2020). Artificial intelligence and the problem of autonomy. *Notre Dame Journal on Emerging Technologies, 1*, 210.

de Oliveira Rodrigues, D. E. F. (2024). Personhood global whispers: Ethical echoes of decentralization? *Procedia Computer Science, 237*, 767–774. DOI: 10.1016/j.procs.2024.05.164

Elliott, A. (2021). *Making sense of AI: Our algorithmic world*. John Wiley & Sons.

Estrada, D. (2020). Human supremacy as posthuman risk. *The Journal of Sociotechnical Critique, 1*(1), 5.

Gaviria, C. I. G. (2022). The role of artificial intelligence in pushing the boundaries of US regulation: A systematic review. *Santa Clara High-Technology Law Journal, 38*, 123.

Graef, I., Petročnik, T., & Tombal, T. (2023). Conceptualizing autonomy in an era of collective data processing: From theory to practice. *Digital Society : Ethics, Socio-Legal and Governance of Digital Technology*, *2*(2), 19. DOI: 10.1007/s44206-023-00045-3

Hansen, A. S. (2022). *Relationships, religion, and robotics: The soul and the ethical implications of AI* (Doctoral dissertation).

Harasimiuk, D., & Braun, T. (2021). *Regulating artificial intelligence: Binary ethics and the law*. Routledge. DOI: 10.4324/9781003134725

Herzfeld, N. L. (2023). *The artifice of intelligence: Divine and human relationship in a robotic age*. Augsburg Fortress Publishers.

Horváth, I. (2022). AI in interpreting: Ethical considerations. *Across Languages and Cultures*, *23*(1), 1–13. DOI: 10.1556/084.2022.00108

Krysa, J., & Impett, L. (2022). *Rethinking curating in the age of artificial intelligence*. Art Books Publishing Ltd.

Lovell, J. (2023). *Legal aspects of artificial intelligence personhood: Exploring the possibility of granting legal personhood to advanced AI systems and the implications for liability, rights, and responsibilities*. Rights and Responsibilities.

Lovell, J. J. (2024).. . *International Journal of Artificial Intelligence and Machine Learning*.

Lyreskog, D. M., Zohny, H., Savulescu, J., & Singh, I. (2023). Merging minds: The conceptual and ethical impacts of emerging technologies for collective minds. *Neuroethics*, *16*(1), 12. DOI: 10.1007/s12152-023-09516-3 PMID: 37009261

Marino, D., Cananzi, D., & Aragona, F. (2024). Artificial intelligence and ethics challenges of technologies. In *Ethics and Artificial Intelligence: Towards a Moral Technology* (pp. 41–55). Springer Nature Switzerland. DOI: 10.1007/978-3-031-50902-5_3

Milic-Frayling, N. (2023). On the cusp: Computing thrills and perils and professional awakening. *Proceedings of the VLDB Endowment International Conference on Very Large Data Bases*, *16*(12), 4152–4159. DOI: 10.14778/3611540.3611640

Murgia, M. (2024). *Code dependent: Living in the shadow of AI*. Henry Holt and Company.

Nersessian, D., & Mancha, R. (2020). From automation to autonomy: Legal and ethical responsibility gaps in artificial intelligence innovation. *Michigan Technology Law Review*, *27*, 55.

Nourbakhsh, I. R., & Keating, J. (2020). *AI and humanity*. MIT Press.

Qorbani, M. (2020). *Humanity in the age of AI: How to thrive in a post-human world*. Bloomsbury.

Stahl, B. C. (2021). *Artificial intelligence for a better future: An ecosystem perspective on the ethics of AI and emerging digital technologies*. Springer Nature. DOI: 10.1007/978-3-030-69978-9

Tsakalogianni, I. (2023). *The twin transition: Mapping the use of artificial intelligence in sustainability—the role of law and ethics* (Master's thesis, Πανεπιστήμιο Πειραιώς).

Ward, H. (2024). *Technologies of the self: Diachronic privacy and self-presentational autonomy* (Doctoral dissertation, UCL).

Wong, A. (2021). Ethics and regulation of artificial intelligence. In *Artificial Intelligence for Knowledge Management: 8th IFIP WG 12.6 International Workshop, AI4KM 2021* (pp. 1-18). Springer International Publishing. DOI: 10.1007/978-3-030-80847-1_1

ADDITIONAL READING

Tariq, M. U. (2025). Unlocking Competitive Edge: Harnessing the Power of Collaborative Innovation. In Martínez-Falcó, J., Sánchez-García, E., & Marco-Lajara, B. (Eds.), *Open Innovation Strategies for Effective Competitive Advantage* (pp. 315–344). IGI Global Scientific Publishing., https://doi.org/10.4018/979-8-3693-9266-9.ch011

Tariq, M. U. (2025). Empowering Patients With AI-Ethical Digital Tools for Health Data Management and Decision. In Pesqueira, A., & de Bem Machado, A. (Eds.), *Navigating Privacy, Innovation, and Patient Empowerment Through Ethical Healthcare Technology* (pp. 197–216). IGI Global Scientific Publishing., https://doi.org/10.4018/979-8-3693-7630-0.ch008

Tariq, M. U. (2025). Leveraging Libraries and Information Centers for Startup Innovation and Growth. In Gupta, V., & Gupta, C. (Eds.), *Navigating Strategic Partnerships for Sustainable Startup Growth* (pp. 1–28). IGI Global Scientific Publishing., https://doi.org/10.4018/979-8-3693-4066-0.ch001

Tariq, M. U. (2025). Revolutionizing Retail: Enhancing Customer Experiences With the Metaverse. In Gonçalves Rodrigues, M., Carvalho, M., & Monteiro Pratas, J. (Eds.), *Cases on Metaverse and Consumer Experiences* (pp. 175–202). IGI Global Scientific Publishing., https://doi.org/10.4018/979-8-3693-7245-6.ch007

Tariq, M. U. (2025). AI-Driven Research Methodologies: Revolutionizing Data-Driven Discoveries in Engineering and Physical Sciences. In Basha, J., Alade, T., Al Khazimi, M., Vasudevan, R., & Khan, J. (Eds.), *Optimizing Research Techniques and Learning Strategies With Digital Technologies* (pp. 97–122). IGI Global Scientific Publishing., https://doi.org/10.4018/979-8-3693-7863-2.ch004

KEY TERMS AND DEFINITIONS

AI Ethics: A set of moral principles and guidelines that govern the design, development, and deployment of AI systems to ensure fairness, transparency, and accountability.

Privacy in AI: The protection of personal data and information from unauthorised access, usage, or surveillance in the context of AI-driven technologies.

Personhood: A concept that explores whether advanced AI systems could possess moral or legal standing, shaping discussions on rights, accountability, and autonomy.

Autonomy in AI: The extent to which AI systems can make decisions independently, raising ethical questions about human oversight, control, and the preservation of individual freedom.

Chapter 4
Evolution of Human Rights Through Literature and Law:
Role of AI in Its Exploration in Modern Ages

Shrabana Chattopadhyay
https://orcid.org/0000-0001-7648-6535
University of Engineering and Management, Kolkata, India

Ria Roy Choudhury
https://orcid.org/0009-0009-3487-7461
University of Engineering and Management, Kolkata, India

Prithwijit Chakraborty
University of Engineering and Management, Kolkata, India

ABSTRACT

AI has touched human rights through literature and law, providing modern society both possibilities and challenges. AI has major human rights issues as it spreads. AI in human rights literature and law shows the necessity for flexible legal frameworks to accommodate technology's difficulties. This conversation promotes ethical AI development and proactive human rights protection in an AI-driven world. This paper examines literature and law on human rights and AI's impact on society. It also defines human rights and their value, then trace their history in literature and law. We examine AI and human rights, including its difficulties and prospects for protecting individual rights. Current AI governance legal frameworks have holes and inconsistencies that require legislative revisions. The paper advises govern-

DOI: 10.4018/979-8-3693-9035-1.ch004

ments, technologists, and human rights advocates to integrate human rights into AI research. Finally, it discusses how literature influences public image and AI ethics, promoting communication to protect human rights.

I. INTRODUCTION

Human rights are the inalienable rights that have been accorded to every person by virtue of being human with humanity as the base of modern society civil justice. Human rights definition is closely connected with human dignity that is reflected both in International Covenant on Economic, Social and Cultural Rights and International Covenant on Civil and Political Rights. These documents state that human rights stem from the worth and sanctity of the human person and pointed out that human rights cannot be implemented without responding to the dignity of the human person. The definition of human rights is deeply intertwined with the concept of human dignity, which is recognized in various international legal frameworks, including the International Covenant on Economic, Social and Cultural Rights and the International Covenant on Civil and Political Rights. These documents assert that human rights derive from the inherent dignity of the human person, emphasizing that respect for dignity is paramount in understanding and applying human rights (Tarasova et al., 2020).

Human rights are not limited to the legal systems of the modern society; they play a major role in the resolution of the conflicts of current concerns including human trafficking, the degradation of the environment, and inequality. (Tangen, 2020). This legal recognition is very significant in fighting the modern slavery and slavery like practices and scholars have established how human rights impacts on public policies for the vulnerable groups.

When speaking about the human rights, the framework now is much wider to cover many ethical issues that go with many more global problems, for instance, the environmental questions and social questions (Twiss, 2011). This intersectionality is important as it provides a deeper appreciation of human rights, with reference to economic, social and environmental domains to enhance inclusiveness of global governance (Kotzé, 2014).

Human rights also help in the accountability especially when dealing with state and non-state actors. (Clapham, 2006). Human rights are incorporated into social work education in the practice of declaring and defending human rights as a principle in social justice (Henry, 2015). This paper will show how, due to their roles as advocates for the vulnerable, human rights knowledge is fundamental to the social work framework, and that social workers who do not receive such formal education must still ensure they are knowledgeable about them. It also significantly disciplines social workers to be educated while at the same putting emphasis a general civic responsibility to respect human rights.

Therefore, human rights, both as the concept defining human beings' rights in modern society and as an actual-element of the modern social reality, are multifaceted, legal-ethical-social. The affirmation of human rights and freedoms as inherent in human being underlines and the principle of their realization irrespective of state and non-state actors.

Bringing together information derived from literature and legal systems present a historical, cultural and politico legal progression that defines the modern concept of human rights. Modern human right was established in the post-World War II period especially after adoption of Universal declaration of Human Rights in 1948. This document set a watershed in the making of human rights where the treatment of individuals in a civilized society, no matter the color of their skin or country of origin, religion, or political status, was formalized (Chilton & Posner, 2016).

Historically, literature has been employed as practical tools in creating awareness of human right abuses and in the process put pressure on those in power. (Bayisa, 2023). The two come closer where authors come out clearly to express their concern towards injustice and advocate for change. (Dawes, 2009).

Aside from literature in human rights, there has been developments on legal frameworks too. Key developments in human rights regime include the two major covenants; The International Covenant on Civil and Political Rights and The International Covenant on Economic, Social and Cultural Rights both of which were signed in 1966. They also provided more definition concerning the subject area of human rights and the duties of the states in the protection of rights (Roberts, 2014).

The formation of human rights has also been informed by social impulses related to gender rights activism, anti-colonialism. For instance, the feminist movement has shifted the focus of human rights through the introduction of women's' rights as part of the human rights approaches (Donert, 2023). The mid twentieth century saw colonized people also repelling their colonizers and also using human rights discourse to give voice to fight, for liberty and freedom, thus showing that human rights and political freedom go hand in hand (Ibhawoh, 2014).

The relationships between these aspects have not only defined what is modern understanding of human rights but also continued activism and commitment to global promotion of human rights. This backup points that as the society advances and faces new issues for example, climate change and technology advancement the debate relating to human rights will similarly change and a new lens has to be developed and taken through the existing lens of including everyone in the rights conversation.

AI has brought about major changes in the many areas of life, and the same is true of human rights. AI and Human Rights – important themes of privacy, discrimination, and how answers to questions of accountability and human dignity in the future world of automation and artificial intelligence are to be found. The use of AI may assist with the detection of examples of human rights violation and allow for quicker responses from officials, and NGOs. This capability of making AI to work for the desired effects is in line with the human rights principles where the demand for information and people's capability are upheld.

But at the same time, the use of artificial intelligence entails considerable threats to human rights. A very valid issue within the concept of AI is that of lack of consciousness, and specifically discrimination that AI systems might contain. This means that algorithms that are built from past data may reinforce the current imbalances meaning more prejudice in places like the police force, employment and social services.

Moreover, with existing use of applications of AI in the surveillance technology it has created concerns over the rights of privacy. This is especially the case with subverting the security/privilege binary: what must remain a complex concern for civil liberties and human rights in the face of AI technologies is the question of how such technologies are mobilized by state actors.

Similarly, human rights have some influence on how the--accountability and what is explained to be legitimate by AI. AI is uniquely poised in human rights discussions for both positive engagement and risks that need to be well managed. Above all, AI has benefits in strengthening rights-for human beings and in individual participation but also threats regarding discriminations, personal and privacy rights, accountability.

II. OBJECTIVE OF THE STUDY

In this article, the author has aimed to analyse the relationship between human rights, literature and a constantly changing field of artificial intelligence. In essence, this article aims to offer a historical figure of the causal development of human rights, taking into account the importance of short and literary fictional works as ways of influencing and informing society, as well as forming legal framework and presenting brief reflection of what artificial intelligence means for human rights.

It starts with a brief history of human rights starting with UDHR before closely looking at how human rights have been presented and advocated for through literature (Leite, 2023). The article will explore the analysis of the narratives of other literature that have depicted injustices and human right violations to raise social awareness to these concerns (Bayisa, 2023).

Besides, the article will also offer a critical analysis in regard to the impacts of AI on the human rights such as privacy, right to free speech, and rights to assembly as well as association. The discussion of the topic will also ponder over the ethics of using AI, especially when recommending for the need to embrace a rights-oriented approach to adopting technologies that is anchored on the UDHR (Cath & Floridi, 2016). It will insist on the possibility of integrating converging approaches to manage the new contexts and the emergence of artificial intelligence on human rights as they arise to meet the opening human rights evolution in the past and the future (Boncheva & Serrano, 2022).

III. HISTORICAL CONTEXT OF HUMAN RIGHTS

In particular, literature has always been effective in raising awareness of different injustice cases and pressing for the acknowledgement of human rights. Those literary stories have been useful in raising awareness of the human rights violation incidents and eliciting positive feelings among the readers. In his article, Bayisa uses Amelmal's "Yältäkoactice Guzo" to show how literature can also be used as an advocacy tool informing society on human rights violations and resulting in advocacy involvement (Bayisa, 2023).

Literature is not only about representation of human rights it also includes an analysis of the ethical concerns about telling stories. Dawes also provides a good account of the problematics of voice in the context of speaking on behalf of the suffering whose speaking may both save and take away their voice (Dawes, 2009).

According to the psychological insight to human rights attitudes provided by Cohrs et al., present study reveal that literature can indeed shape people's perceptive and behavioral patterns towards human rights (Cohrs et al., 2007). As will be argued in this particular paper, human rights advocacy and activism stand to greatly benefit from literature, given that human emotions and ethics can be motivated through personal narrative. With literature we have seen it become an essential tool in campaigning for human rights, promoting understanding and fighting for the change of the social injustices all over the world as it continues to develop.

- **Watershed Moments in the Evolution of Human Rights Legislation**

There are several legal innovations which are achievements in the field of human rights for people's acknowledgement of human values, freedom and justice. The primary and one of the most important achievements in the formation of an individual's liberties was the signing of the "Universal Declaration of Human Rights" (UDHR) in 1948 by the United Nations. After UDHR was came out, more international human rights conventions and covenants were ratified adding more specification in rights and responsibilities for nations. The "International Covenant on Civil and Political Rights" (ICCPR) and the "International Covenant on Economic, Social and Cultural Rights" (ICESCR) which were adopted in 1966 broaden the concept of human rights compared with the UDHR.

Another important achievement also included the creation of regional human rights systems, like the "European Convention on Human Rights" (ECHR) in 1950. In the same way "American Convention on Human Rights" as well as the "African Charter on Human and Peoples' Rights" have provided the regional systems for the protection of the human rights and it has adapted to that culture and laws of the Americas as well as for the Africa respectively.

The "Convention on the Rights of Persons with Disabilities" (CRPD), in 2006, has been the next step to work for people with disabilities' rights and their protection. This brought about the legal recognition of yet a new fundamental individual right known as the right to development ultimately leading to the adoption of the Declaration on the Right to Development in 1986.

However, these are areas of human rights that still have numerous obstacles regarding the organized promotion and protection of human rights. Because there is quite often a difference between international norms and local realities the latter results in violations especially if it is an authoritarian state or a country with an ongoing war.

IV. THE INTERSECTION OF AI AND HUMAN RIGHTS

The role of AI technologies in the development of human rights is therefore not a simple one as the technologies are already widely part of society.

1. The Role of Artificial Intelligence in Human Rights Analysis

AI is broken down into a set of technologies that help in the replication of the human brain when solving a particular task (Hunkenschroer & Kriebitz, 2022). In its observation, Chatterjee et al. state that the use of AI technologies enhances human rights abuses due to the valorisation of AI systems (Chatterjee et al., 2021).

2. Ethics in Artificial Intelligence Laws

Currently, it is crucial to underline that the creators and developers of AI need a proper legalization of the legal rules concerning those challenges. As it was underlined by Xudaybergenov, preventing approach to acknowledge AI as subject of law with limited rights may offer clarification of liability, ownership and agency questions and therefore to facilitate the handling of the conflict connected with AI

(Xudaybergenov, 2023). The creation of more or less ethical standards for artificial intelligence is necessary.

3. Important Moral Factors to Consider When Creating AI

The use of AI is moral and more so in sensitive sectors like the healthcare industry where such technologies are today applied for diagnoses and cures. (Elendu, 2023). Explainable AI (XAI) is already becoming a popular solution for the problems associated with non-transparency and non-accountability of artificial intelligence (Kuźniacki et al., 2022).

4. The Role of Stakeholders in Promoting Human Rights

The implementation of AI with respect to human rights means that multiple actors – governmental and non-governmental – need to cooperate. Civil society organisations have the role of promoting human rights within the AI framework, exemplifying how such protection might be violated and ensuring that governments and corporations are held to account for their actions ((Pizzi et al., 2020)).

5. Case Analyses: Artificial Intelligence in Application

This section provides diverse examples that will demonstrate how AI is connected to human rights. Systemic bias in the algorithms violates the principles of justice and equity for which human rights act and perpetuates the integration of social injustices (Leite, 2023; Rathod, 2023). AI systems in healthcare environment shall be directed within appropriate ethical frameworks and adequate legal regimes to protect patients' interests and prevent social inequity between patient population (Elendu, 2023).

6. Future Directions: Harmonizing Technology and Human Rights

Thus, policymakers need to pay more attention to legal change to reflect the challenges that AI presents for ensuring that human rights are at the heart of the advancing field (Leite, 2023). Instead of a creation of AI in opposition to human rights principles, the book shows that it is possible to apply AI development and reap the benefits of this new technology while preserving human dignity and freedom (Pizzi et al., 2020).

V. LEGAL FRAMEWORKS ADDRESSING AI AND HUMAN RIGHTS

A. Contemporary Legal Frameworks for AI Governance

- **Analysis of Current Regulations and Guidelines**

AI regulation is still an emerging discipline that intersects with human right thus there is need for understanding of existing legal frameworks governing its use. Different frameworks have been developed all over the world as organizations and individuals are slowly becoming aware of the ethical use of AI technologies.

To undertake this, Huriye calls for more cooperation on the side of the stakeholders in refining various AI ethics guidelines which normally do not have a well-established structure (Huriye, 2023). A similar opinion is shared by Bell et al., According to his article, despite the fact that there are many guides that have been established today, what is still lacking are methods for holding the systems to account and for making them more transparent (Bell et al., 2023).

The OECD has been a key player in this landscape, articulating principles guiding the development of national AI strategies and regulatory frameworks. The report by the OECD on the implementation status of its AI principles cites evidence that countries are at very different stages in developing those frameworks, underlining the necessity of international cooperation in those efforts. ("The state of implementation of the OECD AI Principles four years on", 2023)

One of the emerging issues in AI governance is accountability since different applications form high-risk contexts require better stewardship. Hohma et al. also explain the related problem of defining who is to blame for the harms introduced by AI systems, stressing the need for the sound solutions regulating these problems (Hohma et al., 2023).

The global regulation of artificial intelligence involves the cooperation of other nations because even if a country will be able to regulate artificial intelligence on its own, it will still be affected by what happens in other countries, according to Erdélyi and Goldsmith who call for the formation of an international regulating agency to set up systematic measures (Erdélyi & Goldsmith, 2020).

The European Commission's Ethics Guidelines for Trustworthy AI points out that basic human rights form the core basis for the development of AI technologies (Almeida et al., 2020). This is further supported by the notion that even if the ethical guidelines for AI formulated are non-binding, they form the necessary infrastructure or toolkit for developers and organizations creating responsible AI systems (Ronanki, 2023). It is in this context that governance landscapes for AI are characterized by a myriad of guidelines and frameworks as an effort to address the ethical implications of AI technologies.

- **The Function of International Human Rights Legislation in the Administration of Artificial Intelligence**

Research carried out in this area notes that there is growing concern for ethical concerns in AI regulation. Huriye also notes that there is a need to integrate different goals and objectives of the stakeholders and align them to various AI ethics guidelines since the latter lack a coherent structure themselves (Huriye, 2023).

B. Inconsistencies in the Law

- **Different Jurisdictions' Approaches to AI Regulation**

Issues of the regulation of AI involved a vast number of inconsistent situations across jurisdictions. Such differences stem from differences in the legal systems, the culture, and the technological development of the countries involved and, thereby, producing an uncoordinated system to regulating AI. This fragmentation has problems in international cooperation and better uses and regulation of the risks and boon of artificial intelligence.

It indicates that the lack of a coherent set of rules governing the use of AI is one of the essential reasons for its unpredictability. According to Manap and Abdullah, most jurisdictions are without proper legislations addressing AI thereby leaving them with the existing laws which may not be fit for purpose technologies (Manap & Abdullah, 2020). The case is different with the European Union that is working on having a more specific law to regulate AI known as AI Act where, depending on the level of risk of the AI system, it comes with correlative regulatory measures ("The state of implementation of the OECD AI Principles four years on", 2023). However, as a general direction, the EU work may not be relevant to the jurisdictions with different legal frameworks or exposure to the technology.

Currently, AI as a medical device is regulated in the United States through the Food and Drug Administration have drawn a lot of criticism as being more of a follower rather than a leader when it comes to regulating AI technologies (Monteith et al., 2022). On the other hand, the EU issued the Medical Device Regulation (MDR) with requirement regarding clinical evaluation and post-market surveillance of medical AI, which is relatively more precautionary in comparison with other nations (Niemiec, 2022). From this, there may be variations in the safety and effectiveness of the applied AI- technologies between countries resulting in complications affecting patient safety and the general population's trust. This underscores the need for policies that can be adjusted usually to correspond with changes in technological characteristics (Ubaydullayeva, 2023). However, with different rules in place about data protection, like the GDPR in the EU area and a more disassembled system in the USA, it is difficult for agencies to collaborate internationally (Ubaydullayeva, 2023).

In addition, the question of AI regulation ethical has some rather conflicting interpretations at the international level. The guidelines apparently share principal postures globally, including bluntness, equity, and no-do-harm policies, but valuable differences are available concerning how these principal ideas are implemented by assorted countries' guidelines (Jobin & Ienca, 2019).

Therefore, the differences of AI regulation across jurisdictions are tremendous for governing efficiency and global cooperation. Lack of a common regulatory platform along with having different ethical issues, and data protection policies make this picture less clear for AI development and implementation.

- **The Necessity for Dynamic Legal Frameworks to Accommodate Evolving Technology**

This showed that formal legal systems in many instances lack adequate mechanisms to rapidly adapt to rapidly emerging technological developments and this results in the weakening of the legal gap which poses a threat to public confidence, human rights and ethical considerations. This paper seeks to address the issue of legal flexibility in addressing the emerging technologies since their growth requires appropriate legal approach.

As it has been pointed out by Leite human rights could be either positively or negatively influenced by AI technologies, hence existing legal frameworks requires reconsideration (Leite, 2023). Foundational in nature is the UDHR however legal principles do need to evolve to relate to AI in the necessary specific detail.

Similarly, while applying AI to different segments of an economy, including healthcare, finance, and law enforcement, the nature of the laws governing these areas experiences aspects that existing laws fail to cover. If it emerges that more adaptable legal approaches can help integrate ethical principles into the regulatory processes, the responsible development of AI systems would be possible. The above challenges can be managed by having an adaptive legal framework that encouraged the cooperation of nations and commonality of standards (Zheng, 2023).

Because all cultures are constantly evolving, various stakeholders need to take an active part in the regulation process to create pliable legal acts, as Chatterjee and Sreenivasulu have pointed, involvement of various stakeholders, including technologist, ethicist and civil society can create more nuanced rules that respond to societal norms (Chatterjee & Sreenivasulu, 2021).

C. Considerations for the Future of Law in Light of AI Difficulties

As there is a fast pace advancement in AI, current laws are a challenge to develop legal solutions hence there are calls for legal reforms. In the following, this paper presents a roadmap for legal evolution that can improve the capacity of legal systems in addressing new technologies.

There is one suggestion to create separate regulatory authorities to deal with the problems of creating and applying the AI-related legal frameworks. According to Xudaybergenov, such bodies should have participation of legal scholars, AI creators, ethicists, and others so that a more profound appreciation of the concern is achieved regarding the possible effects of AI technologies (Xudaybergenov, 2023).

To this end, Nweke stresses the need for an interdisciplinary approach with a focus on legal, ethical, as well as technological strategies that ought to be adopted as for dealing with challenges of AI (Nweke, 2024). Such cooperation may assist in guaranteeing that legal standards do more than rely on compliance, but also advance ethical AI use.

The idea of adaptive legal frameworks should apply to liability and accountability as well. As Bartlett pointed out: the emergence of mistakes resulting from AI use generates important questions about legal jurisdiction (Bartlett, 2023). It is unimaginable how powerful AI must be, and the international cooperation is required in AI regulation. As discussed by Zheng, since AI technologies are global the legal systems requiring AI must be harmonized (Zheng, 2023).

Last but not the least, raising purely legal awareness, let alone training, remains essential for the consistent promotion of continuous learning related to AI technologies. According to Babikian, it is important for artists, developers, citizens and everyone involved to understand the legal issues surrounding artificial intelligence (Babikian, 2023).

Therefore, the questions AI raises to the law call for approaching legal changes as a process that should be actively driven. Some actions as getting established regulatory agencies, including ethical factors, searching for novel liability frameworks, developing cross-jurisdictional collaborations, and improving legal awareness are the key actions in development of legal resilient frameworks.

- **Approaches for Including Human Rights Concerns in AI Research and Development**

As there is a fast pace advancement in AI, current laws are a challenge to develop legal solutions hence there are calls for legal reforms. In the following, this paper presents a roadmap for legal evolution that can improve the capacity of legal systems in addressing new technologies.

There is one suggestion to create separate regulatory authorities to deal with the problems of creating and applying the AI-related legal frameworks (Xudaybergenov, 2023). According to Xudaybergenov, such bodies should have participation of legal scholars, AI creators, ethicists, and others so that a more profound appreciation of the concern is achieved regarding the possible effects of AI technologies (Xudaybergenov, 2023).

To this end, Nweke stresses the need for an interdisciplinary approach with a focus on legal, ethical, as well as technological strategies that ought to be adopted as for dealing with challenges of AI (Nweke, O., 2024). Such cooperation may assist in guaranteeing that legal standards do more than rely on compliance, but also advance ethical AI use.

The idea of adaptive legal frameworks should apply to liability and accountability as well. As Bartlett pointed out: the emergence of mistakes resulting from AI use generates important questions about legal jurisdiction (Bartlett, 2023). It is unimaginable how powerful AI must be, and the international cooperation is required in AI regulation. As discussed by Zheng, since AI technologies are global the legal systems requiring AI must be harmonized (Zheng, 2023).

Last but not the least, raising purely legal awareness, let alone training, remains essential for the consistent promotion of continuous learning related to AI technologies. According to Babikian, it is important for artists, developers, citizens and everyone involved to understand the legal issues surrounding artificial intelligence (Babikian, 2023).

Therefore, the questions AI raises to the law call for approaching legal changes as a process that should be actively driven. Some actions as getting established regulatory agencies, including ethical factors, searching for novel liability frameworks, developing cross-jurisdictional collaborations, and improving legal awareness are the key actions in development of legal resilient frameworks.

- **The Significance of Interdisciplinary Cooperation in Legal Reforms**

Multidisciplinary collaboration is indeed pivotal in navigating the complexities of legal reform in the context of evolving technologies like artificial intelligence (AI). By bringing together legal experts, technologists, and ethicists, a more nuanced and holistic understanding of the associated challenges can emerge. This approach not only addresses the legal dimensions but also integrates insights from technical and ethical perspectives, ensuring a more robust framework.

As Xudaybergenov aptly highlights, the establishment of regulatory bodies that include a diverse array of stakeholders is critical. These bodies can bridge the gap between legal structures and the practical implications of technological advancements. When regulatory efforts are informed by a blend of technical realities and ethical imperatives, they are better equipped to anticipate potential risks, balance competing interests, and uphold societal values effectively (Xudaybergenov, 2023). Such collaboration underscores the importance of foresight and inclusivity in shaping legal systems that remain agile and equitable amidst rapid technological change.

Multidisciplinary collaboration also plays a crucial role in facilitating the exchange of best practices and lessons learned across jurisdictions. As Pomaza-Ponomarenko notes, international cooperation in legal reform allows countries to learn from one another's experiences, thereby enhancing the effectiveness and resilience of their own legal systems (Pomaza-Ponomarenko, 2023). This cross-pollination of ideas and strategies can help nations address common challenges and avoid redundant efforts in tackling the complexities introduced by AI technologies.

Moreover, such collaboration helps bridge the gap between legal theory and practical application. Babikian emphasizes the importance of creating adaptable legal frameworks that can evolve alongside technological advancements, ensuring laws remain relevant and effective in dynamic environments (Babikian, 2023). By integrating insights from various disciplines, legal systems can strike a balance between flexibility and stability, enabling them to respond to emerging issues without sacrificing foundational principles.

Finally, fostering a culture of dialogue among stakeholders—spanning technologists, legal professionals, policymakers, and the public—can enhance legal literacy and promote a deeper understanding of the rights and responsibilities associated with AI technologies. This awareness is essential for cultivating trust in legal systems and ensuring that AI-driven innovations are deployed ethically and equitably. In sum, multidisciplinary collaboration is not merely beneficial but essential for shaping legal reforms that can withstand the pace and scope of change in the age of AI.

VI. THE ROLE OF LITERATURE IN SHAPING AI AND HUMAN RIGHTS DISCOURSE

Literature has been a testament to the unflinching desire in humans for justice against the unjust and resistance against exploitation of rights. The same has been institutionalized by law as well. Law and literature have been intertwined since ages with the former displaying human demands for equality and rights, while the latter playing a pivotal role in safeguarding the same. Literature has always been the documented and collective voice of the subdued and the subalterns. It has sparked several revolutions while teaching the subalterns to speak. On the other hand, law has codified and enforced what literature speaks for. However, with the on-going development of technology in contemporary times, there emerges a game-changer called Artificial Intelligence, which not only cooperates with the existing legal frameworks and assists in making them more potent, but, contradictory, also proves to be the primary challenge against safeguarding of human rights and liberty.

Several instances of encounters of literature with the evils ravaging the society, have been witnessed in the works such as Maria by Mary Wollstonecraft or Uncle Tom's Cabin by Harriet Beecher Stowe. The plots of such works have revolved around discriminations against protagonists and other characters on the basis of caste, creed, region, religion and sex. Human voice has apparently been suppressed only in presence of a totalitarian regime, often uniting with the misuse of technology, as evident in modern times. Such fettering of human rights and unchecked growth of technology have been critiqued in the works of George Orwell and Mary Shelley. Literary masterpieces such as *1984* and *Frankenstein*, offer a cautionary lens to view the onset of dystopia and dehumanization with absolute powers vested upon the autocrats and scientific innovations. Literature cultivates an intrinsic demand in humans for justice by placing the reader at a juncture wherefrom she/he is able to empathize and live the realities of the characters. Dawes states rightfully that the art of storytelling enables ethical portrayal of suffering. These portrayals are worth empathizing and are often immortalized in the pages of literary contributions.

If literature is the voice of human conscience, law materializes the ethical demands of the voice and enforces the same. A global consensus on equity and dignity, is reflected in the Universal Declaration of Human Rights (UDHR), applicable across diverse environments and culture. As Hendrix and Wong (2012) observe, legal frameworks often interact with societal norms, shaping and being shaped by them.

However, as new challenges are brought forth by the evolving agents of scientific hubris and technological advancements, by Artificial Intelligence, to be more specific, the ability of legal frameworks to regulate AI, has been an issue of major concern. Legal scholars like Marchenko and Энтин (2022) emphasize the need for comprehensive regulatory frameworks that align AI development with fundamental human rights principles.

Malala Yousufzai in *I am Malala* emphasizes upon education being a basic necessity for human survival and a shield against human exploitation. As she says, "We realize the importance of our voices only when we are silenced," she speaks not only for the women confined to the Swat valley of Pakistan, but has also successfully advocated for the rights of women all across the world. Malala's voice reverberates the notion of education being one of the cardinal rights demanded by humans and how it can be a resilient force against tyranny and oppression.

George Orwell's *1984* is a cautionary tale of the strict vigilance of the technologically powered "Big Brother", who can efficiently detect even the slightest of anti-state actions and thoughts that might occur in the minds of the citizens. Such a social condition echoes the contemporary society in which the AI-powered technology can slyly intrude in personal lives, hack profiles, go through search history; where algorithms can read human thoughts, listen to what they speak and deep fakes and impersonations can be as easy as pie. Orwell's far-sightedness in imagining a dystopian backdrop for human slavery to technology, must be applauded and the legal armour to resist such challenges must also be inspected. Scholars like Leite (2023) warn that without proper regulation, AI could violate privacy and fairness principles, much like the totalitarian surveillance Orwell depicted in *1984*.

The aforementioned discourse emphasizing upon the engulfing of human capabilities through scientific overreach, gets further substantiated in the seminal work of Mary Wollstonecraft, *Frankenstein.* As the character of Victor evolves against the concerns of social anxieties and overreaching scientific inquiries, the readers witness his obsession with creation of a new life which exceeds the mere quest for scientific knowledge and runs beyond ethical boundaries and moral and social responsibilities. As Victor says, "I had worked hard for nearly two years, for the sole purpose of infusing life into an inanimate body", the compromise with humanity becomes inevitable as can easily be mirrored in the contemporary times. His scientific innovation, after all, becomes a monster, an idea that resonates in the modern day society at the cost of devouring human rights, liberty and privacy.

Thus, while Malala's memoir reminds us of education being an empowering right, Wollstonecraft and Orwell highlight the darker aspects of the evolving times in which, an ambition-centric progress may lead to a whimpering fall of humans at the feet of unchecked technological prowess. Hammond's interpretation of *Frankenstein* highlights that the novel is not simply an anti-science warning but a call to critically examine "what kind of socio-nature we want produced, by whom, for what purposes, and under what conditions" . But who would ultimately control this unprecedented technological growth and to what extent, is a question that still requires an answer.

The Netflix series, *Black Mirror* also raises a similar concern by analysing our relationship with digital devices and interfaces and the consequent dystopic impacts, thereby generating technophobia. Oliveira (2023) underscores the potential of AI to enhance rights, but only if systems are developed within ethical frameworks that emphasize human dignity. *Frankenstein* and *1984* echo this need, challenging us to remember that technology alone cannot dictate the course of human rights: it is human agency, ethics, and empathy that ultimately determine whether technology serves to elevate or diminish society.

VII. CONCLUSION AND RECOMMENDATIONS

An evolution of human rights through literature and law under the emerging tech of AI emphasizes the importance of an ongoing dialogue between AI developers, legal experts, and human rights advocates. The complexity of issues involving AI technologies and implications for fundamental rights requires collaboration among policymakers, technologists, and human rights advocates to bring about sophisticated regulatory frameworks addressing the ethical, legal, and social challenges posed by the technology in question.

Hence, the intersection of AI and human rights represents both challenges and opportunities. By embracing multidisciplinary collaboration and focusing on human rights in AI development, we can take on the complexities in modern technology and make it an effective tool for good in society. To that end, the following recommendations can be considered into:

i. The Critical Need for Ongoing Dialogue Between AI Developers, Legal Experts, and Human Rights Advocates

This calls for continuous dialogue between AI creators and users, jurists, and human rights defenders in the face of rapidly evolving technologies. According to Bakiner, constant engagement on the lines separating the AI and human rights communities should be maintained to foster awareness and appreciation of how AI affects human rights (Bakiner, 2023).

Furthermore, multidisciplinary teams are proposed by Xudaybergenov to better enable the working of legal frameworks on AI (Xudaybergenov, 2023). As outlined by Begishev et al., several technical control systems lack the ability to keep up with the evolving nature of technological advancements thus creating gaps that can negatively affect human rights protections (Begishev et al., 2019). In relation to AI, therefore, it is necessary to incorporate an ongoing discussion by AI developers, legal experts, and human rights advocates into navigating complexities further defining AI technologies.

ii. Call to Action for a Collaborative Approach to Ensure That AI Serves to Uphold, Rather Than Undermine, Human Rights

Critical to this end is the call for a collaborative approach in development and implementation so that AI promotes protection of human rights rather than degrades them.

According to Pizzi et al., AI-enabled systems have transformative applications in the humanitarian sector but expose human rights to unique risks, thereby calling for a framework based on international human rights law (Pizzi et al., 2020). To work towards an enabling environment, policymakers should be proactive. This involves the setting up of dialogue spaces between AI developers and legal experts, or as much as human rights advocates to put forth different shades of opinion when a decision is to be made. As a matter of fact, Xudaybergenov underlines that not legally recognizing AI triggers deep questions on those problems of accountability and treating highly autonomous AI systems (Xudaybergenov, 2023). It will require collective action by the concerned parties in solving the issues. So, their commitment toward collaboration will allow the ethical deployment of AI with public trust towards such emerging technologies in this self-drive age.

iii. Recommendations for Policymakers, Technologists, and Human Rights Advocates

First, appropriate frameworks of regulation, including human rights, need to be established in AI development. According to Morley et al., discussing the potential impact on ethics of AI issues in a layered manner can divide the complexity into dichotomous debates and effectively lead to rules (Morley et al., 2019).

This should attract more technologists to consider integration ethical notions into their design processes. According to the best practices and ethical codes, there is a proposal as enumerated by Awad et al., integrating machine ethics with human values (Awad et al., 2018).

Thirdly, all stakeholders must acquire constant education and training. In summary, a joint move toward cooperation between policymakers, technologists, and human rights advocates is the way forward in making AI for human rights a reality. Putting these recommendations into practice will aid us to create a future where the use of AI end positively to humankind's benefits while protecting basic rights.

REFERENCES

Almeida, P., Santos, C., & Farias, J. (2020). Artificial intelligence regulation: a meta-framework for formulation and governance.. DOI: 10.24251/HICSS.2020.647

Awad, E., Dsouza, S., Kim, R., Schulz, J., Henrich, J., Shariff, A., Bonnefon, J.-F., & Rahwan, I. (2018). The moral machine experiment. *Nature*, *563*(7729), 59–64. DOI: 10.1038/s41586-018-0637-6 PMID: 30356211

Babikian, S. (2023). The role of legal frameworks in protecting digital artifacts: Artists' and collectors' perspectives. *ISSLP*, *2*(4), 18–24. DOI: 10.61838/kman.isslp.2.4.4

Bakiner, O. (2023). The promises and challenges of addressing artificial intelligence with human rights. *Big Data & Society*, *10*(2), 20539517231205476. Advance online publication. DOI: 10.1177/20539517231205476

Bartlett, B. (2023). The possibility of ai-induced medical manslaughter: Unexplainable decisions, epistemic vices, and a new dimension of moral luck. *Medical Law International*, *23*(3), 241–270. DOI: 10.1177/09685332231193944

Bayisa, G., Dagnew, T., & Gebremeskel, T. (2023). An analysis of the presentation of human and democratic rights abuses in amelmal's yältäkoäche guzo (1974- 2018). *ERJSSH*, *10*(1), 173–184. DOI: 10.4314/erjssh.v10i1.10

Begishev, I., Khisamova, Z., & Gaifutdinov, R. (2019). On methods to legal regulation of artificial intelligence in the world. *International Journal of Innovative Technology and Exploring Engineering*, *9*(1), 5159–5162. DOI: 10.35940/ijitee.A9220.119119

Bell, A., Nov, O., & Stoyanovich, J. (2023). Think about the stakeholders first! toward an algorithmic transparency playbook for regulatory compliance. *Data & Policy*, *5*, e12. Advance online publication. DOI: 10.1017/dap.2023.8

Boncheva, A., & Serrano, R. (2022). Climate change, human rights and sustainability. *Revista Mexicana de Economía y Finanzas*, *17*(4), 1–20. DOI: 10.21919/remef.v17i4.802

Cath, C., & Floridi, L. (2016). The design of the internet's architecture by the internet engineering task force (ietf) and human rights. *Science and Engineering Ethics*, *23*(2), 449–468. DOI: 10.1007/s11948-016-9793-y PMID: 27255607

Chatterjee, S., Sreenivasulu, N., & Hussain, Z. (2021). Evolution of artificial intelligence and its impact on human rights: From sociolegal perspective. *International Journal of Law and Management*, *64*(2), 184–205. DOI: 10.1108/IJLMA-06-2021-0156

Chilton, A., & Posner, E. (2016). Respect for human rights: law and history. SSRN Electronic Journal. DOI: 10.2139/ssrn.2815272

Clapham, A. (2006). Human rights obligations of non-state actors.DOI: 10.1093/acprof:oso/9780199288465.001.0001

Cohrs, J., Maes, J., Moschner, B., & Kielmann, S. (2007). Determinants of human rights attitudes and behavior: A comparison and integration of psychological perspectives. *Political Psychology*, *28*(4), 441–469. DOI: 10.1111/j.1467-9221.2007.00581.x

Dawes, J. (2009). Human rights in literary studies. *Human Rights Quarterly*, *31*(2), 394–409. DOI: 10.1353/hrq.0.0071

Dawes, J. (2009). *That the world may know: Bearing witness to atrocities*. Harvard University Press.

Donert, C. (2023). Women's rights as human rights after the end of history. *Gender & History*, *35*(3), 862–880. DOI: 10.1111/1468-0424.12729

Elendu, C., Amaechi, D. C., Elendu, T. C., Jingwa, K. A., Okoye, O. K., John Okah, M., Ladele, J. A., Farah, A. H., & Alimi, H. A. (2023). Ethical implications of ai and robotics in healthcare: A review. *Medicine*, *102*(50), e36671. DOI: 10.1097/MD.0000000000036671 PMID: 38115340

Erdélyi, O., & Goldsmith, J. (2020). Regulating artificial intelligence: proposal for a global solution. https://doi.org//arxiv.2005.11072DOI: 10.48550

Henry, P. (2015). Analyzing the integration of human rights into social work education. *Issues in Social Science*, *3*(2), 100. DOI: 10.5296/iss.v3i2.8041

Hohma, E., Boch, A., Trauth, R., & Lütge, C. (2023). Investigating accountability for artificial intelligence through risk governance: A workshop-based exploratory study. *Frontiers in Psychology*, *14*, 1073686. Advance online publication. DOI: 10.3389/fpsyg.2023.1073686 PMID: 36760454

Hunkenschroer, A., & Kriebitz, A. (2022). Is ai recruiting (un)ethical? a human rights perspective on the use of ai for hiring. *AI and Ethics*, *3*(1), 199–213. DOI: 10.1007/s43681-022-00166-4 PMID: 35909984

Huriye, A. (2023). The ethics of artificial intelligence: Examining the ethical considerations surrounding the development and use of ai. *American Journal of Technology*, *2*(1), 37–45. DOI: 10.58425/ajt.v2i1.142

Ibhawoh, B. (2014). Testing the atlantic charter: Linking anticolonialism, self-determination and universal human rights. *International Journal of Human Rights*, *18*(7-8), 842–860. DOI: 10.1080/13642987.2014.951340

Jobin, A., Ienca, M., & Vayena, E. (2019). The global landscape of ai ethics guidelines. *Nature Machine Intelligence*, *1*(9), 389–399. DOI: 10.1038/s42256-019-0088-2

Kotzé, L. (2014). Human rights and the environment in the anthropocene. *The Anthropocene Review*, *1*(3), 252–275. DOI: 10.1177/2053019614547741

Kuźniacki, B., Almada, M., Tyliński, K., & Górski, Ł. (2022). Requirements for tax xai under constitutional principles and human rights., 221-238. DOI: 10.1007/978-3-031-15565-9_14

Leite, E., Leite, M., & Leite, A. (2023). Ai's impact on human rights: The need for legal evolution. *JER*, *1*(2), 81–86. DOI: 10.29073/jer.v1i2.16

Manap, N., & Abdullah, A. (2020). Regulating artificial intelligence in malaysia: The two-tier approach. *UUM Journal of Legal Studies*, *11*. Advance online publication. DOI: 10.32890/uumjls.11.2.2020.8779

Monteith, S., Glenn, T., Geddes, J., Whybrow, P., Achtyes, E., & Bauer, M. (2022). Expectations for artificial intelligence (ai) in psychiatry. *Current Psychiatry Reports*, *24*(11), 709–721. DOI: 10.1007/s11920-022-01378-5 PMID: 36214931

Morley, J., Machado, C., Burr, C., Cowls, J., Taddeo, M., & Floridi, L. (2019). The debate on the ethics of ai in health care: a reconstruction and critical review. SSRN Electronic Journal. DOI: 10.2139/ssrn.3486518

Niemiec, E. (2022). Will the eu medical device regulation help to improve the safety and performance of medical ai devices? *Digital Health*, *8*, 205520762210890. DOI: 10.1177/20552076221089079 PMID: 35386955

Nweke, O., & Nweke, G. I. (2024). Legal and ethical conundrums in the ai era: A multidisciplinary analysis. *International Law Research*, *13*(1), 1. DOI: 10.5539/ilr.v13n1p1

Pizzi, M., Romanoff, M., & Engelhardt, T. (2020). Ai for humanitarian action: Human rights and ethics. *International Review of the Red Cross*, *102*(913), 145–180. DOI: 10.1017/S1816383121000011

Pomaza-Ponomarenko, A., Leonenko, N., Cherniahivska, V., Lehan, I., & Puzanova, G. (2023). Legal reform and change: Research on legal reform processes and their impact on society. analysis of factors that facilitate or hinder legal change, including political, social, and economic factors. *Journal of Law and Sustainable Development, 11*(10), e1854. DOI: 10.55908/sdgs.v11i10.1854

Roberts, C. (2014). The contentious history of the international bill of human rights.. DOI: 10.1017/CBO9781139035675

Ronanki, K. (2023). Re-centric recommendations for the development of trustworthy(er) autonomous systems., 1-8. DOI: 10.1145/3597512.3599697

Shelley, M. (1818). *Frankenstein; or, The Modern Prometheus*. Lackington, Hughes, Harding, Mavor & Jones.

Tangen, J. (2020). Timescapes in public policy – constructing the "victim of trafficking". *Journal of Borderlands Studies, 37*(3), 475–492. DOI: 10.1080/08865655.2020.1787189

Tarasova, Y., Bobkova, T., & Bobkov, O. (2020). The contradictions of human rights definition in the russian civil law.DOI: 10.15405/epsbs.2020.04.80

Twiss, S. (2011). Global ethics and human rights: A reflection. *The Journal of Religious Ethics, 39*(2), 204–222. DOI: 10.1111/j.1467-9795.2011.00473.x

Ubaydullayeva, A. (2023). Artificial intelligence and intellectual property: Navigating the complexities of cyber law. *Irshad J. Law and Policy, 1*(4). Advance online publication. DOI: 10.59022/ijlp.57

Wang, J., Mao, W., & Wenjie, W. (2023). The ethics of artificial intelligence: Sociopolitical and legal dimensions. *ISSLP, 2*(2), 27–32. DOI: 10.61838/kman.isslp.2.2.6

Xudaybergenov, A. (2023). Toward legal recognition of artificial intelligence proposals for limited subject of law status. *Irshad J. Law and Policy, 1*(4). Advance online publication. DOI: 10.59022/ijlp.55

Yousafzai, M., & Lamb, C. (2013). *I Am Malala: The girl who stood up for education and was shot by the Taliban*. Little, Brown and Company. • Orwell, G. (1949). *1984*. Secker & Warburg.

Zheng, Y. (2023). Redefining rights in the age of ai: Philosophical perspectives in shaping the future of international human rights law. *Lecture Notes in Education Psychology and Public Media, 16*(1), 124–132. DOI: 10.54254/2753-7048/16/20231124

Chapter 5
AI, Ethics, and Human Rights:
Navigating the Intersection of Technology and Human Dignity

Sakshi
https://orcid.org/0009-0007-8467-8079
Lovely Professional University, India

Shivani Dhand
https://orcid.org/0000-0002-4809-1365
Lovely Professional University, India

ABSTRACT

The rapid integration of artificial intelligence (AI) into critical domains such as healthcare, governance, and education has sparked significant concerns about its implications for human rights. While AI holds the potential to enhance welfare, improve service accessibility, and monitor compliance with human rights standards, its unregulated use poses risks. Privacy invasion, algorithmic bias, and erosion of individual autonomy threaten to undermine fundamental human rights, raising questions about the ethical and legal frameworks governing AI technologies. Addressing this dual potential of AI has become a critical challenge for policymakers and stakeholders worldwide. This paper explores the ethical and legal mechanisms to align AI development and deployment with human rights principles. Consulting more global standards such as the International Covenant on Civil and Political Rights (ICCPR) and the Convention on the Rights of the Child (CRC) with such proposed ethical guidelines as inclusive design, regular system checks, and strict data protection.

DOI: 10.4018/979-8-3693-9035-1.ch005

INTRODUCTION

The emergence of Artificial Intelligence (AI) has altered the dynamics in various industries–healthcare, and education, for instance, as well as finance, security, and public administration, emphasizing a new mode of interaction between businesses and individuals (Noorbakhsh-Sabet et al., 2019; Russell & Norvig, 2016). AI, due to its ability to examine a large pool of data, recognize complicated structures, and mechanize several tasks, has enhanced unimaginable improvements, especially in decision-making and predictive analyses (Binns, 2018). The horizon of the effects of AI is worldwide; hence, as it believably puts in parameters that align it with societal developments, AI in the breasts of this generation has become a game-changer. Nonetheless, given that AI has grown to feature in many spheres of the public, the concerns about how they positively and negatively affect the protection of human rights are difficult to ignore, as well as concerns about fairness, accountability, and transparency (Floridi & Cowls, 2022; Eubanks, 2018). This chapter will examine AI's positive and negative aspects concerning human rights issues and how these challenges play out in migration and refugee management, which is sensitive to biases, surveillance, and informed consent.

There is no doubt that the development of artificial intelligence has increased efficiency and brought about new advancements in technology. For instance, applications that use artificial intelligence have improved medical diagnosis, speeded up finance-related processes, and bettered the management of resources during welfare (Vinuesa et al., 2020). For instance, AI has reduced administrative costs by automating processing, cutting down on queues, and providing better user service. This can help promote human rights by making access to services much easier (Cath, 2018). This efficiency and reach enable AI to be appreciated more in conflict zones where there are extremes in demand and supply, such as in the control of the movement of people. For example, AI tools can facilitate in-course Data Resource Management in which data changes in the migration dynamics are made to track available resources, appreciate losses, and, most importantly, predict intervention in crisis scenarios, thus ensuring governments and organizations act faster and more effectively (Latonero, 2019; Daugherty & Wilson, 2018).

However, as these artificial intelligence systems broaden in scope and depth in their application to dealing with refugees and migration at large, some disagreements on their implementation concerning transparency, bias, and accountability have come to the surface (Eubanks, 2018). Adoption of AI in this field is often characterized by a significant engagement of data collection exercises that include biometrics and geolocation data, which, when applied in unjust ways, are capable of violating people's privacy as well as unduly surveilling at-risk groups (Binns, 2018; Latonero, 2019). Considering migration and refugee borders integrated with such mathematically driven systems are used in the processes of asylum seekers, screening potential migrants creates issues of equity and clarity. All such systems are prone to biases within their algorithms caused by discriminatory and latent training datasets, which may worsen global issues and enable unfair treatment. These could be the same groups that such systems are undermining the objective of proposing them, epitomized by the concern of discrimination in treatment and Availability of service (Verma, 2019).

A fundamental concern regarding the use of AI is its potential for abuse in a manner that contravenes the basic principles of human rights, including dignity, equality, and autonomy (Floridi & Cowls, 2022). For example, many immigration systems with facial recognition technology or other predictive policing ethos now have an o'clock identity verification mechanism. There is no doubt that those tools would help in speeding up the processes involved. However, introducing such technology also poses serious challenges concerning privacy and consent since most individuals would not understand how their information would be collected, manipulated, or even stored. There is no consent from the individual, which is a significant issue, especially when considering refugees, who are often not even consulted about how their information may be used (Cath, 2018). In addition, some issues arise with surveillance AI applications concerning the freedom of individuals from unnecessary supervision, as such an environment leads to monitoring all the time, negatively impacting privacy and personal agency (Zuboff, 2019). The potential for such interference with personal autonomy speaks to ethical concerns or dilemmas around usurping individual liberties and security in the context of AI being incorporated within border control systems (Latonero, 2019).

Algorithmic bias is a significant ethical issue of using AI in refugee and migration management. No machine learning algorithm is free of any bias because any such algorithm has to be trained on data, and that data tends to incorporate, among others, historical prejudices and systemic discrimination (Verma, 2019; Eubanks, 2018). Such biases can extend to the practice of migration management, creating unfounded discriminatory practices such as denying asylum to persons of certain nationalities, ethnicities, or religions. One such case is an AI application in evaluating refugee claims, where there is a chance that such biased algorithms would work against individuals of specific demographics (Binns, 2018). Such biases not only threaten fairness but also the justice and equity that are at the core of human rights. Tackling this entails creating more diverse and representative datasets and ensuring those developing AI solutions respect these issues during model training (Russell & Norvig, 2016).

A further aspect of the ethical concerns with using AI in migration management involves responsibility. Unfortunately, Most AI systems tend to function as "black boxes," making it almost impossible to assign blame when the AI systems cause unexpected effects (Cath, 2018). Especially in the case of decisions influenced by AI, such as allowing asylum seekers, deportation is of high concern as it calls for responsibility. The complexity of burdening accountability comes with the fact that several parties can make such decisions (Floridi & Cowls, 2022). This is worsened by the fact that AI accountability is given that those subjected to AI decisions usually have no clear way of knowing why these aversive outcomes were arrived at, making it difficult for them to challenge or appeal the decisions made. Thus, there is a critical challenge in how AI systems work to ensure that respect for human rights is achieved and that the beneficiaries can question and comprehend the outcomes of the systems' application (Vinuesa et al., 2020).

Given the potential of AI to affect human rights, there is a need for a solid regime to regulate its development and use. In this regard, existing international human rights standards, such as the Universal Declaration of Human Rights (UDHR), offer a starting point for safeguarding individuals from potential abuse in anticipation of technological innovations (Daugherty & Wilson, 2018). While these principles set definite states of play against 'misuse' or 'abuse,' implementing them on AI requires other ethical and legal considerations geared towards the digital space. As there are few if any, comprehensive provisions addressing the regulation of AI, such proportionate and even-handed arguments have led to propositions that call for the protection of fairness, dignity, and responsibility of people in AI, among other issues (Floridi & Cowls, 2022). For example, the GDPR and the proposed Artificial Intelligence Act of the European Union care for provisions that protect society against aggression by AI technologies (Cath, 2018). Suppose ethical dimensions are built into the design and use of AI technologies. In that case, the two forces – humankind and technology will move together towards progress without violating human rights even as the use of technology advances.

Objectives

- To analyze the role of artificial intelligence in advancing and protecting human rights.
- To examine the ethical implications of artificial intelligence in refugee and migration management.
- To assess the legal and ethical frameworks governing the application of artificial intelligence in these contexts.

AI'S ROLE IN PROMOTING AND UNDERMINING HUMAN RIGHTS

As AI starts being applied in different industries, its use in protecting human rights is more accessible and better with improved service delivery and suiting one's needs; on the other hand, there are challenges such as bias, discrimination, and violation of privacy, among others. Even though using AI in migration processes and refugee management is encouraging and can improve processes significantly, it comes with ethical and legal issues that can threaten fundamental human rights. Therefore, one needs to appreciate both the positive and negative aspects of AI application in this field to formulate ethical principles that will correctly guide the use of AI in that area.

Positive Contributions of AI

1. Enhanced Service Delivery

Improvements in service delivery across various sectors have become a reality thanks to the use of AI. One of the notable examples is in the industries where migration and refugee management rely heavily on resource allocation, especially during crises. The role of AI is efficient in collecting, processing, and analyzing large amounts of data geared towards making decisions on how best to distribute resources promptly, devoted to providing aid to people in need. AI, for example, can use data and construct models that show possible areas where crises will arise and the optimal way to deploy resources (Vinuesa et al., 2020). Migration enables the application of resources in the most appropriate parts of the context, thus enhancing response time and improving the adequate provision of humanitarian assistance. Moreover, the role of AI also extends to helping humanitarian institutions manage their logistics so that food, water, and medical aid are appropriately distributed to the regions that are seeing an influx of refugees (Ienca, 2018). Therefore, AI promotes the right to needed supplies and temporary emergency services by becoming more efficient and effective at detecting and managing such situations. Moreover, the predictive aspects of AI assist organizations in anticipating an increase in the number of refugees, thus managing an already strained resource. Smart deployment of AI mechanization makes for better service delivery, especially where upholding human rights such as the right to health and safety is paramount.

2. Personalized Support

However, another meaningful aspect of AI is that it can assist refugees and migrants who encounter various challenges due to language, culture, and laws in an alien environment. These needs can be effectively addressed using applications powered by AI technology. A relevant case in point may be the inmates in the host country, who are supposed to avail of services from service providers and lawyers, among others, to access the information and services they need (McGregor et al., 2019). AI has brought changes to improve such people's language use more effectively concerning citizenship, incorporating rights such as the right to information and fair treatment. Furthermore, such AI tools can lead migrants or refugees to search for adequate legal services concerning their particular situations. For instance, this may even incorporate a new collection of chatbots where users are led through procedural steps along the law and learn about their rights and options: AI-driven patient-participation tools that are increasingly being developed to support mental health will instead be used in assisting refugees in the way that respects their right to receive mental well-being services (Mittelstadt 2019). This type of personalized assistance, which involves providing relevant information on the hierarchy of such services as legal rights and mental health, enhances the dignity of every individual by upholding the differing needs of displaced persons. The Availability of such assistance elevates the role of AI in advancing people's rights by ensuring that needed services are abundant and catered for specific individuals.

Risks to Human Rights

1. Bias and Discrimination

The use of artificial intelligence has come a long way, even with the glaring concessions to human rights, particularly in internalizing and operationalizing discrimination and bias systems. Even on the very second step, most AI systems with erected structures are still exposed to historical data, making them enter inbuilt biases history. This challenge is more evident in migration and refugee systems, and it is caused by the ability of flawed AI systems to determine asylum requests, repartition of refugees, and even service provision (Eubanks, 2018). Suppose an artificial intelligence system receives pre-labeled training data concerning people carrying negative biases towards ethnic groups, for example. In that case, such a system will learn to mimic that exact behavior without their awareness, even though it will produce harmful results for the groups it aims to serve. As a result, certain types of refugees and migrants who belong to certain ethnic, cultural, and socio-economic groups will face prejudice. Several issues have arisen concerning the presence of bias in AI algorithms, and one of them is fairness in these systems when they are used in making decisions that determine people's rights and opportunities. A particular case would be when such an AI system, which is mainly based on nationality or race to evaluate applicants for asylum services, is used, and it will prejudice the outcomes of decision-making in their applications. Research results have indicated that biased algorithms construct an oppressive social system by integrating bias into automatic apparatus, thus creating a right to equality violation (Binns, 2018). To combat such an application, mitigation of biases in AI systems needs to be approached in a more structured manner, which involves sourcing data from multiple sources, performing audits on a routine basis, and Testing systems with all intended users.

2. Privacy Violations

The use of AI systems in multiple aspects, such as controlling and managing human activities, has human rights violations, as outlined by many scholars. There is always the question of privacy, especially with surveillance and data management systems. Certain AI technologies, such as facial recognition and data mining, are increasingly becoming employed to 'watch' over the migrating populations in the name of national security and other regulations, compliance, or enforcement. These technologies, however, can violate the privacy rights of those at risk, which is the case since it allows the use of such data without the individual's consent through mass surveillance and data collection (Zuboff, 2019). Refugees and migrants do not stand much of a chance against the authoritarian use or collection of their data as they are already in a derogatory position about their protection through the law. This raises human rights and human law issues about privacy and personal control. Even if AI is involved, it can also be a severe risk by scrutinizing the monitoring of highly vulnerable people, which could cause further grievous harm because it consists of storing sensitive information. Third-party organizations or private affiliations can access the data collected within an AI surveillance network (Floridi et al., 2018).

Moreover, the problem is complicated when we talk about the inability to track the operations of the AI systems in terms of how information is collected, stored, or processed, putting into the realm of concern regarding how accountable the users of the systems are. For risk cases or privacy violations - especially for diaspora populations - there might not be any pathways to action found, which further amplifies the sense of exposure in such spectacular populations or, at minimum, a loss of faith in government services. Also, it discourages them from accessing critical services because they fear the misuse of their data with surveillance through artificial intelligence. For instance, patients might dodge healthcare or legal aid services and avoid risk factors such as possible surveillance or identification, obstructing their healthcare and legal assistance rights. These concerns can be addressed by implementing appropriate data protection measures and carefully supervising their application. Such restrictions should include limits on sharing data collected for one purpose with other agencies or for different purposes, accountability in the use of data, and transparent standards for seeking and providing consent (Rahwan, 2018). The double-edged sword that is AI in the promotion and violations of human rights highlights the importance of its ethical, legal, and societal implications. AI, for instance, has effectively improved service delivery and provided specialized services to vulnerable groups, facilitating the enjoyment of rights such as access to social services, information, and justice. However, this technology is also prone to bias and risks violating privacy, which diminishes the protection of equality, justice, and privacy. This situation calls for intervention from all stakeholders, including policymakers, technology experts, and human rights campaigners, in developing guidelines to govern AI technologies in line with human rights and social justice. Such adaptive approaches, including but not limited to designing Virtual Agents that function without bias, protecting users' data, and ensuring accountability for services provided, tend to make AI develop and function ethically.

Ethical Considerations in Refugee and Migration Management

The application of artificial intelligence (AI) systems in improving the efficiency of processes and services to manage refugees and migrants has many prospects and ethical concerns that must be addressed. Refugee and migration management is a prime example of many moral dilemmas that lurk because of the many challenges they face. These systems would give more access to all these risk considerations, be it regarding possibilities of fraud or resource management, once they extend into the processing of applications in this activity area. However, such systems raise more severe ethical issues that still need resolution to ensure that refugees and migrants are respected regarding their autonomy and rights. Some of the critical moral issues in this regard include but are not limited to:

1. Transparency

Among the most significant moral issues that arise from using AI to manage the migration process is the issue of transparency. Transparency in AI refers to making the decision-making process of the AI system visible and understandable to relevant persons. This is especially critical in migration, affecting people's rights and safety, especially those already in precarious situations. Refugees and migrants are usually at the mercy of such decisions aimed at resources, state protection, or residency status. In this regard, transparency helps address the issue of system navigation, especially the articulation of reasons for certain decisions. However, in practice, transparency with AI systems is challenging to realize. Most refugees and migration-related AI systems are predictive analytics and machine learning models, the reasoning of which is based on a black box mathematical model (Pasquale, 2015). This complexity becomes detrimental when standing AI systems are implemented to evaluate the potential of individuals or clear them to enter because only a few understand the rationale, primarily what the systems base their decisions on, and many would be unhappy with the outcome (Mittelstadt et al., 2016). For example, AI systems based on historical data may adopt a particular approach to populations, while discrimination exists within the data it ingests without a reasonable or coherent defense (Angwin et al., 2016). Therefore, transparency must be promoted to guarantee justice and avoid reinforcing existing social hierarchies through AI solutions. Also, the need for AI and its transparency extends across the society and the governments. Lastly, those vendors or governments that incorporate AI tools in managing migration must be clear on how data about refugees and migrants is collected, processed, and used to reach certain conclusions (Veale & Binns, 2017). Such transparency aims to enhance the confidence of the stakeholders in the system. It provides a way for other stakeholders, such as NGOs and interested parties, to keep watch over operations that AI drives and ensure that they comply with the morals and the law. In this regard, ethical principles governing the use of AI have to call for processes and building blocks, such as audit trails, that will help shed light on why and how certain conclusions were reached through the application of any AI system (Floridi et al., 2018).

2. Informed Consent

Informed consent is another aspect of ethics that stands out when applying artificial intelligence tools to manage refugees and migration. In principle, informed consent means that an individual is aware, to a reasonable extent, of the purpose for which their data is required and the risks associated with providing it. When it comes to AI systems applied in the management of migration, it is often the case that such consent is not possible because, for instance, the refugees and migrants may be illiterate, speak a different language, or other circumstances arise that limit their freedoms (Binns, 2018). Informed consent is also necessary to honor the rights of migrants and refugees as autonomous agents. Such systems generate claims about a person's eligibility for asylum, residence, or other movement services based on personal information like biometrics, travel data, and even mueve patterns (Leslie, 2019). In many instances, however, the purpose for which such information is collected and for what it is used is not effectively communicated to migrants, or they are made to agree because of the situation. Thus, ethical frameworks for using AI must provide precise mechanisms to ensure that meaningful informed consent is obtainable and that refugees and migrants are made aware of the purpose of their data without fear of reprisals for refusal to participate in the study (Eubanks, 2018). To attain informed consent in managing refugees and migrants, other issues, such as the matter of power, must be sorted. A significant number of these migrants feel compelled to furnish their personal information as they tend to think that this is something that is needed to get help from services or shelters or to obtain some sort of legal protection (Mayer-Schönberger & Cukier, 2013). This disposition can render obtaining consent oppressive as migrants may not have a choice but to agree with the AI-based evaluations, they do not comprehend. It follows that ethical rules should state that consent processes must be developed concerning the position and situation of those groups and that it should be made clear that withdrawal will not have negative consequences on the individuals' cases (Whittaker et al., 2018).

3. Accountability

Accountability is a critical issue regarding the ethical implications of managing migration and refugees in a manner heavily reliant on Artificial Intelligence. In this case, the concern would be who takes responsibility for the consequences of AI-based decisions and whether individuals can appeal against such decisions. The issue of accountability becomes baffling when it involves a decision or recommendation that first magnifies the risk of, say, being granted an asylum, residence, or protection status an individual seeks. The consequences of an algorithm making a wrong turn on an individual can be far-reaching. However, it is usually unclear whom to blame for making such mistakes – is it the software developers, the ones who provided the data, or the institution using the marketed system (Gasser & Almeida, 2017)? In migration management, it is essential to acknowledge that accountability provides redress and appeals mechanisms to individuals who may be disadvantaged by an AI decision affecting their applications. For example, a person whose asylum claim has been rejected by an AI system due to incorrect reasoning or reasoning based on discriminatory trends should have the right to appeal this judgment to another human (Binns, 2018). For AI technology migration management into practice, users should strive towards designing and creating infrastructures with accountability mechanisms such as internal controls and independent scrutiny to ensure all systems are ethical and working as designed (Hagendorff, 2020). Enforcement of duty and accountability for AI-led and oriented decisions, choices, and actions in charters and migration policies should, in an ideal situation, encompass many parties, such as the AI business, the data suppliers, and the enactors of such policies. It is fair and appropriate that no single person bears the totality of blame for reasonably predictable events or failures in their corner; all those people who were involved in creating, executing, and effectively bringing the AI systems to the 'market' should have a certain amount of blame shared among them. Such accountability frameworks must also integrate the periodic and systematic appraisal and review of the artificial intelligence-based systems to correct any inaccuracies or biases that may occur over time Rahwan et al. 2019. In addition to these considerations, a hostile system that prevents access to redress for people harmed by such AI systems should be avoided.

In summary, moral values such as transparency, informed consent, and accountability stand out within migration and refugee management frameworks through artificial intelligence. The importance of trust, as well as its fairness and the understanding of AI in these processes, hinges on transparency. Informed consent gives dignity to refugees and migrants by allowing them control over their data and how it is used. Accountability, on its part, ensures that there is a person or organization who will be responsible and provide an explanation for the usage of AI in decision-making. These three moral values present a solid basis for the use of AI technology that defends the rights of refugees and migrants and promotes migration management that is just and fair.

LEGAL AND ETHICAL FRAMEWORKS FOR HUMAN RIGHTS IN AI DEVELOPMENT

With the advancement of technology, artificial intelligence (AI) has become part of every facet of life, particularly in businesses and governments. Nevertheless, with the new technologies utilizing AI or its development in systems come some legal and ethical problems, especially concerning human rights. In this regard, it becomes challenging to find the corresponding policies that have now gone into near-exponential development progress advancements in AI today. When considering the possibility of new policies, it has become imperative to internalize international human rights standards into the local milieu. Highlights in various legal and political regimes require that artificial intelligence technology development processes should also consider fundamental human freedoms such as respect for privacy, equality, and freedom of choice, among others. This section explains the contemporary international human rights norms on AI and the ethics framework proposed to ensure that AI is beneficial and fair to society.

1. Existing Legal Standards

International human rights law is very effective in promoting and protecting rights and freedoms in the context of AI. Specifically, these include international covenants, particularly the International Covenant on Civil and Political Rights (ICCPR) and the Convention on the Rights of the Child (CRC), which are very significant limitations on the matters concerned with AI and its impact on the world. These internationally binding contracts are signed by countries that commit them to certain unavoidable rights in designing and deploying AI systems.

The ICCPR, which was adopted by the United Nations in in1966, provides a basis of human rights law where it defends such liberties as the right to believe what they want, the right to control the information about oneself, and the right to be free from being discriminated against (United Nations, 1966). These rights are of utmost importance in the context of AI. For instance, it is known that in very many cases, large amounts of data regarding personal habits, behavior, and backgrounds of individuals are often gathered and processed for purposes of crime prediction – a form of analysis aimed at forecasting which groups of people are more prone to specific criminal activities (Brundage et al., 2018). The erosion of privacy that AI technologies can cause intersects with the ICCPR's guarantees of freedom of expression and personal dignity – both highly valued in society. Therefore, advanced technology such as AI must be implemented and controlled to uphold those rights.

The impact of AI on vulnerable groups, particularly children, as surmised by the Convention of the Rights of the Child (CRC), also ratified in 1989, is worth mentioning. For example, in social networking and online gaming, children are targeted by advertisements and other content by the AI capability. Children have exercised the right not to be economically exploited and shielded from detrimental content. This should also be true for educational and other applications incorporating new technologies. Children, as a sensitive group whose data is highly prone to abuse, cannot afford to have any AI systems that assist in the gathering or processing of their data without due care taken (Lemaignan et al., 2021).

These international human rights standards set the groundwork for how AI should be regulated to avoid infringing on individuals' rights. Nevertheless, as essential as they are, these frameworks are not exceptionally tailored to the challenge of artificial intelligence. With the advancing landscape of artificial intelligence, there is an increasing demand for a more relevant ethical framework to deal with issues beset with AI.

2. Proposed Ethical Guidelines

Considering the imperfections of present human rights theories, researchers and administrators have fashioned ethical standards to help guarantee the compliance of AI applications with human rights. The principles of inclusive design, regular auditing, and data protection are the main ethical principles that govern these standards. These principles together form a basis for designing AI systems that are just, open, and observant of human rights.

2.1. Inclusive Design

The first and perhaps the most fundamental moral consideration in creating AI is inclusive design, which supports and promotes the participation of all appropriate groups in creative and operational activities involving AI. Many sources of AI algorithms are trained, considering the social prejudices that may be reflected in the data and, unaddressed, contribute to discriminate results. Participatory design also includes those from dwelling or marginalized groups, ethnicities, and verticals who are designing AI systems. It helps to address inequality in developed systems, such as the development of AI systems (Nissenbaum, 2019). For instance, facial recognition systems have often been discriminatory towards ethnic minority groups (Buolamwini & Gebru, 2018). Suppose the source data used for training the models is not diverse. In that case, the models trained on such data may lead to a dangerous discrimination risk where individuals are misidentified, which may breach an individual's human rights. This way, considering inclusive design principles will allow AI practitioners to eliminate biases and contribute to a more developed and ethical AI ecosystem. Policymakers can enhance inclusive design by prescribing accountability for processes and outputs and forcing the developers to provide a record of the data, methods, and relevant sections of the general population identified in their work (Binns, 2018).

2.2. Regular Auditing

Routine auditing of AI systems is essential to any ethical strategy designed for AI development. Audits are aimed at determining if AI systems respect human rights and do not incorporate negative stereotypes or discriminative attitudes on any basis. AI algorithms are such that algorithms' sophistication hides their decision-making processes; hence, errors or biases created in their workings become very hard to trace and rectify. Regular audits, in turn, help to systematically address the issues regarding such processes and the ethical robustness of the pursuable AIs by the responsible parties (Cath, 2018). Independent oversight is needed to ensure that auditing is done well and achieves its intended objective. Typically, ethical audits carried out by external bodies, as opposed to their organizations' units, help provide neutral evaluations of an AI system's adherence to its ethical code. This also aids in the improvement of the public's confidence in the system, where the independent evaluation assures the people that the AI systems in question are not being used for the wrong reasons and are maintained within the ethical boundaries that are intended(L Floridi & Taddeo, 2016). Besides external audits, the developers should perform internal evaluations and develop measures that would assist them in finding, reporting, and fixing biases or mistakes in their systems. One of the main drawbacks of conducting regular audits for AI systems in development and deployment is the lack of motivation on the part of the developers and the AI companies to conduct such assessments on their own since the regulatory body does not have the means to compel them. One option that has worked in the past could be implementing an incentive program for those who base their activities on ethics. These certifications would ensure some level of accountability, motivating developers towards ethics, therefore increasing the adoption rate of regular audits (Whittaker et al., 2018).

3. Data Protection

The significance of protecting human rights in AI development must include data protection. This implies that personal information must be guarded and protected from unauthorized access to prevent individuals' unethical misuse of such information, as AI systems depend on loads of data. The use of data by such applications raises issues of privacy where the data is used, for instance, in AI applications without proper security, leading to a possibility of a given individual being a victim of identity theft, surveillance, or exploitation (Mayer-Schönberger & Cukier, 2013). The General Data Protection Regulation (GDPR), a law that was enacted in the European Union, serves to demonstrate the requirement for such a high level of data protection prescribed by the law that personal data should be processed legally, only for legitimate purposes as stipulated in the law (Vogit & Von dem Bussche, 2017). Such principles must be embraced and adopted as a matter of importance for all who are embracing AI technology to protect people's privacy and empower them to manage their data. In AI technologies, data protection cannot be limited to building practices that ensure compliance with the GDPR. Using such technologies poses certain privacy risks that may not yet be covered by law. When it comes to AI development, it is essential that the developers practice data minimization and only collect the necessary data for a specific purpose.

Moreover, data anonymization techniques help lessen the chances that individuals in a particular dataset can be located, reducing privacy concerns (Zarsky, 2016). There should be practical ethics that also insist on showing how there should be strong measures to protect data from unwanted intrusion or abuse of sensitive information and why the users should be informed on the projected uses of their data. In addition to those mentioned above, data security in AI development should prevent data loss. Without solid protection, malicious entities may carry out cyber warfare against the AI system, compromising the private information of the service users. Because of this, ethical considerations should include implementing effective cybersecurity measures to avoid such access, especially physical access, whereby intrusion-tolerant measures such as data encryption are employed (Leslie, 2019).

It is vital to consider human rights issues in the design and development of artificial intelligence systems to ensure that such systems are built per international laws and for the greater good of society. For example, the existing legal instruments such as ICCPR, CRC, and so on serve as the baseline; however, these are not designed to address the intricacies posed by artificial intelligence. Hence, ethical principles are necessary to fill that gap. All interested parties need to be consulted to achieve unbiasedness and equity. It has to be mentioned that most AI-powered systems are being reviewed now and then to ensure ethicality and fairness regarding the processes concerning these systems. Finally, these laws and principles prevent an unwarranted invasion of the private lives of individuals and misuse of their information. Such laws and ethical grounds are for building artificial intelligence systems that respect human dignity and rights.

CONCLUSION

Artificial intelligence has become a pretty powerful force that has the potential to protect and, at the same time, threaten human rights in a variety of spheres. AI has a highly transformative potential to improve human welfare, widen access to services, and even support the protection of human rights through quickly available tools for monitoring, analyzing, and intervening. Conversely, the deep penetration of AI into human life, coupled with its dubious or sometimes harmful property, raises questions about potential privacy violations, discrimination, and erosion of human autonomy. Both of these forces indicate the urgency of addressing the influence of AI on human rights in a balanced, proactive manner. An extensive ethical guide for developing and deploying AI systems must be established to ensure that the systems uphold human dignity and fundamental rights. Transparency, accountability, inclusivity, and respect for privacy should be the foundation principles of this framework. Technologists will be enabled to divert eventualities and engineer technologies that respect the values and rights central to human society through ethical practices of AI within human welfare priorities. Just like legal standards will evolve with technological progress in their relevance, so will ethical guidelines. Ensuring that AI is aligned with human rights objectives will require a cooperative effort between all relevant stakeholders, technologists, and civil society organizations to create regulations, standards, and spaces for public input. Policymakers play a pivotal role in setting ethical principles for AI development by drafting and enforcing laws within clear boundaries. Concerning this, technologists convert such values into the design and operation of AI systems. Civil society organizations: Their contribution is fundamental as they bring perspectives on social justice, human rights, and ethical accountability into the discourse, engaging on behalf of those who will be most affected by the developments of AI. Hence, the collaborative interdisciplinary effort will make AI a resource and a weapon for social justice and human rights. In so doing, stakeholders will create a sophisticated AI environment in which technology advances in innovation while upholding and advancing human rights interests. AI is seen as aligned with equity, fairness, and dignity principles.

REFERENCES

Angwin, J., Larson, J., Mattu, S., & Kirchner, L. (2022). Machine bias. In *Ethics of data and analytics* (pp. 254–264). Auerbach Publications. DOI: 10.1201/9781003278290-37

Binns, R. (2018, January). Fairness in machine learning: Lessons from political philosophy. In Conference on fairness, accountability, and transparency (pp. 149-159). PMLR.

Brundage, M., Avin, S., Clark, J., Toner, H., Eckersley, P., Garfinkel, B., . . . Amodei, D. (2018). The malicious use of artificial intelligence: Forecasting, prevention, and mitigation. *arXiv preprint arXiv:1802.07228.*

Buolamwini, J., & Gebru, T. (2018, January). Gender shades: Intersectional accuracy disparities in commercial gender classification. In *Conference on fairness, accountability and transparency* (pp. 77-91). PMLR.

Cath, C. (2018). Governing artificial intelligence: Ethical, legal and technical opportunities and challenges. *Philosophical Transactions. Series A, Mathematical, Physical, and Engineering Sciences, 376*(2133), 20180080. DOI: 10.1098/rsta.2018.0080 PMID: 30322996

Daugherty, P. R., & Wilson, H. J. (2018). *Human+ machine: Reimagining work in the age of AI.* Harvard Business Press.

Eubanks, V. (2018). *Automating inequality: How high-tech tools profile, police, and punish the poor.* St. Martin's Press.

Floridi, L., & Cowls, J. (2022). A unified framework of five principles for AI in society. *Machine learning and the city: Applications in architecture and urban design*, 535-545.

Floridi, L., Cowls, J., Beltrametti, M., Chatila, R., Chazerand, P., Dignum, V., Luetge, C., Madelin, R., Pagallo, U., Rossi, F., Schafer, B., Valcke, P., & Vayena, E. (2018). AI4People—an ethical framework for a good AI society: Opportunities, risks, principles, and recommendations. *Minds and Machines, 28*(4), 689–707. DOI: 10.1007/s11023-018-9482-5 PMID: 30930541

Floridi, L., Cowls, J., King, T. C., & Taddeo, M. (2021). How to design AI for social good: Seven essential factors. *Ethics, Governance, and Policies in Artificial Intelligence*, 125-151.

Floridi, L., & Taddeo, M. (2016). What is data ethics? *Philosophical Transactions. Series A, Mathematical, Physical, and Engineering Sciences, 374*(2083), 20160360. DOI: 10.1098/rsta.2016.0360 PMID: 28336805

Gasser, U., & Almeida, V. A. (2017). A layered model for AI governance. *IEEE Internet Computing, 21*(6), 58–62. DOI: 10.1109/MIC.2017.4180835

Hagendorff, T. (2020). The ethics of AI ethics: An evaluation of guidelines. *Minds and Machines, 30*(1), 99–120. DOI: 10.1007/s11023-020-09517-8

Latonero, M. (2018). Governing artificial intelligence: Upholding human rights & dignity. *Data & Society, 38.*

Lemaignan, S., Newbutt, N., Rice, L., Daly, J., & Charisi, V. (2021). UNICEF guidance on AI for children: Application to the design of a social robot for and with autistic children. *arXiv preprint arXiv:2108.12166.*

Leslie, D. (2019). Understanding artificial intelligence ethics and safety. *arXiv preprint arXiv:1906.05684.*

Mayer-Schönberger, V., & Cukier, K. (2013). *Big data: A revolution that will transform how we live, work, and think.* Houghton Mifflin Harcourt.

Mittelstadt, B. D., Allo, P., Taddeo, M., Wachter, S., & Floridi, L. (2016). The ethics of algorithms: Mapping the debate. *Big Data & Society, 3*(2), 2053951716679679. DOI: 10.1177/2053951716679679

Nissenbaum, H. (2019). Contextual integrity up and down the data food chain. *Theoretical Inquiries in Law, 20*(1), 221–256. DOI: 10.1515/til-2019-0008

Noorbakhsh-Sabet, N., Zand, R., Zhang, Y., & Abedi, V. (2019). Artificial intelligence transforms the future of health care. *The American Journal of Medicine, 132*(7), 795–801. DOI: 10.1016/j.amjmed.2019.01.017 PMID: 30710543

Pasquale, F. (2015). *The black box society: The secret algorithms that control money and information.* Harvard University Press. DOI: 10.4159/harvard.9780674736061

Rahwan, I., Cebrian, M., Obradovich, N., Bongard, J., Bonnefon, J. F., Breazeal, C., Crandall, J. W., Christakis, N. A., Couzin, I. D., Jackson, M. O., Jennings, N. R., Kamar, E., Kloumann, I. M., Larochelle, H., Lazer, D., McElreath, R., Mislove, A., Parkes, D. C., Pentland, A. S., & Wellman, M. (2019). Machine behavior. *Nature, 568*(7753), 477–486. DOI: 10.1038/s41586-019-1138-y PMID: 31019318

Russell, S. J., & Norvig, P. (2016). *Artificial intelligence: a modern approach.* Pearson.

United Nations. (1966). *International Covenant on Civil and Political Rights.* ICCPR.

United Nations. (1989). *Convention on the Rights of the Child*. CRC.

Veale, M., & Binns, R. (2017). Fairer machine learning in the real world: Mitigating discrimination without collecting sensitive data. *Big Data & Society*, *4*(2), 2053951717743530. DOI: 10.1177/2053951717743530

Verma, S. (2019). Weapons of math destruction: How big data increases inequality and threatens democracy. *Vikalpa*, *44*(2), 97–98. DOI: 10.1177/0256090919853933

Vinuesa, R., Azizpour, H., Leite, I., Balaam, M., Dignum, V., Domisch, S., Felländer, A., Langhans, S. D., Tegmark, M., & Fuso Nerini, F. (2020). The role of artificial intelligence in achieving the Sustainable Development Goals. *Nature Communications*, *11*(1), 1–10. DOI: 10.1038/s41467-019-14108-y PMID: 31932590

Voigt, P., & Von dem Bussche, A. (2017). The eu general data protection regulation (gdpr). *A Practical Guide, 1st Ed., Cham: Springer International Publishing, 10*(3152676), 10-5555.

Whittaker, M., Crawford, K., Dobbe, R., Fried, G., Kaziunas, E., Mathur, V., & Schwartz, O. (2018). *AI now report 2018*. AI Now Institute at New York University.

Zuboff, S. (2019). *The Age of Surveillance Capitalism: The Fight for a Human Future at the New Frontier of Power, edn*. Public Affairs.

Chapter 6
A Machine Learning Approach to Social Equity Analysis

Usharani Bhimavarapu
https://orcid.org/0000-0002-0246-1420
Department of Computer Science and Engineering, Koneru Lakshmaiah Education Foundation, Vaddeswaram, India

ABSTRACT

Social justice and social equity are critical frameworks for ensuring fairness and inclusivity in societal systems. This study employs Support Vector Machines (SVM) to classify and analyze social equity across diverse domains, including access to education, healthcare, and resources. By leveraging a comprehensive dataset, features representing equity metrics were extracted and preprocessed using statistical and natural language processing techniques. The SVM model, optimized with kernel functions to handle non-linear data patterns, effectively categorized data into equity-related classes, enabling the identification of disparities. The findings highlight the potential of machine learning in promoting evidence-based decision-making to address inequities and advance social justice initiatives. This study underscores the importance of integrating computational tools with social equity research to create more inclusive and equitable societies.

INTRODUCTION

Social equity refers to the fair and just distribution of resources, opportunities, and access to services across all societal groups. It emphasizes addressing systemic inequalities that disproportionately affect marginalized communities, ensuring that

DOI: 10.4018/979-8-3693-9035-1.ch006

every individual has the tools and resources to succeed. As Guy and McCandless (2012) noted, social equity is both a legacy of social justice movements and a promise for future progress. It extends beyond formal equality, focusing on substantive equity by recognizing historical disadvantages and taking proactive measures to level the playing field.

Transportation systems and environmental conditions are central to social equity concerns. Bullard et al. (2004) highlighted how transportation systems have historically marginalized low-income and minority communities through practices like "transportation racism," where resources are disproportionately allocated to wealthier areas. Similarly, Braubach and Fairburn (2010) emphasized that housing-related environmental risks, such as exposure to pollution, disproportionately impact marginalized populations. Addressing these inequities requires reevaluating infrastructure investments and ensuring that marginalized groups are prioritized in urban planning and environmental policymaking.

As technology becomes integral to public services, algorithmic bias presents a significant challenge to social equity. Bozdag (2013) highlighted how algorithmic filtering, and personalization can inadvertently reinforce existing biases, excluding underrepresented communities from opportunities. Similarly, Eubanks (2018) demonstrated how high-tech tools used to manage public services often profile and punish low-income communities. To promote equity in the digital age, it is essential to design transparent, inclusive algorithms and actively monitor their impact on marginalized groups.

Advanced technologies like drones and autonomous vehicles hold potential to improve social equity in urban environments but also present challenges. Reed et al. (2022) noted that while autonomous vehicles can enhance access to last-mile deliveries for underserved areas, they may exacerbate inequities if not deployed thoughtfully. Similarly, Golub et al. (2013) discussed how new mobility solutions often ignore the unique needs of African American communities in urban settings. Integrating equity considerations into the deployment of these technologies ensures that their benefits are shared across all demographic groups.

Achieving social equity requires deliberate policy interventions and inclusive decision-making processes. Litman (2017) advocated for comprehensive transportation equity evaluations to assess how policies impact different social groups. Policies must not only address immediate disparities but also consider long-term systemic changes, such as promoting access to education, healthcare, and sustainable infrastructure. By combining ethical governance, community engagement, and innovative solutions, societies can fulfill the promise of social equity and create more inclusive and just systems for all.

Artificial Intelligence (AI) has the potential to address systemic inequalities and promote social equity by providing data-driven insights, improving access to services, and enabling more inclusive decision-making. By analyzing complex datasets, AI can identify patterns of inequity and recommend targeted interventions, empowering policymakers to address disparities in education, healthcare, transportation, and other critical sectors. When designed and implemented responsibly, AI can serve as a powerful tool to reduce social inequities and create more inclusive societies.

AI-powered tools can improve access to essential services for underserved populations. For instance, in education, AI-driven platforms provide personalized learning experiences that cater to diverse needs, including those of students with disabilities or limited access to traditional resources. Similarly, in healthcare, AI applications such as telemedicine and AI-enhanced diagnostics improve access to quality care in remote or economically disadvantaged areas, bridging the healthcare gap and addressing inequities in medical outcomes.

Although algorithms have been criticized for perpetuating biases, AI can also be used to mitigate them. By employing fairness-aware machine learning techniques, AI systems can identify and correct biases in decision-making processes, such as hiring, lending, and criminal justice. This ensures that decisions are based on objective criteria rather than discriminatory factors. For example, AI tools can audit hiring algorithms to ensure equitable opportunities for all candidates, irrespective of race, gender, or socioeconomic status.

AI is revolutionizing urban planning by optimizing resource allocation and improving service delivery. For example, AI-driven analysis of transportation data can identify underserved areas and suggest equitable solutions, such as expanded public transit networks or more efficient routes. Similarly, AI can model environmental risks, such as pollution or climate change impacts, to prioritize interventions in vulnerable communities. These applications enable planners to address social and environmental inequities more effectively.

AI-powered platforms can amplify the voices of marginalized communities by analyzing public opinion and feedback at scale. Natural language processing (NLP) tools can process large volumes of data from social media, surveys, and public forums, helping policymakers understand the unique needs of different groups. This ensures that marginalized populations are included in decision-making processes and that their concerns are addressed in public policies.

While AI offers immense potential for promoting social equity, it also comes with challenges, such as ensuring transparency, avoiding new forms of bias, and maintaining accountability. To harness AI for equity, it is essential to adopt ethical AI practices, prioritize fairness in algorithm design, and involve diverse stakeholders in its development. By addressing these challenges, AI can be a transformative force in creating a more equitable and inclusive society.

Algorithmic personalization models (Bozdag, 2013) often introduce biases that can lead to echo chambers and lack of exposure to diverse perspectives, while social media-based opinion-mining approaches (Han et al., 2020; Li et al., 2020) can struggle with data reliability due to misinformation and biased content. Models for transportation equity (Litman, 2017) and social inequities in environmental risks (Braubach & Fairburn, 2010) face challenges in capturing nuanced disparities across populations. Similarly, transportation models addressing racial and social inequities (Bullard et al., 2004; Golub et al., 2013) often fail to comprehensively evaluate systemic barriers affecting marginalized communities. Advanced routing algorithms for urban logistics (Cattaruzza et al., 2017) lack scalability when applied to large, complex networks. Drone-based delivery models (Chiang et al., 2019; Khan et al., 2019; Park et al., 2018; Moshref-Javadi & Winkenbach, 2021) face limitations in terms of regulatory barriers, public acceptance, and energy efficiency, while their environmental impact models fail to consider broader ecological consequences (Torija et al., 2020). AI-driven medical drones (Damoah et al., 2021) present ethical concerns and dependency on data quality, whereas UAV-based logistics models (Li et al., 2023; Mendonca et al., 2022) lack comprehensive frameworks for integrating ground-based systems. Last-mile delivery models using autonomous vehicles (Reed et al., 2022) remain challenged by urban congestion and safety concerns. Latent Dirichlet Allocation (LDA) models (Jelodar et al., 2019; Maier et al., 2018) struggle with interpretability and scalability in large datasets. Public acceptance studies for aviation technologies (Tepylo et al., 2023; Zhu, 2019) often fail to account for cultural and regional differences in attitudes. Moreover, environmental justice models in transportation (Schweitzer & Valenzuela, 2004) lack robust mechanisms to quantify long-term health impacts. Drone-centric supply chain models (Rejeb et al., 2023; Jahani et al., 2024) face implementation challenges in densely populated urban areas and are hindered by insufficient legal frameworks. Furthermore, critiques of high-tech automation (Eubanks, 2018) reveal inherent risks of perpetuating inequalities, while reviews on Google Wing's delivery drones (Zenz & Powles, 2024) underscore the public's resistance to monopolized airspace control and technological determinism.

The integration of PCA, TF-IDF, and SVM can effectively address many limitations of existing models by combining dimensionality reduction, feature extraction, and robust classification capabilities. Principal Component Analysis (PCA) helps reduce the dimensionality of complex datasets while preserving essential information, allowing it to overcome scalability issues seen in models like Latent Dirichlet Allocation (Jelodar et al., 2019; Maier et al., 2018) and advanced routing algorithms (Cattaruzza et al., 2017). PCA also enhances computational efficiency, making models more adaptable to large-scale applications such as urban logistics and UAV-based systems (Chiang et al., 2019; Li et al., 2023). TF-IDF excels in extracting meaningful textual features, addressing the interpretability and bias issues associated with algorithmic personalization (Bozdag, 2013) and social media opinion mining (Han et al., 2020; Li et al., 2020). By identifying significant terms and reducing noise from unimportant words, TF-IDF enhances the reliability of content analysis in areas like transportation equity (Litman, 2017) and public perception of technologies (Tepylo et al., 2023).

Support Vector Machines (SVM) bring robust classification capabilities, enabling effective handling of imbalanced datasets and reducing bias. This is particularly important for addressing systemic disparities in environmental justice (Schweitzer & Valenzuela, 2004) and transportation equity (Bullard et al., 2004; Golub et al., 2013). SVM's ability to work well with sparse, high-dimensional data complements TF-IDF and PCA, providing accurate predictions in scenarios like last-mile delivery optimization (Reed et al., 2022) and drone-based logistics (Moshref-Javadi & Winkenbach, 2021). Furthermore, the combination of these methods allows for better scalability and generalization, addressing public acceptance challenges for drones and aviation technologies (Zhu, 2019; Zenz & Powles, 2024). Together, PCA, TF-IDF, and SVM form a synergistic framework that not only enhances model performance but also mitigates ethical concerns, scalability limitations, and data reliability issues prevalent in existing methodologies.

MATERIALS AND METHODOLOGY

The methodology begins with data collection, encompassing diverse social equity indicators such as access to education, healthcare, and resources. The data is preprocessed using techniques like normalization, encoding, and handling missing values to ensure quality and consistency. Feature extraction is performed using statistical and NLP methods to derive meaningful equity metrics. A Support Vector Machine (SVM) model is then employed, utilizing kernel functions to handle both linear and non-linear data patterns. The model is trained and validated using cross-validation to optimize performance. Finally, the SVM classifier predicts equity-related categories, enabling the identification of disparities and patterns in the data.

Materials

We collected a comprehensive dataset to analyze academic outcomes and evaluate social media data to understand public concerns and perceptions regarding social equity in urban UAV logistics. The academic outcomes dataset consists of performance metrics, student demographics, and institutional resources, enabling a detailed assessment of educational achievements and disparities. Concurrently, the social media data was gathered from multiple platforms, focusing on public discussions related to urban UAV logistics, such as delivery reliability, environmental impact, accessibility, and inclusivity. By employing advanced data processing techniques, we extracted meaningful insights from these datasets to bridge the gap between academic performance and societal perceptions. This dual approach not only sheds light on the challenges and opportunities in urban UAV logistics but also fosters a deeper understanding of social equity concerns, ensuring that technological advancements benefit diverse communities equitably.

Preprocessing

Data preprocessing is a crucial step in preparing textual data for analysis or modeling. It involves a series of systematic transformations to clean, structure, and standardize the raw text. Common techniques include tokenization, which breaks text into individual words or sentences, and stopword removal to eliminate commonly used words that add little analytical value, such as "is" or "the." Lemmatization and stemming are applied to reduce words to their root forms, ensuring uniformity across similar terms. Text normalization is performed by converting text to lowercase, removing punctuation, numbers, and special characters, and handling contractions. Additionally, more advanced methods like named entity recognition (NER) can identify and classify entities such as names, locations, or dates. For noisy or domain-specific text, such as social media data, spelling correction and slang normalization may also be incorporated. These preprocessing techniques ensure that the textual data is clean, consistent, and meaningful, enabling more accurate and effective NLP modeling and analysis.

Topic Modeling

Topic modeling using Principal Component Analysis (PCA) is an effective dimensionality reduction approach to identify latent topics within a large corpus of text data. Unlike traditional topic modeling methods like Latent Dirichlet Allocation (LDA), PCA focuses on transforming high-dimensional word frequency or term-document matrices into a lower-dimensional space that captures the most significant variance in the data. Initially, text data is preprocessed using NLP techniques, such as tokenization, stopword removal, and vectorization (e.g., TF-IDF or Bag of Words). PCA is then applied to the resulting term-document matrix to reduce dimensionality, allowing the identification of principal components that correspond to dominant themes or topics. These components are linear combinations of words that capture patterns and relationships across the corpus. By examining the highest-weighted words within each principal component, researchers can interpret and label the underlying topics. PCA-based topic modeling is particularly useful for exploring large datasets efficiently, uncovering hidden structures, and enabling visualizations like scatter plots to show document clusters based on shared themes.

Feature Extraction

Feature extraction using NLP techniques is a critical step in transforming raw text into structured data suitable for machine learning or statistical analysis. It involves identifying and extracting meaningful attributes from textual data to represent its semantic and syntactic properties. One widely used technique is the Bag of Words (BoW), which converts text into a numerical format by counting word occurrences, although it disregards word order and context. TF-IDF (Term Frequency-Inverse Document Frequency) builds on this by assigning weights to words based on their importance in a document relative to the corpus, highlighting significant terms. Advanced methods like word embeddings, such as Word2Vec, GloVe, and FastText, generate dense vector representations of words, capturing their contextual and semantic relationships. Techniques like part-of-speech (POS) tagging and dependency parsing provide grammatical and syntactic features, while named entity recognition (NER) identifies specific entities such as names, dates, and locations. N-grams extract sequences of 'n' words to capture context and word order, while topic modeling methods like Latent Dirichlet Allocation (LDA) uncover latent themes in text. Sentiment analysis features capture the emotional tone of the text, and custom domain-specific features, such as keywords or patterns, add task-specific insights. These techniques transform unstructured text into meaningful features, enabling more accurate and insightful modeling for applications like classification, clustering, and predictive analytics.

Social Equity Classification

Social equity classification using Support Vector Machines (SVM) is a powerful approach to analyze and categorize data related to fairness and inclusivity in various domains. SVM, a supervised machine learning algorithm, is particularly well-suited for handling complex and high-dimensional data, making it ideal for social equity studies involving diverse factors like income, access to resources, healthcare, and education. The process begins with collecting and preprocessing data, such as normalizing values, handling missing entries, and encoding categorical variables. Relevant features representing equity metrics are then extracted using statistical or NLP techniques, depending on the data type. The SVM algorithm separates data points into classes by finding the optimal hyperplane that maximizes the margin between them. Kernel functions, such as linear, polynomial, or radial basis function (RBF), are often used to handle non-linear relationships within the data. The model is trained and validated using techniques like cross-validation to ensure robust performance. Once trained, the SVM classifier can effectively predict and classify new data into equity-related categories, such as equitable or inequitable. This approach provides valuable insights for policymakers and organizations to identify disparities and implement targeted interventions to promote social equity.

EXPERIMENTAL RESULTS

The experimental results demonstrate the effectiveness of the Support Vector Machine (SVM) model in classifying social equity data across various domains. After preprocessing the dataset and extracting relevant features, the SVM classifier was trained and evaluated using cross-validation to ensure robust performance. The results revealed significant distinctions between equity-related categories, highlighting patterns of disparity in areas such as healthcare access, education, and resource allocation. The model's accuracy, precision, recall, and F1 score were analyzed, showing promising results in identifying inequities and predicting outcomes related to social justice. These findings underscore the potential of machine learning techniques in advancing social equity analysis and decision-making.

Table 1 highlights the distribution of frequencies for various social equity factors, providing an overview of their prevalence or focus in the data. The most frequently discussed factor is Income Inequality (200 occurrences), followed by Healthcare Access (150) and Employment Opportunities (180). Other significant factors include Food Security (160), Social Mobility (140), and Access to Education (120). Factors such as Disability Access (85), Environmental Justice (75), and Public Safety and Policing (95) appear less frequently, indicating relatively lower emphasis. Overall, the table reflects diverse areas of focus within social equity, emphasizing both systemic and resource-based issues.

Table 1. Frequencies of social equity factors

Social Equity Factor	Frequency
Access to Education	120
Healthcare Access	150
Income Inequality	200
Employment Opportunities	180
Housing Affordability	130
Racial Equality	110

continued on following page

Table 1. Continued

Social Equity Factor	Frequency
Gender Equality	90
Disability Access	85
Environmental Justice	75
Public Safety and Policing	95
Social Mobility	140
Access to Clean Water	105
Food Security	160
Civic Participation	115
Digital Equity	125

Table 2 ranks social equity factors by their relative importance in contributing to predictive models, using importance scores. Access to Education emerges as the most influential factor (0.22), followed by Healthcare Access (0.18) and Income Inequality (0.15). Lower-ranked factors include Public Safety and Policing, Social Mobility, and others with equal scores (0.01), such as Civic Participation and Digital Equity. The ranking illustrates the varying significance of factors in shaping equitable outcomes and informs priority areas for intervention.

Table 2. Feature importance

Feature	Importance Score
Access to Education	0.22
Healthcare Access	0.18
Income Inequality	0.15
Employment Opportunities	0.12
Housing Affordability	0.10
Racial Equality	0.08
Gender Equality	0.05
Disability Access	0.04
Environmental Justice	0.03
Public Safety and Policing	0.02
Social Mobility	0.01
Access to Clean Water	0.01
Food Security	0.01
Civic Participation	0.01
Digital Equity	0.01

Table 3 compares the performance of various models based on accuracy, precision, recall, and F1 score. The Proposed Model demonstrates superior performance across all metrics, achieving 92.5% accuracy and high precision (91.0%) and recall (94.2%). Other models, such as SVM (88.7% accuracy) and XGBoost (90.2%), perform well but fall short of the proposed approach. Models like Logistic Regression (80.2%) and Decision Tree (78.3%) show moderate performance, reflecting varying efficacy in feature extraction.

Table 3. Feature extraction performance comparison

Model	Accuracy (%)	Precision (%)	Recall (%)	F1 Score (%)
Proposed Model	92.5	91.0	94.2	92.5
SVM	88.7	85.6	90.0	87.7
Random Forest	86.5	83.2	89.3	86.2
Logistic Regression	80.2	78.5	82.1	80.3
Decision Tree	78.3	76.2	79.0	77.6
KNN	84.5	82.3	85.7	83.9
Naive Bayes	79.8	77.0	81.5	79.1
AdaBoost	81.4	79.8	83.0	81.3
XGBoost	90.2	89.5	91.8	90.6
MLP	87.3	84.7	88.4	86.5

Table 4 evaluates topic modeling approaches using coherence scores, topic diversity, interpretability, and model complexity. The Proposed Model excels with the highest coherence score (0.75) and high interpretability and diversity, coupled with low complexity. Competing methods, such as BERTopic and Word2Vec + LDA, also perform well, while models like Latent SVM and K-means lag in coherence and diversity, indicating limited utility in identifying equitable topics effectively.

Table 4. Topic modeling performance comparison

Model	Coherence Score	Topic Diversity	Interpretability	Model Complexity
Proposed Model	0.75	High	High	Low
LDA	0.68	Medium	Medium	Medium
NMF	0.72	Medium	Medium	Medium
Latent SVM	0.65	Low	Low	High
LDA + VAE	0.70	High	High	High

continued on following page

Table 4. Continued

Model	Coherence Score	Topic Diversity	Interpretability	Model Complexity
BERTopic	0.73	Medium	High	Medium
K-means	0.60	Low	Low	Low
DeepWalk	0.66	Medium	Low	High
Doc2Vec	0.69	Medium	Medium	High
Word2Vec + LDA	0.74	High	Medium	Medium

Table 5 assesses model performance for social equity classification based on metrics like accuracy, precision, recall, F1 score, and AUC. The Proposed Model consistently outperforms others, achieving 92.5% accuracy and an AUC of 95.0%. XGBoost follows closely with 90.2% accuracy and 93.0% AUC. Models like Decision Tree and Naive Bayes show lower performance, underscoring the importance of sophisticated methods in capturing complex social equity patterns.

Table 5. Social equity classification performance comparison

Model	Accuracy (%)	Precision (%)	Recall (%)	F1 Score (%)	AUC (%)
Proposed Model	92.5	91.0	94.2	92.5	95.0
SVM	88.7	85.6	90.0	87.7	91.5
Random Forest	86.5	83.2	89.3	86.2	90.0
Logistic Regression	80.2	78.5	82.1	80.3	84.0
Decision Tree	78.3	76.2	79.0	77.6	83.5
KNN	84.5	82.3	85.7	83.9	88.4
Naive Bayes	79.8	77.0	81.5	79.1	84.7
AdaBoost	81.4	79.8	83.0	81.3	85.0
XGBoost	90.2	89.5	91.8	90.6	93.0
MLP	87.3	84.7	88.4	86.5	89.8

Table 6 compares models by training time, complexity, and memory usage. The Proposed Model balances low training time (15 minutes) and low complexity with moderate memory usage (350 MB). In contrast, models like MLP and XGBoost are computationally intensive, requiring significant memory and time. Naive Bayes is the most resource-efficient, with the shortest training time (5 minutes) and lowest memory usage (100 MB).

Table 6. Model comparison based on training time and computational complexity

Model	Training Time (Minutes)	Model Complexity	Memory Usage (MB)
Proposed Model	15	Low	350
SVM	30	Medium	600
Random Forest	45	High	750
Logistic Regression	10	Low	200
Decision Tree	25	Medium	500
KNN	35	Medium	550
Naive Bayes	5	Low	100
AdaBoost	40	High	700
XGBoost	50	High	800
MLP	60	High	900

Table 7 examines how models classify specific equity categories, including education, healthcare, income inequality, housing affordability, and gender equality. The Proposed Model demonstrates consistently high classification accuracy across all categories, with the highest accuracy in Education Equity (94.5%). Other models, such as XGBoost, perform well but show slight variability across categories. Logistic Regression and Decision Tree have moderate accuracy, particularly in more nuanced categories like Gender Equality.

Table 7. Classification performance across different equity categories

Model	Education Equity (%)	Healthcare Access (%)	Income Inequality (%)	Housing Affordability (%)	Gender Equality (%)
Proposed Model	94.5	92.0	90.0	91.5	93.2
SVM	89.2	87.8	85.4	86.5	88.9
Random Forest	85.5	83.6	80.2	82.3	84.1
Logistic Regression	79.1	77.3	74.8	76.1	78.4
Decision Tree	77.5	75.0	72.1	73.9	76.0
KNN	83.4	81.0	79.0	80.2	82.5
Naive Bayes	76.0	74.2	72.5	74.0	75.1
AdaBoost	78.9	76.7	75.0	76.5	78.2
XGBoost	91.2	89.5	87.0	88.7	90.0
MLP	88.1	85.4	83.2	84.9	86.7

Table 8 analyzes models based on topic coherence, average topic size, and the number of topics identified. The Proposed Model achieves the highest coherence score (0.75) with moderate topic size (12 terms) and identifies eight topics, balancing granularity and coherence. Other models, such as Word2Vec + LDA and BERTopic, also perform well, while models like Naive Bayes and AdaBoost identify more or fewer topics with lower coherence

Table 8. Topic coherence and topic size comparison

Model	Topic Coherence Score	Average Topic Size (Terms)	Number of Topics Identified
Proposed Model	0.75	12	8
SVM	0.68	15	10
Random Forest	0.72	13	9
Logistic Regression	0.65	14	7
Decision Tree	0.70	11	6
KNN	0.73	12	8
Naive Bayes	0.60	16	12
AdaBoost	0.66	10	5
XGBoost	0.69	14	9
MLP	0.74	13	8

Table 9 categorizes model performance by feature types, including demographic features, access to resources, economic equity, and social mobility. The Proposed Model achieves the highest accuracy across all feature categories, with 93.2% for demographic features and 92.0% for social mobility. XGBoost and SVM follow as strong performers, while simpler models like Naive Bayes and Decision Tree show limited classification accuracy.

Table 9. Social equity classification by feature type

Model	Demographic Features (%)	Access to Resources (%)	Economic Equity (%)	Social Mobility (%)	Overall Classification Accuracy (%)
Proposed Model	93.2	91.8	89.4	92.0	92.5
SVM	87.1	84.5	83.2	86.3	88.7
Random Forest	84.8	82.6	80.9	82.7	86.5

continued on following page

Table 9. Continued

Model	Demographic Features (%)	Access to Resources (%)	Economic Equity (%)	Social Mobility (%)	Overall Classification Accuracy (%)
Logistic Regression	77.4	75.9	73.5	74.0	80.2
Decision Tree	75.3	73.4	71.2	72.1	78.3
KNN	82.0	79.7	78.1	80.0	84.5
Naive Bayes	74.5	72.3	70.4	71.6	79.8
AdaBoost	76.8	74.6	73.5	75.2	81.4
XGBoost	90.1	88.0	85.4	87.5	90.2
MLP	86.2	83.9	82.1	83.7	87.3

Table 10 evaluates models' abilities to detect disparities in income inequality, healthcare, and education. The Proposed Model achieves the highest overall detection accuracy (92.5%), with exceptional performance in educational inequities (94.0%). XGBoost is a close second, while models like Logistic Regression and Decision Tree exhibit lower accuracy, underscoring their limited capability to address complex disparity detection challenges effectively.

Table 10. Comparison of equity disparity detection

Model	Detection of Income Inequality	Detection of Healthcare Disparities	Detection of Educational Inequities	Overall Disparity Detection (%)
Proposed Model	93.0	91.5	94.0	92.5
SVM	88.3	86.0	89.5	88.7
Random Forest	85.2	83.4	86.5	86.5
Logistic Regression	79.1	77.2	80.4	80.2
Decision Tree	76.4	74.3	78.9	77.5
KNN	82.0	79.3	82.7	83.4
Naive Bayes	75.9	73.5	76.1	76.0
AdaBoost	77.8	75.6	78.3	78.9
XGBoost	89.7	87.8	90.0	90.2
MLP	85.6	83.0	86.1	87.3

CONCLUSION

This research demonstrates the effectiveness of Support Vector Machines (SVM) in classifying and analyzing social equity, providing a data-driven approach to identify and address disparities in critical domains. The use of SVM, combined with robust feature extraction and preprocessing techniques, enabled the creation of a reliable classification model that can support policymakers and stakeholders in making informed decisions. The results reveal that machine learning can play a pivotal role in advancing social justice by offering actionable insights into inequities and fostering targeted interventions. Future work should explore integrating additional data sources and advanced algorithms to enhance classification accuracy and expand the scope of social equity analysis. This study reaffirms the importance of leveraging computational tools to promote fairness, inclusivity, and justice in societal systems.

REFERENCES

Bakshy, E., Messing, S., & Adamic, L. A. (2015). Exposure to ideologically diverse news and opinion on Facebook. *Science*, *348*(6239), 1130–1132. DOI: 10.1126/science.aaa1160 PMID: 25953820

Bozdag, E. (2013). Bias in algorithmic filtering and personalization. *Ethics and Information Technology*, *15*(3), 209–227. DOI: 10.1007/s10676-013-9321-6

Braubach, M., & Fairburn, J. (2010). Social inequities in environmental risks associated with housing and residential location—A review of evidence. *European Journal of Public Health*, *20*(1), 36–42. DOI: 10.1093/eurpub/ckp221 PMID: 20047933

Bullard, R. D., Johnson, G. S., & Torres, A. O. (Eds.). (2004). *Highway robbery: Transportation racism & new routes to equity*. South End Press.

Cattaruzza, D., Absi, N., Feillet, D., & Gonzalez-Feliu, J. (2017). Vehicle routing problems for city logistics. *EURO Journal on Transportation and Logistics*, *6*(1), 51–79. DOI: 10.1007/s13676-014-0074-0

Chiang, W. C., Li, Y., Shang, J., & Urban, T. L. (2019). Impact of drone delivery on sustainability and cost: Realizing the UAV potential through vehicle routing optimization. *Applied Energy*, *242*, 1164–1175. DOI: 10.1016/j.apenergy.2019.03.117

Damoah, I. S., Ayakwah, A., & Tingbani, I. (2021). Artificial intelligence (AI)-enhanced medical drones in the healthcare supply chain (HSC) for sustainability development: A case study. *Journal of Cleaner Production*, *328*, 129598. DOI: 10.1016/j.jclepro.2021.129598

Eubanks, V. (2018). *Automating inequality: How high-tech tools profile, police, and punish the poor*. St. Martin's Press.

Golub, A., Marcantonio, R. A., & Sanchez, T. W. (2013). Race, space, and struggles for mobility: Transportation impacts on African Americans in Oakland and the East Bay. *Urban Geography*, *34*(5), 699–728. DOI: 10.1080/02723638.2013.778598

Guy, M. E., & McCandless, S. A. (2012). Social equity: Its legacy, its promise. *Public Administration Review*, *72*(s1, Suppl. S1), S5–S13. DOI: 10.1111/j.1540-6210.2012.02635.x

Han, X., Wang, J., Zhang, M., & Wang, X. (2020). Using social media to mine and analyze public opinion related to COVID-19 in China. *International Journal of Environmental Research and Public Health*, *17*(8), 2788. DOI: 10.3390/ijerph17082788 PMID: 32316647

Jahani, H., Khosravi, Y., Kargar, B., Ong, K. L., & Arisian, S. (2024). Exploring the role of drones and UAVs in logistics and supply chain management: A novel text-based literature review. *International Journal of Production Research*, ●●●, 1–25.

Jelodar, H., Wang, Y., Yuan, C., Feng, X., Jiang, X., Li, Y., & Zhao, L. (2019). Latent Dirichlet Allocation (LDA) and topic modeling: Models, applications, a survey. *Multimedia Tools and Applications*, *78*(11), 15169–15211. DOI: 10.1007/s11042-018-6894-4

Khan, R., Tausif, S., & Javed Malik, A. (2019). Consumer acceptance of delivery drones in urban areas. *International Journal of Consumer Studies*, *43*(1), 87–101. DOI: 10.1111/ijcs.12487

Li, J., Xu, Q., Cuomo, R., Purushothaman, V., & Mackey, T. (2020). Data mining and content analysis of the Chinese social media platform Weibo during the early COVID-19 outbreak: Retrospective observational infoveillance study. *JMIR Public Health and Surveillance*, *6*(2), e18700. DOI: 10.2196/18700 PMID: 32293582

Li, X., Tupayachi, J., Sharmin, A., & Martinez Ferguson, M. (2023). Drone-aided delivery methods, challenges, and the future: A methodological review. *Drones (Basel)*, *7*(3), 191. DOI: 10.3390/drones7030191

Litman, T. (2017). *Evaluating transportation equity*. Victoria Transport Policy Institute.

Maier, D., Waldherr, A., Miltner, P., Wiedemann, G., Niekler, A., Keinert, A., Pfetsch, B., Heyer, G., Reber, U., Häussler, T., Schmid-Petri, H., & Adam, S. (2018). Applying LDA topic modeling in communication research: Toward a valid and reliable methodology. *Communication Methods and Measures*, *12*(2-3), 93–118. DOI: 10.1080/19312458.2018.1430754

Mendonca, N., Murphy, J. R., Patterson, M. D., Alexander, R., Juárez, G., & Harper, C. (2022). *Advanced air mobility vertiport considerations: A list and overview*. American Institute of Aeronautics and Astronautics, Inc.

Moshref-Javadi, M., & Winkenbach, M. (2021). Applications and research avenues for drone-based models in logistics: A classification and review. *Expert Systems with Applications*, *177*, 114854. DOI: 10.1016/j.eswa.2021.114854

Park, J., Kim, S., & Suh, K. (2018). A comparative analysis of the environmental benefits of drone-based delivery services in urban and rural areas. *Sustainability (Basel)*, *10*(3), 888. DOI: 10.3390/su10030888

Reed, S., Campbell, A. M., & Thomas, B. W. (2022). The value of autonomous vehicles for last-mile deliveries in urban environments. *Management Science*, *68*(1), 280–299. DOI: 10.1287/mnsc.2020.3917

Rejeb, A., Rejeb, K., Simske, S. J., & Treiblmaier, H. (2023). Drones for supply chain management and logistics: A review and research agenda. *International Journal of Logistics*, *26*(6), 708–731. DOI: 10.1080/13675567.2021.1981273

Schweitzer, L., & Valenzuela, A.Jr. (2004). Environmental injustice and transportation: The claims and the evidence. *Journal of Planning Literature*, *18*(4), 383–398. DOI: 10.1177/0885412204262958

Tepylo, N., Straubinger, A., & Laliberte, J. (2023). Public perception of advanced aviation technologies: A review and roadmap to acceptance. *Progress in Aerospace Sciences*, *138*, 100899. DOI: 10.1016/j.paerosci.2023.100899

Torija, A. J., Li, Z., & Self, R. H. (2020). Effects of a hovering unmanned aerial vehicle on urban soundscapes perception. *Transportation Research Part D, Transport and Environment*, *78*, 102195. DOI: 10.1016/j.trd.2019.11.024

Zenz, A., & Powles, J. (2024). Resisting technological inevitability: Google Wing's delivery drones and the fight for our skies. *Philosophical Transactions. Series A, Mathematical, Physical, and Engineering Sciences*, *382*(2285), 20240107. DOI: 10.1098/rsta.2024.0107 PMID: 39533913

Zhu, X. (2019). Segmenting the public's risk beliefs about drone delivery: A belief system approach. *Telematics and Informatics*, *40*, 27–40. DOI: 10.1016/j.tele.2019.05.007

Chapter 7
Addressing the Critical Challenge of AI Bias Through Empathy:
The Role of Empathizing–Systemizing Theory in AI Teams

Rita Tegon
https://orcid.org/0000-0002-1225-1806
University of Salamanca, Spain

ABSTRACT

AI is increasingly shaping critical sectors such as healthcare, education, commerce, and governance, offering advancements in efficiency, predictive accuracy, and personalised solutions. However, its rapid adoption raises ethical concerns, particularly regarding biases embedded in AI systems. These biases stem from data used to train machine learning models, reflecting implicit ontological, psychological, cognitive, and cultural assumptions. As such, AI biases perpetuate structural inequalities and reinforce historical discrimination. This chapter explores the urgent need for global frameworks to mitigate AI bias, highlighting policy initiatives to address these challenges. It then shifts to an interdisciplinary perspective, proposing empathy as a tool for tackling AI bias. Drawing on Baron-Cohen's Empathizing-Systemizing theory, the chapter argues that fostering empathy in AI development strengthens accountability and trust, essential for ethical AI deployment. Gender-balanced teams, with greater empathy emphasis, can offer solutions to reduce bias, supporting equitable AI systems.

DOI: 10.4018/979-8-3693-9035-1.ch007

INTRODUCTION

Artificial Intelligence (AI) has become a transformative and pervasive force in modern society, fundamentally reshaping critical domains such as healthcare, education, financial services, commerce, employment, housing, transportation, and governance. Its applications promise significant advancements in efficiency, predictive accuracy, and personalised solutions, addressing societal challenges on an unprecedented scale. However, these capabilities are accompanied by profound challenges, particularly ethical concerns arising from biases embedded within AI systems. Such concerns are evident in numerous reports, including the key findings outlined in the *AI Index 2024 Annual Report*, recently released by an independent initiative under the Stanford Institute for Human-Centered Artificial Intelligence (Maslej et al., 2024). These biases, often systemic rather than incidental, are not merely technical anomalies; they reflect broader socio-political issues that perpetuate structural inequalities, reinforce historical discrimination, and threaten individual and collective freedoms. Bias in AI typically originates from the datasets used to train machine learning models, which are shaped by implicit and explicit ontological, psychological, cognitive, and cultural assumptions. Such biases can manifest in various forms, including gender, racial, and socio-economic discrimination, which in turn amplify inequities in critical areas like hiring, lending, and law enforcement (Eubanks, 2018). This growing awareness has catalysed interdisciplinary efforts to address AI bias, encompassing technical, ethical, and policy-oriented strategies.

This chapter addresses these challenges, proposing actionable strategies to mitigate AI bias through a holistic and interdisciplinary lens. A central argument is that technical solutions must be underpinned by a robust socio-psychological foundation, with empathy playing a pivotal role. Empathy, the capacity to understand and respond to the emotions and perspectives of others, directly influences mechanisms of accountability and trust—two elements consistently highlighted in legal frameworks and scholarly literature as essential for ethical AI development. The absence of empathy in AI development not only exacerbates bias but also undermines efforts to build systems that respect human rights and align with societal expectations of fairness and transparency.

The chapter commences by reviewing global official documents, emphasising that, despite differences in regulatory development, both existing and proposed frameworks consistently identify bias as a critical issue closely linked to accountability and trust. Building on this foundation, the discussion transitions to an in-depth analysis of how bias operates and undermines both accountability and trust within AI systems. By linking these concepts to empathy, the chapter argues that a lack of empathetic understanding during the design and implementation phases of AI perpetuates discriminatory outcomes, as developers fail to anticipate the lived experiences of diverse user groups. Empathy, therefore, emerges as a foundational element in counteracting bias by fostering a deeper awareness of its origins and implications.

Empathy's relevance is further explored through the lens of *Empathizing-Systemizing (E-S) theory* (Baron-Cohen, Knickmeyer, & Belmonte, 2005) which differentiates between two cognitive dimensions: empathising, associated with emotional understanding, and systemising, linked to analytical reasoning. While systemising has traditionally dominated AI development, the chapter contends that integrating empathising perspectives can enhance ethical design and decision-making. Gendered dimensions of empathy, as evidenced in *E-S theory*, underscore the importance of diversity in development teams; empirical studies suggest that mixed-gender teams, characterised by cognitive diversity, are better equipped to identify and mitigate biases embedded in AI systems.

The methodology employed adopts an interdisciplinary approach, integrating insights from diverse academic fields such as neuroscience, social and gender studies, ethics, and AI development, offering both practical and theoretical insights. By drawing on these varied perspectives, the paper offers a nuanced examination of how biases are embedded in AI systems and their far-reaching implications for societal equity and justice. At the core of the research methodology is a qualitative analysis that synthesises global perspectives on AI regulation and the ethical challenges posed by AI bias. This approach includes a comprehensive review of relevant policy documents and scholarly literature to provide contextual insights, alongside the application of a theoretical framework—most notably the *E-S theory*—to explore the pivotal role of empathy in AI development.

Thus, the chapter outlines two principal research objectives. The first is to elucidate a global consensus on the critical challenges posed by AI bias, as demonstrated in policy documents and academic research. The second objective, which is closely linked to the first, is to explore the *E-S theory* as a foundational framework for designing AI systems that prioritise empathy and fairness. Given that AI bias presents a significant challenge, the chapter proposes empathy, grounded in the *E-S theory*, as a key approach to mitigating this issue. By integrating these interconnected objectives, the aim is to contribute to the broader discourse on developing AI systems that are genuinely equitable, inclusive, and accountable, aligning with the demands of an increasingly interconnected and ethically aware global landscape.

BACKGROUND

The evolving landscape of AI regulation reflects growing global recognition of the challenges posed by unregulated AI development. A multitude of key policy frameworks and initiatives have emerged globally to ensure that AI technologies are developed and deployed responsibly. These frameworks vary significantly depending on regional contexts, societal values, and technological priorities, but they share the common goal of addressing ethical concerns, including AI's potential to exacerbate bias, infringe on privacy, and undermine individual rights. Authoritative platforms such as the *AI Policy Tracker* (Stanford University, 2024), the OECD.AI *Policy Observatory Dashboard* (OECD, 2024a), the UNESCO's *Global AI Ethics and Governance Observatory* (UNESCO, 2024), and the *AI Readiness Index* (Oxford Insights, 2024) monitor the development, implementation, and updates of AI-related legislation worldwide, offering a global perspective on regulatory progress and enabling stakeholders to stay informed about the latest developments in AI oversight.

The *AI Policy Tracker* is an independent, real-time monitoring platform that tracks AI-related policy developments, regulatory actions, and legislative initiatives at a global level. Developed and maintained by the *Future of Life Institute (FLI)*, a non-profit organisation, it provides detailed, actionable views of specific regulatory actions and trends across different jurisdictions, allowing users to explore AI policies by country, region, sector, and policy type. This makes it particularly valuable for those seeking immediate insights into policy shifts as they unfold.

In contrast, the OECD.AI *Policy Observatory Dashboard*, financed through contributions from OECD member countries and other stakeholders, including public and private entities that support the organisation's goals, offers a more analytical framework, providing in-depth data on AI strategies, national policies, and governance models from OECD member countries and beyond. The platform is designed to facilitate comparisons between different national approaches to AI governance, offering robust visualisations and datasets that reflect AI's economic, social, and ethical impacts.

UNESCO's *Global AI Ethics and Governance Observatory* provides a platform for tracking and promoting ethical AI development worldwide. It brings together research, policies, and initiatives to foster responsible AI practices that address societal and ethical challenges. The observatory aims to support the alignment of AI advancements with human rights and social justice, offering resources for governments, researchers, and industry stakeholders to navigate the complex landscape of AI ethics.

The *AI Readiness Index* by Oxford Insights evaluates the ability of countries (up from 193 in 2023) to integrate artificial intelligence into public service delivery. Using 39 indicators across 10 dimensions grouped under three pillars—Government,

Technology Sector, and Data and Infrastructure—it provides a detailed framework for assessing AI preparedness globally, highlighting disparities in AI readiness and offering best practices to support equitable and innovative AI implementation.

Based on information gathered from these platforms, though not limited to them, the following provides a review of the most recent global AI legislation, highlighting the diverse regulatory approaches designed to address associated risks while promoting innovation.

Canada was one of the first countries to develop a national AI strategy, introducing the *Pan-Canadian Artificial Intelligence Strategy* in 2017 (Canadian Institute for Advanced Research, 2017). This initiative, led by the *Canadian Institute for Advanced Research*, aims to promote AI innovation, strengthen research, and advance ethical practices in AI development. The *Artificial Intelligence and Data Act* (Government of Canada, 2023), focuses on ensuring trust in AI technologies by addressing risks such as algorithmic bias, transparency, and misuse of data. Canada also plays a significant role on the global stage, co-founding the *Global Partnership on AI* in 2020: it seeks to foster international collaboration on responsible AI deployment, making Canada a leader in aligning AI ethics and governance globally (Barrett et al., 2023).

The European Union is at the forefront of AI regulation, with the Artificial Intelligence Act (European Parliament & Council of the European Union, 2024) being a cornerstone of its strategy. It proposes a risk-based framework that categorises AI applications into four tiers based on the potential risk they pose to individuals' rights and safety. High-risk AI systems, such as those used in healthcare, transportation, and law enforcement, face stringent requirements, including transparency, accountability, and robust safeguards against bias and discrimination. The EU's regulatory approach aims to balance technological innovation with public protection, ensuring that AI systems are both trustworthy and aligned with fundamental rights. This approach is part of a broader EU initiative to create a unified regulatory environment for digital technologies, which also includes data protection regulations like the *General Data Protection Regulation* (GDPR) (European Parliament & Council of the European Union, 2016).

The UK's AI policy framework also centres around fostering innovation while addressing ethical risks. Its *National AI Strategy*, released in 2021 (UK Government, 2021), provides a comprehensive vision for AI governance, focusing on three key areas: strengthening AI research and development, improving AI skills and infrastructure, and addressing public trust concerns. Unlike the EU's approach, the UK favours sector-specific guidelines rather than overarching legislation, enabling flexibility in addressing AI's diverse applications.

China's AI regulatory framework reflects its top-down governance model, focusing on both the rapid advancement of AI technologies and strict control over their ethical implications. The country has enacted a series of regulations that emphasise

AI's role in national development, including the *New Generation Artificial Intelligence Development Plan* (State Council of the People's Republic of China, 2017), which outlines a strategic roadmap for becoming the global leader in AI by 2030. Chinese regulations prioritise innovation but have begun to incorporate more ethical considerations, especially regarding data privacy, algorithmic transparency, and bias. Notably, the *Measures for the Administration of Algorithmic Recommendation Services* (Translate, 2022) aim to curb the harmful effects of biased or discriminatory algorithmic decision-making, particularly in areas like e-commerce and social media. China's model contrasts with that of the EU, focusing on centralised control while attempting to balance AI's risks and benefits through government oversight (Rolf, 2023).

In contrast, the United States has adopted a fragmented approach to AI regulation, unlike the more unified frameworks in the EU and China. There is currently no single, comprehensive AI law in the United States. Instead, various federal agencies have issued guidelines for specific AI applications, such as facial recognition, autonomous vehicles, and health AI. One notable legislative effort is the *Algorithmic Accountability Act* (U.S. Congress, 2023), introduced in Congress. This bill is aimed at increasing transparency and accountability in AI systems, particularly those used in high-risk sectors like healthcare and criminal justice. The bill mandates that companies assess the impacts of their algorithms and ensure they do not disproportionately harm marginalised groups. Alongside this document, the United States recently introduced the *Eliminating Bias in Algorithmic Systems (BIAS) Act*, a bill focusing specifically on algorithmic bias and discrimination, proposing that each federal agency utilising, funding, or overseeing AI establish a Civil Rights Office (U.S. Congress, 2024). The Civil Rights Offices are required to submit biennial reports to Congress, outlining the risks associated with algorithmic systems, the measures implemented to address these risks, and proposals for additional legislative or administrative actions. The *BIAS Act* establishes an interagency working group, led by the Department of Justice, to coordinate efforts and share best practices across federal agencies to protect civil rights in AI systems. It aims to address algorithmic bias in critical sectors like healthcare, finance, and law enforcement, where AI systems have disproportionately harmed marginalised communities, such as through discriminatory facial recognition or risk assessment tools.

In Southeast Asia, the approach to AI regulation emphasises fostering innovation while addressing risks in a more flexible manner.

Countries like Singapore have developed proactive AI strategies that include ethics guidelines and an emphasis on AI governance through international collaborations. The *National Artificial Intelligence Strategy* (NAIS) of Singapore, as outlined in the 2019 and 2023 documents, reflects the country's commitment to being a global leader in AI development while ensuring responsible use. The 2019 document focuses on establishing Singapore as an AI hub by investing in talent, infrastructure, and international partnerships. Key priorities include the development of AI in sectors such as healthcare, transport, and finance, alongside promoting ethical governance (Singapore Government, 2019). It aims to build a trusted AI ecosystem through frameworks that ensure transparency, fairness, and accountability. In the 2023 update, the strategy introduces more specific measures for AI governance, particularly fostering public trust through improved standards and risk management. This strategy places greater emphasis on AI's role in enhancing public sector efficiency and addressing challenges such as the digital divide (Singapore Government, 2023). Together, these documents position Singapore as a leader in integrating AI into society, with a robust focus on regulation, collaboration, and ethical considerations.

India, similarly, is making strides in establishing a comprehensive AI framework through its National Strategy for Artificial Intelligence (NITI Aayog, 2018). This strategy outlines AI's potential to drive economic growth and societal benefits, particularly in sectors such as healthcare, agriculture, education, smart cities, and smart mobility. The Indian government is also grappling with issues of data privacy, infrastructure, and accountability, aiming to prevent the misuse of AI while ensuring fairness. National discussions are increasingly concerned with AI-driven inequality and data sovereignty, as India seeks to regulate the technology while also fostering its widespread use.

AI regulation in Africa is guided by the *African Union's (AU) Continental AI Strategy* (2024), which advocates for a people-centric, development-oriented, and inclusive approach. This strategy ensures that AI technologies are used to benefit African societies while addressing global risks such as data privacy and algorithmic bias. The strategy is closely linked to the *AU's Agenda 2063*, which focuses on sustainable development across the continent. Notably, countries such as Kenya and Ghana are aligning their national AI policies with the AU's framework, ensuring that the broader continental strategy influences national regulatory practices.

New Zealand's approach to AI regulation is defined by its recently launched *AI Strategy Blueprint* (New Zealand Government, 2023) which prioritises transparency, inclusivity, and trust. The strategy is in its early stages and emphasises collaboration between the government, academia, and industry. New Zealand focuses on harnessing AI for social and economic benefits while addressing risks like algorithmic bias and privacy concerns. Its frameworks stress the ethical deployment of AI without stifling innovation, and the government actively consults stakeholders to refine its AI governance.

Australia's approach to artificial intelligence regulation emphasises transparency, accountability, and trust in the public sector. The Australian Government has developed the *National Framework for Assurance of Artificial Intelligence*, aiming to provide clear guidelines for the ethical and responsible use of AI within government operations. This framework is designed to ensure AI systems are implemented in ways that are transparent, accountable, and align with ethical standards, particularly in addressing the challenges of bias, discrimination, and privacy concerns. It outlines steps for assessing AI risks, ensuring the protection of public data, and fostering public trust in the deployment of AI technologies. The framework is intended to guide the development of AI systems that are both effective and aligned with public welfare, while minimising potential social and economic harms (Australian Government, 2024).

In Latin America, AI regulation is gradually taking shape as various countries in the region recognise the importance of establishing frameworks that balance innovation with responsible AI use.

Argentina's approach to AI regulation is emerging through the proposed Bill 2505-D-2023 that outlines key terms, principles, and prohibitions, with the aim of ensuring ethical and responsible AI development (Morales, 2023). This bill includes provisions to penalise AI-related crimes, particularly those that impact sexual integrity. A notable component of the proposal is the establishment of a national registry of AI systems, designed to conduct impact assessments and risk classifications. Additionally, the bill emphasises the development of AI systems that respect fundamental rights and democratic values, alongside a transparency and data protection programme to safeguard against misuse of AI technologies.

Brazil is in the process of developing a bill aimed at creating a regulatory framework for AI development and use, which will focus on protecting citizens' rights, categorising risks, and establishing governance mechanisms and sanctions. While the bill is still under consideration, it is expected to include a multistakeholder approach, involving experts from various sectors, including industry and civil society. A key element of the regulatory framework will be ensuring that AI systems comply with data protection laws. This is already being managed by the *National Data Protection Authority*, which was established in 2018 under the *General Data Protection Law*, ensuring that AI systems are aligned with data protection principles. Compliance measures for AI systems will likely include fines, penalties, incident reporting, and audits (Brazilian Government, 2018).

The *Política Nacional de Inteligencia Artificial* of Chile, developed under the Ministry of Science, Technology, Knowledge and Innovation, is structured around three main pillars: enabling factors, AI development and adoption, and ethics and regulation (Ministry of Science, Technology, Knowledge and Innovation. (2021). The policy aims to empower citizens in the development and application of AI, while encouraging active participation in discussions on the social, ethical, and economic implications of AI technologies. Its objectives include developing human capital, technological infrastructure, and data availability to enable AI progress. The policy also prioritises fostering AI research, development, and innovation, and ensuring its adoption in both the public and private sectors. Additionally, the policy focuses on creating a consensus on the ethical, regulatory, and socio-economic impacts of AI, including issues such as privacy, transparency, and gender equality.

In November 2019, the Colombian government adopted the *Digital Transformation and Artificial Intelligence policy* (Consejo Nacional de Política Económica y Social, República de Colombia, & Departamento Nacional de Planeación, 2019). This policy aims to create socio-economic conditions conducive to AI development in Colombia, focusing on the adoption of an ethical framework for responsible and inclusive AI development, the utilisation of data as a fundamental pillar for AI development, and the establishment of a competitive AI-driven market. In November 2023, the Colombian Senate passed Law 059/23, which sets public policy guidelines for AI development, use, and implementation (Senado de la República, 2023).

In 2018, Mexico launched its AI strategy, aiming to enhance competitiveness, foster inclusive development, and reduce inequality. The initiative was supported by the IA2030Mx citizen coalition, consisting of nine institutions from various sectors (IA2030Mx, 2018). The strategy focuses on using AI as a tool for social and economic progress, while emphasising ethical development and mitigating both current and future social and environmental risks.

In February 2024, Peru's Congress introduced a bill (BoL 07033/2023) to regulate AI use, aimed at safeguarding citizens' rights while encouraging ethical AI development (Madariaga Zeballos, 2023). Modelled after the *EUs AI Act*, the bill defines key terms, establishes risk categories, and holds developers accountable for AI system errors. It also bans manipulative AI practices and mandates verification, certification, and civil liability insurance.

The United Arab Emirates (UAE) has implemented a comprehensive national AI strategy, which focuses on positioning AI as a key driver of its future development. The UAE's strategy, which is aligned with its vision to become a global leader in AI, is built around several pillars, including the creation of a robust regulatory framework, investment in AI research, and the promotion of AI adoption across both public and private sectors. The strategy emphasises the ethical development and deployment of AI technologies, ensuring that these systems are designed to enhance societal welfare, foster innovation, and address global challenges. The UAE government has established initiatives to boost AI talent, support AI-driven businesses, and integrate AI solutions across various sectors such as healthcare, transportation, and government services. This commitment to fostering an AI-enabled future underscores the nation's focus on trust, accountability, and transparency in AI applications, aiming for responsible innovation in line with international standards (UAE Artificial Intelligence Strategy, 2024).

Russia's national AI strategy, adopted in 2019 and aligned with its broader national development goals, also focuses on establishing the country as a global leader in artificial intelligence technologies by 2030. The strategy includes six key objectives: fostering a competitive AI industry, cultivating talent through education and training, developing data and computational infrastructure, enhancing AI-related research, ensuring ethical standards and transparency, and integrating AI solutions across sectors such as healthcare, transport, and manufacturing. A notable emphasis is placed on the secure and responsible use of AI, particularly in state and national security applications. By prioritising transparency and ethical considerations, the strategy aims to build trust in AI technologies while advancing innovation and economic growth in line with international standards (Decree of the President of the Russian Federation No. 490, 2019).

Alongside these significant documents, which represent only the tip of the iceberg in the vast array of global regulatory efforts, it is noteworthy that during the G7 Summit held in Cernobbio, Italy, on 15 October 2024, G7 nations reaffirmed their commitment to transparency, ethical standards, and accountability in AI. This declaration builds on the momentum of earlier discussions, particularly those at the 2023 Hiroshima Summit. The *2024 Ministerial Declaration*, a key component of the Hiroshima AI Process, underscores the importance of ensuring that AI systems are developed and implemented responsibly, with a pronounced emphasis on human rights and fairness (G7 Italy, 2024).

After reviewing key legislative documents regulating AI globally, it is also essential to examine recent contributions that provide a deeper understanding of AI governance. These contributions include comprehensive frameworks and platforms that address governance models, risk assessments, and ethical considerations. Examples include the *AI Index 2024 Annual Report* by Stanford HAI (Maslej et al., 2024), the *OECD Report on AI Risks, Benefits, and Policy Imperatives* (OECD, 2024b) and the *AI risk repository* (MIT, 2024).

The top ten key takeaways of the *AI Index 2024 Annual Report* by Stanford HAI, reveal both advancements and significant challenges, especially regarding biases, trust, and accountability. While AI excels in specific tasks, such as image classification and language understanding, it still struggles with more complex tasks like advanced mathematics and commonsense reasoning. This highlights the limitations of AI systems and the importance of ensuring that these technologies are not over-relied upon in high-stakes contexts. Moreover, the rapid pace of AI development is largely driven by industry, with major corporations, such as OpenAI and Google, dominating the creation of large models. These models come with staggering costs, raising concerns about the accessibility and accountability of AI technologies. As AI models become more complex, the costs and risks associated with their use escalate, further complicating the trust and accountability issues surrounding them. The lack of standardised evaluations for AI responsibility is another critical issue. The *AI Index 2024 Annual Report* reports that responsible AI benchmarks vary widely between companies, making it difficult to compare the ethical risks and limitations of different models. This inconsistency undermines trust in AI and raises questions about how accountable developers are for the systems they create. Without a clear framework for evaluating responsibility, biases embedded in AI models—such as those related to race, gender, and socio-economic status—can go unchecked, perpetuating inequality and reinforcing societal discrimination. At the same time, generative AI investment has soared, suggesting an increased focus on AI systems that produce content, but also raising questions about the ethics of their use, especially when these models may inadvertently reinforce biases. The public is increasingly aware of these risks, with surveys showing rising concern about the societal impacts of AI, which only reinforces the need for ethical oversight and transparent practices.

The *OECD Report on AI Risks, Benefits, and Policy Imperatives*, a detailed analysis of the risks and benefits of AI, while emphasising the dual nature of AI technologies, it synthesises research and expert perspectives on the potential benefits, risks, and necessary policy actions regarding AI. It highlights ten key benefits, including the acceleration of scientific progress, enhanced productivity, and improved decision-making and forecasting. It also outlines ten major risks, such as the enabling of advanced cyberattacks, the spread of disinformation, threats to democracy, the concentration of power, incidents within critical systems, and the exacerbation of inequality and poverty. The document also evaluates current public policy and governance initiatives while identifying remaining gaps in regulation and oversight. Alongside, the report highlights the risks related to bias, discrimination, and algorithmic transparency. The document stresses that robust regulatory frameworks are essential to mitigate these risks while ensuring equitable distribution of AI's benefits. Lastly, the report suggests ten policy priorities, such as defining clearer liability frameworks, establishing AI *red lines*, investing in safety measures, and ensuring effective risk management protocols.

The *AI Risk Domain Taxonomy* developed by the Massachusetts Institute of Technology (MIT) categorises and assesses risks associated with AI technologies. It identifies various domains of risk, such as technical, ethical, and social, and provides frameworks to evaluate potential harm or challenges posed by AI systems. This resource aims to guide researchers, policymakers, and developers in managing and mitigating AI-related risks through comprehensive, structured analysis.

Thus far, we have reviewed an extensive array of materials, including approved and proposed regulations, analytical reports, metrics, and risk assessment tools, which collectively underscore the paramount importance of trust in AI systems. These findings consistently highlight bias as a pervasive and significant threat to trust, reinforcing the necessity for comprehensive strategies aimed at mitigating its effects and fostering equitable, reliable AI outcomes. This analysis sets the stage for a deeper examination of the nature and implications of AI bias, addressing both the moral imperatives and practical considerations required to prevent discriminatory practices in the design and application of AI. By framing bias as an issue that transcends technical dimensions and extends into ethical and societal contexts, we recognise the broader responsibilities involved in AI development. Central to this discourse is the pivotal role of empathy as a conceptual bridge between accountability and trust. Empathy offers a multidimensional framework for understanding and addressing the human impact of biased systems, encouraging practices that align technological innovation with ethical soundness. As we proceed, we will delve into the definitions and manifestations of bias within AI systems and explore how integrating empathy into the AI development process—particularly through the formation of gender-balanced teams—can contribute to the creation of systems that are not only technically proficient but also inherently just and trustworthy.

BIAS IN AI SYSTEMS: THE ROLE OF EMPATHY FOR ACCOUNTABILITY AND TRUST

Technical issues within AI systems not only have technological implications but also raise profound ethical concerns, with one of the most significant being the potential for bias, as highlighted in the previous section. However, it is crucial to recognise that the biases embedded in AI are those present in humans as described by the *Manoogian Codex* (Manoogian and Benson, 2017). This assumption underscores a critical point: biases, in both human and machine-generated content, are not completely removable. Nevertheless, they can be reduced and controlled, although there is no universal point at which to stop mitigating them (Gonen & Goldberg, 2019).

Daniel Kahneman and Amos Tversky are foundational figures in the study of cognitive biases, positing that such biases arise from heuristics—mental shortcuts that simplify decision-making but often lead to systematic errors. Through their development of *Prospect theory*, they demonstrated how people deviate from rational decision-making models due to biases, leading to predictable irrational behaviour (Kahneman & Tversky, 1979). Kahneman, in *Thinking, Fast and Slow*, distinguishes between two systems of thinking: the fast, automatic, and often biased System 1, and the slower, more deliberate System 2, which can help counter biases when applied. This research highlights the systematic nature of biases, which are not random errors but predictable patterns in judgment (Kahneman, 2011). In contrast, Gerd Gigerenzer and Wolfgang Gaissmaier critique the traditional view of cognitive biases as errors in judgment. Instead, they suggest that these biases are rational deviations from formal logic, not flaws in reasoning, and serve as functional strategies for effective decision-making in complex situations (Gigerenzer & Gaissmaier, 2011).

AI systems operate by processing large datasets that may reflect ingrained biases—gender, racial, and cultural stereotypes—that could be inadvertently embedded within the system. When AI systems are trained on such biased datasets, they risk perpetuating and amplifying harmful stereotypes or unintentionally silencing marginalised voices. Nevertheless, in the context of AI systems, biases are not confined to the data used for training; they also extend to the algorithms and models that are built upon such data. For instance, machine learning models often perform better when trained on large, rich datasets that favour well-represented groups, while marginalising those groups with limited representation. Furthermore, spurious correlations—non-linear relationships that do not necessarily have causal significance—can also amplify biases. For example, an AI model might learn to associate the presence of certain objects or contexts in an image with gender characteristics, even when those associations are purely coincidental. This highlights the challenges that arise when bias becomes deeply embedded in AI systems, making it difficult to disentangle and address. According to Barocas and Selbst (2016), biases in data can be exacerbated by skewed sampling, tainted examples, and limited feature selection, which reinforce existing prejudices. As highlighted by Suresh and Guttag (2021) six different types of bias may arise within AI systems: historical bias stems when the state of the world in which the data was generated is flawed; representation bias emerges when building datasets for training a model, if those datasets poorly represent the people that the model will serve; measurement bias occurs when the accuracy of the data varies across groups; aggregation bias results from improperly combining groups, leading to models that may fail to perform effectively for specific populations, favouring the majority instead; evaluation bias occurs when the benchmark data used to assess a model's performance—by comparing it to other models with similar tasks—fails to adequately represent the population the model is intended to serve; deployment bias, on the other hand, arises when the intended problem the model was designed to solve differs from its actual application. When end users apply the model in unintended ways, its performance and reliability cannot be assured.

Bias in AI-supported decision-making systems can lead to significant harms across critical domains, particularly in areas like healthcare, where the promise of personalised and precise interventions often collides with the risks of perpetuating existing inequalities. Without the deliberate inclusion of diverse and representative data, AI systems risk reinforcing harmful biases (Vicente & Matute, 2023; Van Kolfschooten, 2023). These biases manifest in both individual and collective social dimensions, showcasing the pervasive and multifaceted nature of their impact. At the individual level, biases in algorithms can result in illegal discrimination and unfair practices across critical sectors such as hiring, insurance, housing, and education. Such outcomes often exacerbate systemic inequities, leading to a profound loss of opportunities for marginalised groups. For instance, algorithms trained on skewed datasets may perpetuate racial, gender, or socioeconomic disparities, effectively excluding certain individuals from accessing vital resources (Buolamwini & Gebru, 2018). Economic harms are also prominent; algorithmic decisions that govern credit allocation and pricing strategies often disadvantage vulnerable populations, deepening financial inequities. Beyond individual effects, algorithmic bias also inflicts harm on a societal scale. Surveillance technologies disproportionately target specific demographics, resulting in a loss of liberty for certain groups. Additionally, biased AI systems often reinforce harmful stereotypes, fostering social stigmatisation that undermines societal cohesion. These outcomes erode trust and fairness, thereby perpetuating systemic inequities. The concept of dignitary harm further underscores the damaging societal impact of biased algorithms. This form of harm highlights the erosion of recognition, respect, and dignity afforded to marginalised communities. For instance, Hofmann, Kalluri, Jurafsky, and others (2024) detail how algorithmic systems have generated discriminatory outcomes, such as racist decisions based on dialect, amplifying the marginalisation of affected groups.

The presence of bias, therefore, undermines not only fairness but also accountability and trust in AI systems. The erosion of trust in these systems is intrinsically linked to the presence of bias, as algorithms imbued with skewed data or flawed methodologies can subvert principles of equity and impartiality. Trust serves as a cornerstone for the legitimacy of AI systems, particularly given the expectation that decisions rendered by these technologies will be objective and equitable, devoid of preferential treatment towards any specific group (Eubanks, 2018; Noble, 2018). However, when biases manifest—whether using non-representative datasets or poorly constructed algorithms—they harm individuals and perpetuate systemic inequities, further corroding public confidence in these technologies (O'Neil, 2016). This scenario underscores the critical need for the incorporation of empathy into the design and development of AI systems. Research suggests that empathy can play a pivotal role in enhancing accountability by fostering systems that are sensitive to the diverse needs and experiences of their users (Srinivasan & González, 2022). By embedding empathy as a foundational principle in the AI development process, it becomes possible to address the detrimental impacts of bias, mitigate the resulting inequalities, and ultimately rebuild public trust in these technologies. Such an approach not only aims to ensure technical proficiency but also aligns with the ethical imperative to create AI systems that promote fairness, equity, and inclusivity.

Empathy is a multifaceted concept that has been extensively explored across disciplines such as psychology, neuroscience, and philosophy. The term was first introduced over a century ago by Titchener, adapting the German word *Einfühlung*, while discussions surrounding empathy can be traced back to the earliest origins of philosophical thought (Stotland et al., 1978). Despite this long history, empathy remains a complex and often ambiguous concept with numerous interpretations. Cuff et al. (2016) note that there are over forty-three different definitions of empathy in the literature. Across various fields, the common thread among these definitions is the ability to understand and share the emotions or perspectives of others. These varying perspectives contribute to the continuing complexity of empathy as a concept in both theoretical and applied contexts. In psychology and neuroscience, empathy is defined as the capacity to grasp and resonate with another's emotional state or thoughts. It comprises cognitive empathy, which involves understanding another's perspective, and affective empathy, which allows for sharing emotional experiences (Davis, 1983; Decety & Jackson, 2004; Decety & Ickes, 2009). Empathic processes are tied to specific brain regions, such as the anterior insula and the anterior cingulate cortex, which neuroscientific studies identify as crucial to emotional resonance (Singer & Lamm, 2009). In the following section, a deeper examination of empathy from a neuroscientific perspective will be provided, exploring how these neural mechanisms underpin the ability to share and understand emotions. From a philosophical perspective, empathy is not merely a cognitive or emotional phenomenon but also a moral and relational practice. Philosophers emphasise recognising the significance of others' subjective experiences and argue that empathy forms the ethical foundation for social justice and moral responsibility (Coplan & Goldie, 2011). Finally, in the social sciences, empathy is studied as a critical factor for enhancing social cohesion, communication, and conflict resolution. By fostering a better understanding of diverse perspectives, empathy serves as a vital tool for navigating multicultural and complex societal structures (Batson, 2009). According to Matthew Cartabuke, empathy plays a crucial role in moral recognition and ethical decision-making, particularly in contexts that require an understanding of the broader social implications of actions. His research, situated at the intersection of social justice and business ethics, emphasises how empathy leads to a deeper understanding of stakeholders' needs and enhances overall success by fostering responsibility and care. Matthew Cartabuke describes empathy as a multifaceted process involving social, affective, somatic, cognitive, and even spiritual dimensions. These processes are integral to human experiences such as understanding, emotional states, caring, socialisation, intentions, and, importantly, responsibility. In his work, responsibility emerges as a key concept, acting as a common thread that links both accountability and empathy, suggesting that fostering a sense of responsibility is critical to advancing both ethical behaviour and effective decision-making in organisational and societal contexts (Cartabuke et al. 2019).

Despite extensive research on empathy and accountability in AI, these concepts are often treated in isolation, missing the crucial interdependence between them. This chapter aims to highlight this connection, emphasising how the integration of human empathy into AI systems is key to fostering both ethical responsibility and technological effectiveness. Importantly, the focus here is on human empathy, rather than the simulation of empathy by AI, which is a growing area of research but not the focus of this discussion. In the next section, the *Empathising-Systemising theory* will be explored, illustrating how human cognitive and emotional processes, when properly integrated into AI development, can enhance trust, accountability, and foster a more equitable technological landscape. By incorporating empathetic individuals into AI development teams, it is possible to create systems that not only uphold ethical standards but also restore trust and ensure fairness and justice in both design and application.

THE *EMPATHISING-SYSTEMISING THEORY*'S ROLE IN AI TRUST AND ACCOUNTABILITY: A GENDER PERSPECTIVE

After exploring the interplay between empathy and accountability and their role in fostering trust among end-users, this section introduces a neuroscientific framework that defines, measures, and emphasises the relevance of empathy in developing trustworthy AI systems. The focus is on Simon Baron-Cohen's *Empathising-Systemising (E-S) theory* (Baron-Cohen, Knickmeyer, & Belmonte, 2005), a psychological framework that offers valuable insights for assembling gender-balanced AI teams, ultimately promoting trust, accountability, and inclusivity in AI design and implementation.

Empathy is a multifaceted cognitive and emotional process, underpinned by neural mechanisms such as the mirror neuron system, which is located in the premotor cortex and parietal lobe. Discovered by Rizzolatti and Luppino (2001) in primates, mirror neurons activate when an individual performs an action or when they observe the same action being performed by others. This dual activation provides a biological basis for understanding the emotions and intentions of others, facilitating both cognitive and emotional aspects of empathy (Rizzolatti & Craighero, 2004). The ability to understand and predict others' mental states is central to social interaction, and this is where the Theory of Mind (ToM) plays a critical role. ToM allows individuals to attribute beliefs, intentions, and emotions to others, forming the foundation for empathy. Neural regions like the medial prefrontal cortex, temporoparietal junction, and superior temporal sulcus are involved in processing ToM, providing the cognitive scaffolding necessary for empathy-driven interactions (Frith & Frith, 2006).

The concept of ToM, first systematically explored by Premack and Woodruff (1978), suggested that humans and some non-human animals share the ability to attribute mental states to themselves and others. This insight opened a broader understanding of social cognition, influencing developmental psychology and cognitive science. Simon Baron-Cohen expanded on this foundational work through his research on autism and empathy. His work, particularly in the context of *mindblindness*, posits that an innate cognitive mechanism for understanding others' mental states exists, yet in individuals with autism spectrum disorders (ASD), this system is impaired or differently functioning (Baron-Cohen, 1995).

Building on the foundational work of the ToM, Simon Baron-Cohen from the University of Cambridge developed the E-S theory, which we will further explore because it offers a robust psychological framework, supported by empirical tools such as the Empathy Quotient (EQ) and Systemising Quotient (SQ). These tools have been instrumental in advancing research on gender differences, particularly in STEM fields, and their application holds significant promise in AI and technology development.

The *E-S theory* offers a psychological framework that further elaborates on how empathy and systemising tendencies relate to one another, particularly in the context of gender differences and neurodiversity. While ToM focuses on the ability to attribute mental states to others, the E-S theory distinguishes between two key cognitive approaches: empathising, which involves understanding and responding to emotions, and systemising, which focuses on understanding systems, structures, and patterns (Baron-Cohen, 2003).

The E-S theory classifies individuals based on abilities in empathic thinking (E) and systematic thinking (S). It uses an Empathy Quotient (EQ) and a Systemising Quotient (SQ) to assess skills and tries to explain social and communication symptoms in autism spectrum disorders as empathy deficits and delays paired with intact or superior systemising. The E-S theory has been further developed into the Extreme Male Brain (EMB) theory, which associates autism and Asperger syndrome with lower-than-average levels of empathy as conceptualised within the framework of the E-S theory. Simply expressed, E and S denote a basic inclination toward people or things, respectively.

Baron-Cohen suggests that, on average, females tend to achieve higher scores on measures of empathy, while males typically score higher on measures of systemising. This has been discovered using the child and adolescent versions of EQ and SQ. While experience and socialisation have a role in the observed sex differences in empathy and systemising, Baron-Cohen suggests that a contribution is given also by biology. Fetal testosterone (FT) is identified as a potential biological factor influencing empathising (E) and systemising (S). Research indicates that FT levels show a positive correlation with Systemising Quotient (SQ) scores and a negative correlation with Empathy Quotient (EQ) scores. Despite the criticisms received, among others from Fine (2008), who considered the theory a form of Neurosexism, i.e. a bias in the neuroscience of sex differences towards reinforcing harmful gender stereotypes, the theory can be considered to have strong support.

In particular, the importance of the systemasing pathway for science learning has been discussed in detail by Zeyer (2018): gender is much less important for motivation to learn science than a systemising cognition, he argues. In the same study, he observes that, while systemising has a significant impact on motivation to learn science, the influence of empathising is consistently minimal and statistically insignificant. This can be due to the fact that science is mostly about physical things, prediction, and control. In a large-scale 2018 study with over 500,000 participants, the E-S theory was validated, further cementing its importance in understanding gender differences in cognition (Greenberg, Warrier, Allison, & Baron-Cohen, 2018). This study confirmed that typical females are generally more empathic, while typical males are more system-oriented, validating the theory's scientific legitimacy.

For AI development, the E-S theory offers crucial insights. As AI systems become increasingly complex, ensuring their fairness and inclusivity requires teams that bring diverse cognitive perspectives. The recognition that empathy is more commonly associated with females suggests that fostering gender-balanced teams can lead to more empathetic and ethical AI systems. Women, with their higher empathic tendencies, can contribute to AI development by identifying and addressing biases that disproportionately affect women and other marginalised groups. Additionally, empathy can help AI systems better understand and respond to human emotions, improving user trust and satisfaction.

Despite these promising implications, AI development remains gender-imbalanced, with women significantly underrepresented in AI and STEM fields. According to the UNESCO (Global Education Monitoring Report Team, 2024), this gender disparity limits the diversity of perspectives in AI research and development, perpetuating biases that affect the inclusivity and fairness of AI systems. Addressing this underrepresentation is essential for mitigating biases and promoting the development of AI technologies that are more equitable and trustworthy. By promoting gender-balanced teams in AI, it is possible to enhance the design of inclusive and ethical AI systems that better serve diverse populations and address the challenges of bias and accountability.

CONCLUSION

This chapter has explored the pressing issue of AI bias, underscoring the critical role of empathy in mitigating such biases and fostering accountability and trust within AI systems. It has argued that addressing AI bias requires more than just technical interventions; it demands a robust socio-psychological foundation, with empathy as a core element. Empathy—defined as the capacity to understand and respond to others' emotions and perspectives—directly influences mechanisms of trust and accountability, which are vital for the ethical development of AI systems. Without empathy, AI systems risk perpetuating existing biases, undermining efforts to create systems that align with human rights and societal standards of fairness and justice.

The chapter began by examining global policy frameworks, illustrating the increasing recognition of the risks posed by unregulated AI development. Various international guidelines have emerged to ensure responsible AI deployment, focusing on ethical concerns such as bias, privacy, and individual rights. Key platforms, such as the AI Policy Tracker and the OECD AI Policy Observatory Dashboard, provide valuable tools for tracking and analysing AI-related legislation across jurisdictions, thereby enabling global oversight of AI regulation. Although regulatory efforts differ across countries in terms of pace and approach, these frameworks consistently emphasise the need to mitigate AI bias, reflecting its universal recognition as a critical concern intertwined with issues of accountability and trust in AI development.

Following this, the chapter has examined how bias operates within AI systems: bias in AI stems from various sources, particularly the data and algorithms used in machine learning. Biased datasets reflect societal stereotypes, such as gender or racial biases, which AI systems can inadvertently perpetuate if the data is flawed or underrepresentative. Algorithmic bias can emerge even with unbiased data, especially when models perform better on well-represented groups, leading to inequality for underrepresented ones. Additionally, historical bias in society can be embedded in AI systems, reinforcing past discrimination. Representation bias occurs when datasets fail to capture the diversity of the population the AI serves, often seen in healthcare models. Measurement bias happens when data collection methods skew results for certain groups, affecting accuracy. Evaluation bias arises when AI models are tested on unrepresentative data, distorting assessments of fairness. Finally, deployment bias occurs when AI systems are used in contexts different from their training environment, leading to disparities. These biases pose significant ethical challenges in AI development and deployment.

Moreover, the chapter argued that empathy is a critical countermeasure, as it fosters a deeper understanding of the lived experiences of diverse user groups, which is often overlooked in AI design and development.

To deepen the understanding of empathy's role in AI development, the chapter engaged with Simon Baron-Cohen's Empathizing-Systemizing (E-S) theory, which differentiates between two key cognitive dimensions: empathising, associated with emotional understanding, and systemising, linked to analytical reasoning. While AI development has traditionally favoured systemising—focused on understanding structures and patterns—the chapter posited that the integration of empathising perspectives is essential for ethical AI design. The theory suggests that while females tend to score higher in empathizing and males in systemising, these tendencies, while shaped by biology and socialisation, have significant implications for AI development. Gender-balanced teams, characterised by cognitive diversity, are better positioned to identify and mitigate biases, particularly those embedded in AI systems.

Despite these promising insights, AI development remains gender-imbalanced, with women significantly underrepresented in AI and STEM fields. According to the UNESCO (Global Education Monitoring Report Team, 2024), this gender disparity limits the diversity of perspectives in AI research and development, thus perpetuating biases that undermine the inclusivity and fairness of AI systems. Addressing this underrepresentation is essential to mitigate biases and promote the development of AI technologies that are more equitable and trustworthy. The chapter advocates for the promotion of gender-balanced teams in AI to enhance the design of inclusive and ethical systems that better serve diverse populations.

The interdisciplinary methodology employed in this chapter—drawing on insights from neuroscience, social and gender studies, ethics, and AI development—aimed to provide a rich theoretical foundation for understanding how biases are embedded in AI systems. The findings underscore that empathy is essential in AI development, both as a tool to counteract bias and as a means to foster inclusivity. Furthermore, gender-balanced and diverse teams are shown to be more effective in addressing bias and inequality in AI systems, ensuring that AI technologies align with ethical standards and societal values.

By proposing empathy as a cornerstone of AI development, the chapter contributes to the broader discourse on AI ethics and social justice. It suggests that fostering empathy within AI development teams—particularly through gender balance and cognitive diversity—will enhance the ethical design and implementation of AI systems. Such an approach ensures that AI systems are more equitable, transparent, and aligned with the diverse needs of society. As global debates on AI regulation and ethical standards continue to evolve, this interdisciplinary framework may provide a valuable model for creating more inclusive and accountable AI systems that reflect the complexities of human society, ultimately fostering trust and fairness in AI technologies.

REFERENCES

Aayog, N. I. T. I. (2018). *National strategy for artificial intelligence*. Retrieved November 27, 2024 from https://www.niti.gov.in/sites/default/files/2023-03/National-Strategy-for-Artificial-Intelligence.pdf

African Union. (2024). *Continental artificial intelligence strategy*. Retrieved March 15, 2025 from https://au.int/sites/default/files/documents/44004-doc-EN-_Continental_AI_Strategy_July_2024.pdf

Artificial Intelligence Strategy, U. A. E. (2024). *Artificial Intelligence in the UAE*. Retrieved November 27, 2024, from https://ai.gov.ae/strategy/

Australian Government. (2024). National Framework for Assurance of Artificial Intelligence. *Department of Finance*. Retrieved March 15, 2025 from https://www.finance.gov.au/government/public-data/data-and-digital-ministers-meeting/national-framework-assurance-artificial-intelligence-government

Barocas, S., & Selbst, A. D. (2016). Big data's disparate impact. *SSRN*, *104*(3), 671–732. Retrieved November 27, 2024, from. DOI: 10.2139/ssrn.2477899

Baron-Cohen, S. (1995). *Mindblindness: An essay on autism and theory of mind*. MIT Press. Retrieved March 15, 2025 from https://doi.org/DOI: 10.7551/mitpress/4635.001.0001

Baron-Cohen, S. (2003). *The essential difference: The truth about the male and female brain*. Basic Books.

Baron-Cohen, S., Knickmeyer, R. C., & Belmonte, M. K. (2005). Sex differences in the brain: Implications for explaining autism. *Science*, *310*(5749), 819–823. Retrieved March 15, 2025 from. DOI: 10.1126/science.1115455 PMID: 16272115

Barrett, S., & al. (2023). *Canada's Pan-Canadian AI Strategy: Progress and Prospects*. CIFAR. Retrieved March 15, 2025 from https://cifar.ca

Batson, C. D. (2009). These things called empathy: Eight related but distinct phenomena. In J. Decety & W. Ickes (Eds.), *The social neuroscience of empathy* (pp. 3–15). Boston Review. Retrieved March 15, 2025 from https://doi.org/DOI: 10.7551/mitpress/9780262012973.003.0002

Buolamwini, J., & Gebru, T. (2018). Gender Shades: Intersectional Accuracy Disparities in Commercial Gender Classification. *Proceedings of the 1st Conference on Fairness, Accountability and Transparency, Proceedings of Machine Learning Research,* 81, 77–91. Retrieved March 15, 2025 from https://proceedings.mlr.press/v81/buolamwini18a/buolamwini18a.pdf

Canadian Institute for Advanced Research. (2017). *Pan-Canadian Artificial Intelligence Strategy.*

Cartabuke, M., Westerman, J. W., Bergman, J. Z., Whitaker, B. G., Westerman, J., & Beekun, R. I. (2019). Empathy as an antecedent of social justice attitudes and perceptions. *Journal of Business Ethics*, *157*(3), 605–615. Retrieved March 15, 2025 from. DOI: 10.1007/s10551-017-3677-1

CIFAR. Retrieved March 15, 2025 from https://cifar.ca/ai/ai-strategy/

Consejo Nacional de Política Económica y Social. República de Colombia, & Departamento Nacional de Planeación. (2019). *Política Nacional de Transformación Digital e Inteligencia Artificial* [National Council for Economic and Social Policy. Republic of Colombia, & National Planning Department. (2019). National Policy on Digital Transformation and Artificial Intelligence] (CONPES No. 3975). Retrieved March 15, 2025 from https://colaboracion.dnp.gov.co/CDT/Conpes/Econ%C3%B3micos/3975.pdf

Coplan, A., & Goldie, P. (Eds.). (2011). *Empathy: Philosophical and psychological perspectives*. Oxford University Press. Retrieved March 15, 2025 from https://doi.org/DOI: 10.1093/acprof:oso/9780199539956.001.0001

Cuff, B. M. P., Brown, S. J., Taylor, L., & Howat, D. J. (2016). Empathy: A review of the concept. *Emotion Review*, *8*(2), 144–153. Retrieved March 15, 2025 from. DOI: 10.1177/1754073914558466

Davis, M. H. (1983). Measuring individual differences in empathy: Evidence for a multidimensional approach. *Journal of Personality and Social Psychology*, *44*(1), 113–126. Retrieved March 15, 2025 from. DOI: 10.1037/0022-3514.44.1.113

Decety, J., & Ickes, W. (Eds.). (2009). *The social neuroscience of empathy*. MIT Press. Retrieved March 15, 2025 from https://doi.org/DOI: 10.7551/mitpress/9780262012973.001.0001

Decety, J., & Jackson, P. L. (2004). The functional architecture of human empathy. *Behavioral and Cognitive Neuroscience Reviews*, *3*(2), 71–100. Retrieved March 15, 2025 from. DOI: 10.1177/1534582304267187 PMID: 15537986

Decree of the President of the Russian Federation No. 490. (2019). *On the development of artificial intelligence in the Russian Federation*. Retrieved March 15, 2025 from http://www.kremlin.ru/acts/bank/44731

Eubanks, V. (2018). *Automating Inequality: How High-Tech Tools Profile, Police, and Punish the Poor*. New York: Picador, St. Martin's Press. Retrieved March 15, 2025 from DOI: 10.5204/lthj.v1i0.1386

European Parliament & Council of the European Union. (2016). *Regulation (EU) 2016/679 of the European Parliament and of the Council of 27 April 2016 on the protection of natural persons with regard to the processing of personal data and on the free movement of such data*. Official Journal of the European Union. Retrieved March 15, 2025 from https://eur-lex.europa.eu/legal-content/EN/TXT/?uri=CELEX%3A32016R0679

European Parliament & Council of the European Union. (2024). *Regulation (EU) 2024/1689 of the European Parliament and of the Council of 13 June 2024 laying down harmonised rules on artificial intelligence*. Official Journal of the European Union. Retrieved March 15, 2025 from https://artificialintelligenceact.eu/the-act

Fairly.ai. (2024). *Map of global AI regulations*. Retrieved November 27, 2024, from https://www.fairly.ai/blog/map-of-global-ai-regulations

Fine, C. (2008). Will working mothers' brains explode? The popular new genre of neurosexism. *Neuroethics*, *1*(1), 69–72. Retrieved March 15, 2025 from. DOI: 10.1007/s12152-007-9004-2

Frith, C. D., & Frith, U. (2006). The neural basis of mentalizing. *Neuron*, *50*(4), 531–534. Retrieved March 15, 2025 from. DOI: 10.1016/j.neuron.2006.05.001 PMID: 16701204

Gigerenzer, G., & Gaissmaier, W. (2011). Heuristic decision making. *Annual Review of Psychology*, *62*(1), 451–482. Retrieved March 15, 2025 from. DOI: 10.1146/annurev-psych-120709-145346 PMID: 21126183

Global Education Monitoring Report Team. (2024). *Gender equality in STEM education: Breaking down the barriers*. UNESCO. Retrieved March 15, 2025 from https://doi.org/DOI: 10.54676/WVCF2762

Gonen, H., & Goldberg, Y. (2019). Lipstick on a pig: Debiasing methods cover up systematic gender biases in word embeddings but do not remove them. *ArXiv*. Retrieved March 15, 2025 from https://doi.org//arXiv.1903.03862DOI: 10.48550

Government of Canada. (2023). *Artificial Intelligence and Data Act.* Retrieved March 15, 2025 from https://ised-isde.canada.ca/site/innovation-better-canada/en/artificial-intelligence-and-data-act-aida-companion-document

Greenberg, D. M., Warrier, V., Allison, C., & Baron-Cohen, S. (2018). Testing the Empathizing-Systemizing theory of sex differences and the Extreme Male Brain theory of autism in half a million people. *Proceedings of the National Academy of Sciences of the United States of America, 115*(48), 12152–12157. Retrieved March 15, 2025 from. DOI: 10.1073/pnas.1811032115 PMID: 30420503

Hofmann, V., Kalluri, P. R., Jurafsky, D., & King, S. (2024). AI generates covertly racist decisions about people based on their dialect. *Nature, 633*(8028), 147–154. Retrieved March 15, 2025 from. DOI: 10.1038/s41586-024-07856-5 PMID: 39198640

IA2030Mx. (2018). *Agenda Nacional Mexicana de IA 2030.* Retrieved March 15, 2025 from https://wp.oecd.ai/app/uploads/2022/01/Mexico_Agenda_Nacional_Mexicana_de_IA_2030.pdf

Kahneman, D. (2011). *Thinking, fast and slow.* Farrar, Straus and Giroux.

Kahneman, D., & Tversky, A. (1979). Prospect theory: An analysis of decision under risk. *Econometrica, 47*(2), 263–291. Retrieved March 15, 2025 from. DOI: 10.2307/1914185

Madariaga Zeballos, J. (2023). *Perú: Proyecto de ley para la regulación de la inteligencia artificial.* Congreso de la República del Perú. Retrieved March 15, 2025 from https://wb2server.congreso.gob.pe/spley-portal-service/archivo/MTYzNzgw/pdf

Manoogian, J., & Benson, B. (2017). *Cognitive Bias Codex.* Retrieved November 27, 2024 from https://commons.wikimedia.org/wiki/File:Cognitive_Bias_Codex_-_180%2B_biases,*designed_by_John_Manoogian_III*(jm3).jpg

Maslej, N., Fattorini, L., Perrault, R., Parli, V., Reuel, A., Brynjolfsson, E., Etchemendy, J., Ligett, K., Lyons, T., Manyika, J., Niebles, J. C., Shoham, Y., Wald, R., & Clark, J. (2024). *The AI Index 2024 Annual Report.* AI Index Steering Committee, Institute for Human-Centered AI, Stanford University. Retrieved March 15, 2025 from https://aiindex.stanford.edu

Ministry of Science. Technology, Knowledge and Innovation. (2021). *Política Nacional de Inteligencia Artificial.* Retrieved March 15, 2025 from https://www.minciencia.gob.cl/uploads/filer_public/bc/38/bc389daf-4514-4306-867c-760ae7686e2c/documento_politica_ia_digital_.pdf

MIT. (2024.). *AI risk domain taxonomy.* Retrieved March 15, 2025 from https://airisk.mit.edu/#Domain-Taxonomy-of-AI-Risks

Morales, V. (2023). *Proyecto de ley: Régimen de regulación de la inteligencia artificial* (Exp. 2505-D-2023). Cámara de Diputados de la Nación Argentina. Retrieved March 15, 2025 from https://www4.hcdn.gob.ar/dependencias/dsecretaria/Periodo2023/PDF2023/TP2023/2505-D-2023.pdfhttps://www.diputados.gov.ar/diputados/vmoralesg/proyecto.html?exp=2505-D-2023

New Zealand Government. (2023). *AI Strategy Blueprint: Trust and Innovation*. Ministry of Innovation and Employment. Retrieved March 15, 2025 from https://www.govt.nz

Noble, S. U. (2018). *Algorithms of oppression: How search engines reinforce racism*. NYU Press. DOI: 10.18574/nyu/9781479833641.001.0001

O'Neil, C. (2016). *Weapons of math destruction: How big data increases inequality and threatens democracy*. Crown Publishing Group.

OECD. (2024a). *OECD AI Policy Observatory: Overview*. Retrieved November 27, 2024, from https://oecd.ai/en/dashboards/overview

OECD (2024b). *Assessing potential future artificial intelligence risks, benefits, and policy imperatives, OECD Artificial Intelligence Papers*, No. 27, OECD Publishing, Paris. Retrieved March 15, 2025 from DOI: 10.1787/dee339a8-en

Oxford Insights. (2024.). *AI readiness index*. Retrieved March 15, 2025 from https://oxfordinsights.com/ai-readiness/ai-readiness-index/

Premack, D., & Woodruff, G. (1978). Does the chimpanzee have a theory of mind? *Behavioral and Brain Sciences*, *1*(4), 515–526. Retrieved March 15, 2025 from. DOI: 10.1017/S0140525X00076512

Rizzolatti, G., & Craighero, L. (2004). The mirror-neuron system. *Annual Review of Neuroscience*, *27*(1), 169–192. Retrieved March 15, 2025 from. DOI: 10.1146/annurev.neuro.27.070203.144230 PMID: 15217330

Rizzolatti, G., & Luppino, G. (2001). The cortical motor system. *Neuron*, *31*(6), 889–901. Retrieved November 27, 2024, from. DOI: 10.1016/S0896-6273(01)00423-8 PMID: 11580891

Rolf, S. (2023). *China's regulations on algorithms: Context, impact, and comparisons with the EU*. Friedrich-Ebert-Stiftung. Retrieved March 15, 2025 from https://library.fes.de/pdf-files/bueros/bruessel/19904.pdf

Senado de la República. (2023). *Por medio de la cual se establecen los lineamientos de política pública para el desarrollo, uso e implementación de inteligencia artificial y se dictan otras disposiciones* (Proyecto de Ley 059/23). Retrieved March 15, 2025 from https://leyes.senado.gov.co/proyectos/index.php/proyectos-ley/cuatrenio-2022-2026/2023-2024/article/59-por-medio-de-la-cual-se-establecen-los-lineamientos-de-politica-publica-para-el-desarrollo-uso-e-implementacion-de-inteligencia-artificial-y-se-dictan-otras-disposiciones

Singapore Government. (2019). *National Artificial Intelligence Strategy (NAIS) 2019*. Retrieved March 15, 2025 from https://file.go.gov.sg/nais2019.pdf

Singapore Government. (2023). *National Artificial Intelligence Strategy (NAIS) 2023*. Retrieved March 15, 2025 from https://file.go.gov.sg/nais2023.pdf

Singer, T., & Lamm, C. (2009). The social neuroscience of empathy. *Annals of the New York Academy of Sciences*, *1156*(1), 81–96. Retrieved March 15, 2025 from. DOI: 10.1111/j.1749-6632.2009.04418.x PMID: 19338504

Srinivasan, R., & González, B. (2022). The role of empathy for artificial intelligence accountability. *Journal of Responsible Technology*, *9*, 100021. Retrieved March 15, 2025 from. DOI: 10.1016/j.jrt.2021.100021

Stanford University. (2024). *AI policy tracker*. Center for Research on Foundation Models. Retrieved March 15, 2025 from https://aipolicy.stanford.edu/

State Council of the People's Republic of China. (2017). *New generation artificial intelligence development plan* (No. 2017-35). State Council of the People's Republic of China. Retrieved March 15, 2025 from http://fi.china-embassy.gov.cn/eng/kxjs/201710/P020210628714286134479.pdf

Stotland, E., Matthews, K. E., Sherman, S., Hansson, R. O., & Richardson, B. Z. (1978). Empathy, fantasy, and helping. *Sage (Atlanta, Ga.)*.

Suresh, H., & Guttag, J. (2021). A framework for understanding sources of harm throughout the machine learning life cycle. In *Equity and Access in Algorithms, Mechanisms, and Optimization (EAAMO '21)* (Article 17, pp. 1–9). Association for Computing Machinery. Retrieved March 15, 2025 from DOI: 10.1145/3465416.3483305

Translate, C. L. (2022). Provisions on the management of algorithmic recommendations in internet information service. *China Law Translate*. Retrieved March 15, 2025 from https://www.chinalawtranslate.com/en/algorithms/(open in a new window)

UK Government. (2021). *National AI Strategy: Leading the future of artificial intelligence*. Retrieved March 15, 2025 from https://www.gov.uk/government/publications/national-ai-strategy

UNESCO. (2024). *Global AI ethics and governance observatory*. Retrieved March 15, 2025 from https://www.unesco.org/ethics-ai/en/global-hub

U.S. Congress. (2023). *Algorithmic Accountability Act of 2023*, S. 2892, 118th Congress. Retrieved March 15, 2025 from https://www.congress.gov/bill/118th-congress/senate-bill/2892

U.S. Congress. (2024). *H.R.10092 - Eliminating Bias in Algorithmic Systems Act*. Congress.gov. Retrieved March 15, 2025 from https://www.congress.gov/bill/118th-congress/house-bill/10092/text

Van Kolfschooten, H. (2023). The AI cycle of health inequity and digital ageism: Mitigating biases through the EU regulatory framework on medical devices. *Journal of Law and the Biosciences*, *10*(2), lsad031. Advance online publication. Retrieved March 15, 2025 from. DOI: 10.1093/jlb/lsad031 PMID: 38075950

Vicente, L., & Matute, H. (2023). Humans inherit artificial intelligence biases. *Scientific Reports*, *13*(1), 15737. Retrieved March 15, 2025 from. DOI: 10.1038/s41598-023-42384-8 PMID: 37789032

Zeyer, A. (2018). Gender, complexity, and science for all: Systemizing and its impact on motivation to learn science for different science subjects. *Journal of Research in Science Teaching*, *55*(2), 147–171. Retrieved March 15, 2025 from. DOI: 10.1002/tea.21413

KEY TERMS AND DEFINITIONS

Accountability in AI: The principle ensuring that developers, organisations, and stakeholders involved in AI development and deployment are responsible for the ethical, equitable, and transparent use of AI systems. Accountability requires mechanisms to identify and mitigate biases, ensure fairness, and uphold societal and legal standards, fostering trust and minimising unintended harms. It is foundational for maintaining integrity in systems where decisions significantly affect individuals and communities.

AI Bias: A systematic deviation in AI outputs resulting from imbalances or inaccuracies in training datasets, algorithms, or decision-making frameworks. These biases often mirror societal stereotypes and inequalities, leading to discrimination across gender, race, or socio-economic lines. They pose ethical challenges and necessitate robust mitigation strategies to achieve fairness and equity in AI applications.

Dignitary Harm: A social and ethical phenomenon where individuals or groups experience a loss of respect, recognition, or dignity due to discriminatory practices or systemic biases embedded in AI systems. This harm often manifests in marginalised communities, particularly when algorithms reinforce stereotypes or exclude individuals based on race, gender, or socio-economic status. The concept underscores the need for accountability and fairness in technological design.

Empathising-Systemising (E-S) Theory: A psychological framework distinguishing two dimensions: Empathising (E), which is the capacity to identify, understand, and respond to others' emotional states, and Systemising (S), the ability to analyse and predict patterns within systems. The theory suggests that, on average, females excel in empathising, while males tend to perform better in systemising. It has been applied to understanding neurodiversity, particularly autism spectrum conditions, where individuals often show a stronger inclination towards systemising, sometimes at the expense of empathising.

Extreme Male Brain (EMB) Theory: A hypothesis extending the E-S Theory, proposing that autism reflects an extreme manifestation of typical male cognitive traits, characterised by enhanced systemising and diminished empathising abilities. The EMB theory has implications for understanding social and communication challenges in autism and remains a topic of debate within neuroscience and gender studies due to its reliance on binary gender comparisons and biological determinism.

Fetal Testosterone (FT): A prenatal hormone hypothesised to influence cognitive and behavioural development, including empathising and systemising tendencies. Elevated levels of fetal testosterone have been correlated with stronger systemising abilities and reduced empathising skills, contributing to discussions on gender differences and neurodevelopmental conditions such as autism. The role of FT highlights the interplay between biological and environmental factors in shaping cognition.

Mirror Neurons: Specialised neurons discovered in the premotor cortex and parietal lobe that activate both when an individual performs an action and when observing the same action performed by others. These neurons are considered a biological basis for understanding others' intentions, emotions, and behaviours, bridging cognitive and affective dimensions of empathy. They are pivotal to social learning and the development of Theory of Mind (ToM).

Theory of Mind (ToM): A cognitive ability enabling individuals to attribute mental states such as beliefs, intentions, and emotions to themselves and others. ToM is essential for understanding and predicting social behaviours and is considered a foundation for empathy. Research in neuroscience associates ToM with activity in brain regions like the medial prefrontal cortex and temporoparietal junction. Impairments in ToM are frequently observed in autism spectrum conditions.

Trust in AI: A multidimensional concept referring to the confidence of users and society in the reliability, fairness, and transparency of AI systems. Trust is undermined by biases, opaque algorithms, and unethical deployments, while strategies such as ethical guidelines, inclusive design practices, and regulatory oversight aim to rebuild and maintain trust in AI technologies.

Chapter 8
Building Ethical Foundations:
Educating Researchers for Responsible Use of AI

Debajit Karmakar
https://orcid.org/0000-0001-9272-0627
Lakshmibai National Institute of Physical Education, India

Anindita Das
https://orcid.org/0000-0002-6502-9619
Lakshmibai National Institute of Physical Education, India

Awesh Khati
https://orcid.org/0009-0009-9090-1593
Lakshmibai National Institute of Physical Education, India

ABSTRACT

This aims to establish a comprehensive framework for integrating ethics into AI research education. The mission of the chapter is to guide the development of responsible AI researchers by embedding ethical considerations at the core of their training. It explores key ideas such as the necessity of fostering a deep understanding of AI's societal impacts, the importance of promoting transparency, fairness, and accountability, and the challenges posed by biases and unintended consequences. The chapter will outline innovative methods for teaching ethics in AI, including case studies, interdisciplinary collaboration, and the integration of ethical AI design principles into research practices. Additionally, it will discuss recent advancements in AI ethics, changes in academic curricula, and the need for continuous dialogue between researchers, ethicists, and policymakers. Ultimately, this chapter seeks

DOI: 10.4018/979-8-3693-9035-1.ch008

to prepare the next generation of AI researchers to navigate the complex ethical landscape with integrity and responsibility.

INTRODUCTION

Responsible AI practices encompass principles, methods, and frameworks aimed at ensuring artificial intelligence is developed, deployed, and used ethically, transparently, and in alignment with human values. These practices prioritize fairness, accountability, transparency, and privacy (Ouchchy et al., 2020). Fairness seeks to reduce biases within AI systems, especially in sensitive contexts like hiring, healthcare, or law enforcement, to prevent discrimination. Accountability requires developers, companies, and stakeholders to take ownership of AI decisions, explaining and justifying outcomes to build trust, especially in areas impacting public life or individual well-being (Ng et al., 2021). Transparency involves clear communication about how AI functions, the data it uses, and its limitations, allowing users and policymakers to make informed decisions and promoting explainability, so complex models remain interpretable to stakeholders (Huang et al., 2021). Privacy underscores the importance of protecting personal data and complying with data protection regulations like GDPR and CCPA, alongside implementing security measures to guard against malicious use or cyber threats. Ethical design, prioritizing humanity's best interests, requires interdisciplinary collaboration among ethicists, technologists, sociologists, and policymakers to mitigate potential harms proactively (Dhirani et al., 2023). Continuous monitoring and updating of AI systems help maintain ethical standards as they evolve. Educating AI practitioners in these principles fosters a trustworthy, inclusive, and sustainable AI ecosystem, establishing the ethical groundwork for responsible AI in diverse sectors (Huang et al., 2021).

THE NEEDS OF ETHICAL USE OF AI FOR RESEARCHERS

As artificial intelligence (AI) rapidly integrates into nearly every domain of research, the ethical use of AI by researchers becomes paramount. AI offers unprecedented opportunities to analyse vast datasets, discover patterns, and generate insights that were previously unimaginable. The sheer power of AI also presents a range of ethical dilemmas that researchers must address to ensure that the technology is employed responsibly (Hallamaa & Kalliokoski, 2022). These dilemmas encompass issues of privacy, bias, transparency, accountability, and the overall societal impact of AI-driven research. Educating researchers to navigate these ethical challenges is essential for fostering responsible AI use and maintaining public trust in both science and technology.

- **Respect for Privacy and Data Security:** AI systems are often fuelled by vast amounts of data, much of which may be personal or sensitive (Ng et al., 2021). As researchers increasingly rely on AI to analyse datasets involving personal information, such as health records or behavioural data, there is an ethical responsibility to ensure that privacy is protected (Hallamaa & Kalliokoski, 2022). Data anonymization, encryption, and secure storage practices are critical components of ethical AI research, but even these safeguards can be insufficient if not carefully managed (Ouchchy et al., 2020).

One major concern is that AI can sometimes “re-identify” individuals from supposedly anonymized data, especially when combined with other datasets. Researchers must be vigilant about this risk and employ advanced techniques to mitigate it. Moreover, informed consent from participants is crucial, particularly in studies where sensitive personal data is used (Hallamaa & Kalliokoski, 2022). Participants should be made aware of how their data will be utilized by AI systems and what potential risks are involved. The lack of transparency in AI's decision-making processes, often referred to as the “black box” problem, complicates this further, as it can be difficult to explain to participants exactly how their data will be processed and what implications this may have (Ng et al., 2021).

- **Mitigating Algorithmic Bias:** Bias in AI systems is one of the most pressing ethical challenges in modern research. AI models learn from the data they are fed, and if that data is biased, the model will inevitably reflect and potentially amplify those biases. This is particularly concerning in fields like healthcare, criminal justice, and hiring, where biased algorithms can perpetuate discrimination and inequity (Dhirani et al., 2023).

For researchers, addressing algorithmic bias is a critical ethical obligation. They must ensure that the data used to train AI systems is representative and free from biases that could harm specific groups (Ouchchy et al., 2020). This requires not only careful dataset curation but also the development of methodologies to detect and correct bias within AI models. Additionally, researchers must be transparent about the limitations of their AI systems, acknowledging where biases may still exist and how they could impact the research findings (Dhirani et al., 2023).

The responsibility does not end with the technical aspects of bias mitigation. Researchers must also engage in interdisciplinary dialogue, working with ethicists, social scientists, and community stakeholders to fully understand the social implications of biased AI systems. Failing to address these issues could lead to mistrust in AI-based research, particularly among marginalized groups who may be disproportionately affected by biased outcomes (Ouchchy et al., 2020).

- **Ensuring Transparency and Accountability:** One of the core ethical principles in AI research is transparency. Researchers must be open about the methods, data, and algorithms they use to generate results. This transparency is essential not only for reproducibility, a cornerstone of scientific research, but also for building public trust in AI systems (Benefo et al., 2022). When AI systems are deployed in critical areas such as healthcare, public policy, or finance, the lack of transparency can lead to harmful outcomes that go unchecked.

However, ensuring transparency in AI research is no small task. Many AI algorithms, particularly those based on deep learning, are so complex that even their creators may not fully understand how they arrive at certain decisions. This "black box" nature of AI complicates the ethical imperative for transparency. Researchers must find ways to explain AI processes to stakeholders in understandable terms, without oversimplifying the potential risks and limitations (Benefo et al., 2022). Efforts to develop explainable AI (XAI) are a step in the right direction, as they aim to make AI systems more interpretable without sacrificing performance.

Accountability is another vital aspect of ethical AI use. Researchers must take responsibility for the outcomes of their AI systems, particularly when these systems are used to inform real-world decisions. In cases where AI models produce incorrect or harmful results, there must be clear mechanisms for addressing these issues and preventing future occurrences (Stahl et al., 2022). This may require researchers to advocate for regulatory frameworks or institutional policies that ensure AI systems are held to high ethical standards.

- **Fostering Inclusivity and Fair Access:** The development and deployment of AI should benefit all sectors of society, not just those with access to advanced technologies or substantial financial resources. Ethical AI research must consider the inclusivity of its applications and ensure that AI-driven advancements do not widen existing disparities (Hallamaa & Kalliokoski, 2022). For instance, AI in education, healthcare, or job markets has the potential to offer enormous benefits, but only if it is designed to be accessible and fair.

Researchers have a responsibility to consider the broader societal impact of their AI systems, ensuring that marginalized or underrepresented groups are not excluded from the benefits. This includes thinking critically about the long-term social and economic effects of AI adoption (Trocin et al., 2023). Will AI replace jobs disproportionately in certain communities? Will it reinforce existing inequalities in healthcare access or education? Addressing these questions requires researchers to look beyond the technical aspects of AI and engage with social, ethical, and political issues (Floridi et al., 2018).

Additionally, fostering fair access to AI research tools and results is crucial. Researchers must ensure that their findings are shared in open, accessible ways that allow a wide range of individuals and institutions to benefit from them. This includes promoting open-source AI tools, sharing data responsibly, and publishing research in open-access platforms. The democratization of AI knowledge and tools is essential for creating a future where AI benefits are more evenly distributed.

- **Balancing Innovation with Ethical Responsibility:** AI research is at the forefront of innovation, with new breakthroughs happening at an accelerating pace. However, this rapid innovation also requires researchers to balance the excitement of discovery with the ethical responsibilities that come with it. As AI systems become more powerful, the potential for both positive and negative societal impacts grows. Ethical considerations should not be seen as a barrier to innovation, but rather as a guiding framework that ensures AI advancements contribute to the greater good (Ouchchy et al., 2020).

Researchers must be mindful of the potential consequences of their work, particularly when it involves sensitive areas such as autonomous weapons, facial recognition, or predictive policing. These technologies have far-reaching implications for privacy, civil liberties, and human rights (Trocin et al., 2023). Ethical AI research requires foresight, with researchers considering not only the immediate applications of their work but also the broader societal and global impact.

Furthermore, researchers must engage in continuous ethical reflection throughout the AI development process. This means regularly revisiting ethical considerations as new data, tools, and methods emerge (Benzinger et al., 2023). Ethical guidelines should be dynamic, adapting to the rapidly changing landscape of AI research. By fostering a culture of ethical mindfulness, researchers can ensure that their work remains aligned with societal values and contributes positively to the world.

- **The Need for Ethical Education and Training:** Given the complexity of the ethical challenges associated with AI, it is crucial to educate researchers on how to navigate these issues effectively. Ethical education and training should be an integral part of AI research programs, equipping researchers with the knowledge and tools needed to make responsible decisions (Floridi et al., 2018). This includes not only understanding the technical aspects of

AI ethics, such as bias mitigation and data privacy but also fostering a broader ethical awareness that encompasses social justice, human rights, and the long-term societal implications of AI.

Ethical education should be interdisciplinary, drawing on fields such as philosophy, law, sociology, and public policy. By exposing researchers to diverse perspectives, institutions can create a generation of AI researchers who are not only skilled in their technical fields but also deeply attuned to the ethical dimensions of their work (Karimian et al., 2022). This holistic approach to ethical education is essential for building a strong foundation of responsible AI use.

KEY ETHICAL ISSUES IN RESEARCH BY THE USE OF AI

As artificial intelligence (AI) continues to permeate various domains of research, its integration into scientific inquiry brings forth profound ethical challenges. These concerns are particularly significant as AI systems become critical tools for data analysis, decision-making, and even hypothesis generation. This section explores the core ethical issues associated with the use of AI in research, with a focus on bias, privacy, transparency, accountability, and fairness (D'Antonoli, 2020). By addressing these concerns, researchers can be educated on how to responsibly employ AI while fostering ethical integrity in their work.

- **Bias in AI Systems:** One of the most prominent ethical concerns in AI-driven research is the risk of bias. AI systems, particularly those relying on machine learning algorithms, are heavily dependent on the data used to train them (Floridi et al., 2018). If the training data is skewed or contains inherent biases—whether in terms of race, gender, socioeconomic status, or geographical location—the resulting AI models are likely to perpetuate and even amplify these biases.

In research contexts, biased AI models can lead to flawed conclusions, discriminatory practices, and inequitable outcomes. For instance, biased AI systems in medical research could misdiagnose certain populations based on underrepresentation in the training datasets, or in social science research, biased algorithms could reinforce stereotypes. Such outcomes not only undermine the validity of research but also have far-reaching social and ethical consequences (Ng et al., 2021).

Addressing bias requires careful attention to the selection, cleaning, and preprocessing of datasets, as well as the continuous monitoring of AI models to ensure that they perform equitably across different populations (Karimian et al., 2022).

Moreover, researchers must remain vigilant about their own biases in framing research questions and interpreting AI-generated results.

- **Privacy Concerns:** The issue of privacy becomes paramount in AI research, especially when AI systems are used to analyse large datasets that include personal or sensitive information. AI algorithms, particularly those in fields like healthcare, psychology, and behavioural sciences, often rely on vast amounts of data to produce accurate and meaningful insights (D'Antonoli, 2020). This data may include highly sensitive personal details such as medical records, financial information, or behavioural patterns (Ng et al., 2021).

The risk lies in how this data is collected, stored, and processed. Researchers using AI must ensure that data is anonymized, securely stored, and handled in accordance with relevant data protection regulations such as the General Data Protection Regulation (GDPR) in the European Union (Biller-Andorno et al., 2022). Furthermore, researchers must consider the ethical implications of data sharing, particularly when datasets are used across institutions or international borders where varying privacy laws may apply.

The tension between the need for large datasets and the protection of individual privacy underscores the importance of developing robust ethical frameworks for AI research. Informed consent, clear data usage policies, and methods for de-identifying data are critical strategies for addressing privacy concerns in AI-driven research (D'Antonoli, 2020).

- **Transparency in AI Systems:** Transparency is another crucial ethical issue in AI research. AI systems, particularly those involving deep learning models, often operate as "black boxes" where even the developers may not fully understand how the system arrived at a particular decision or prediction. This opacity can be problematic in a research context where the rationale behind findings must be clear and justifiable (Biller-Andorno et al., 2022).

The lack of transparency raises concerns about the interpretability and reproducibility of AI-based research. If researchers cannot explain the decision-making process of their AI tools, the validity of their findings may be questioned (D'Antonoli, 2020). Additionally, stakeholders—including research participants, funding bodies, and the wider scientific community—may be wary of adopting AI-driven findings without a clear understanding of how those findings were generated (Ng et al., 2021).

To mitigate these concerns, researchers must prioritize the development and use of interpretable AI models where possible. Efforts to create explainable AI (XAI) are crucial in ensuring that AI systems can be audited and that their decision-making processes can be understood and scrutinized by humans (Karimian et al., 2022). Transparent communication of how AI systems work, their limitations, and the potential sources of error is essential for maintaining trust in AI-driven research (Paraman & Anamalah, 2023).

- **Accountability in AI-Driven Research:** Accountability is a foundational principle of ethical research, and it takes on added complexity in the context of AI. When AI systems are used to analyse data, make decisions, or even conduct experiments, the question of who is responsible for the outcomes becomes complicated.

In cases where AI systems produce erroneous results, perpetuate biases, or cause harm, researchers must be prepared to take responsibility (Paraman & Anamalah, 2023). This requires a clear understanding of the AI system's capabilities and limitations, as well as the establishment of mechanisms for auditing and correcting mistakes. Ethical AI research demands that researchers be accountable not only for their findings but also for the tools and methodologies they use to arrive at those findings (Floridi et al., 2018).

Moreover, accountability in AI-driven research also involves considering the broader societal implications of the research (Morley et al., 2020). Researchers must ask themselves who benefits from the research and who may be adversely affected. Ensuring accountability means anticipating the potential risks and negative externalities of AI-driven research and taking steps to mitigate them (Shneiderman, 2020).

- **Fairness in AI Applications:** Fairness is closely linked to the issues of bias and accountability, and it plays a critical role in ensuring that AI-driven research does not exacerbate existing inequalities. AI systems, when improperly designed or trained, can disproportionately affect marginalized or vulnerable groups (Vainio-Pekka et al., 2023). Ensuring fairness means that AI models must perform equally well across different demographic groups and should not disproportionately disadvantage any particular population.

Fairness in AI research also extends to the equitable distribution of the benefits and burdens of AI innovations. Researchers must be aware of how their AI applications may affect different populations and whether these impacts are just and fair (Paraman & Anamalah, 2023). For instance, in the realm of public health research, AI models that are used to allocate resources or predict disease outbreaks must be designed to avoid favouring wealthier or more privileged communities at the expense of underserved populations (Shneiderman, 2020).

Ensuring fairness requires researchers to engage in constant dialogue with stakeholders, including those who may be affected by the research outcomes. Participatory approaches, where affected communities have a voice in the design and implementation of AI systems, are essential for fostering fairness in AI-driven research (Biller-Andorno et al., 2022).

INCORPORATING ETHICS INTO RESEARCH USING AI

Incorporating ethics into research using AI is a crucial endeavour, particularly as AI technologies increasingly influence diverse fields of study. Ethical considerations must form the cornerstone of AI-driven research, guiding the responsible development, deployment, and application of AI systems (Vainio-Pekka et al., 2023). This process involves ensuring that AI tools are designed with fairness, transparency, and accountability at their core, thus safeguarding against biases, data misuse, and unintended consequences. Researchers must be trained not only in the technical aspects of AI but also in the ethical dimensions that accompany its use, including issues related to privacy, informed consent, and the potential impact of AI on vulnerable populations (Shneiderman, 2020). By embedding ethics into every stage of AI research—from data collection and algorithm design to interpretation and dissemination—researchers can mitigate risks associated with AI systems, such as reinforcing societal inequalities or perpetuating discrimination. A focus on ethical AI research also requires fostering interdisciplinary collaboration, drawing insights from fields like philosophy, law, and social sciences to anticipate and address the broader societal implications of AI technologies (Vainio-Pekka et al., 2023). Moreover, researchers should engage in continuous ethical reflection and accountability, adapting their approaches as AI technologies evolve and as new ethical challenges emerge (Floridi et al., 2018). Institutions, too, play a vital role in building this ethical foundation by integrating ethical AI research training into academic curricula, supporting the development of ethical guidelines, and fostering a culture of responsible innovation. Ultimately, incorporating ethics into AI research is not a one-time effort but an ongoing commitment to responsible practice, ensuring that AI benefits society while upholding principles of justice, equity, and human dignity (Shneiderman, 2020).

PRACTICAL TOOLS AND RESOURCES FOR ETHICAL DECISION-MAKING IN RESEARCH

As the integration of Artificial Intelligence (AI) into research processes accelerates, the ethical implications of these technologies become increasingly complex. Researchers must navigate a myriad of ethical considerations to ensure that their work is not only scientifically sound but also morally responsible. This landscape necessitates a robust framework for ethical decision-making, underscored by a variety of practical tools and resources that guide researchers in their endeavours (Biller-Andorno et al., 2022).

The **Helsinki Declaration**, formulated by the World Medical Association, provides a foundational ethical framework for conducting biomedical research involving human subjects. This declaration emphasizes the necessity of obtaining informed consent, prioritizing participant welfare, and ensuring that research benefits outweigh risks. Its principles extend beyond medical research, offering a universal ethical compass for researchers across disciplines (Stahl, 2022). The Helsinki Declaration underscores the importance of transparency, requiring researchers to disclose potential conflicts of interest and to submit their protocols for review. Such requirements instil confidence in the research process, ensuring that participants are treated with respect and dignity. For researchers integrating AI, adherence to the Helsinki Declaration is crucial, as it informs the ethical deployment of AI technologies in studies that involve human data (Trocin et al., 2023).

In addition to international declarations, **ethics committees** serve as critical entities that provide oversight and guidance in research. These committees, often referred to as Institutional Review Boards (IRBs) or Research Ethics Committees (RECs), are responsible for evaluating the ethical implications of proposed research projects. They assess the potential risks to participants, the validity of informed consent processes, and the overall integrity of the research design. Ethics committees ensure that researchers comply with established ethical standards and regulatory requirements (Biller-Andorno et al., 2022). For researchers utilizing AI, these committees can evaluate the implications of algorithmic bias, data privacy concerns, and the ethical treatment of data. Engaging with ethics committees during the research design phase allows for proactive identification and mitigation of ethical dilemmas, fostering a culture of responsibility and accountability (Trocin et al., 2023).

Moreover, the **Committee on Publication Ethics (COPE)** provides essential guidelines and resources for maintaining ethical standards in research publication. COPE aims to promote integrity in scholarly publishing by offering resources that help researchers, editors, and publishers navigate ethical issues related to authorship, plagiarism, peer review, and data sharing. The guidelines set forth by COPE emphasize the importance of transparency and ethical considerations in the publication process, ensuring that research findings are disseminated responsibly (Shneiderman, 2020). For researchers employing AI in their methodologies, COPE's resources help address issues such as data provenance, algorithmic transparency, and the ethical use of AI-generated content. By adhering to COPE's guidelines, researchers can cultivate an ethical publishing environment, safeguarding the integrity of their work and that of their peers (Trocin et al., 2023).

In addition to these established frameworks, various **practical tools** are available to aid researchers in ethical decision-making. For instance, ethical decision-making frameworks and models, such as the **Utilitarianism**, **Deontological Ethics**, and **Virtue Ethics** approaches, provide researchers with structured methodologies for evaluating the ethical implications of their work (Shneiderman, 2020). These frameworks encourage critical reflection on the consequences of research actions, obligations to stakeholders, and the character traits that underpin ethical behaviour. Utilizing these models can enhance researchers' ability to navigate complex ethical dilemmas, especially when working with AI, where traditional ethical considerations may require re-evaluation in the context of new technologies (Bakiner, 2023).

Furthermore, **educational resources** such as workshops, online courses, and training programs can significantly bolster researchers' understanding of ethical practices. Many institutions offer training sessions that focus on ethical considerations in research, including the ethical use of AI and data privacy (Trocin et al., 2023). These programs equip researchers with the knowledge necessary to recognize ethical challenges and develop appropriate responses. Engaging with interdisciplinary perspectives on ethics, particularly in the context of AI, fosters a more nuanced understanding of the implications of technology in research (Lund et al., 2023).

Fostering a culture of open dialogue and collaboration within research teams is essential for ethical decision-making. Establishing a forum for discussing ethical concerns encourages researchers to share their experiences and insights, leading to collective problem-solving (Bouhouita-Guermech et al., 2023). By promoting a culture of ethical awareness and accountability, research institutions can create an environment where ethical considerations are prioritized and integrated into every stage of the research process.

STRATEGIES FOR SECURING SENSITIVE DATA AND PREVENTING ETHICAL BREACHES IN RESEARCH USING AI

In the context of artificial intelligence (AI) and research, securing sensitive data and preventing ethical breaches are paramount concerns. As researchers leverage AI to gather, analyze, and interpret vast amounts of data, the ethical implications of their methodologies come under scrutiny (Bouhouita-Guermech et al., 2023).

The following strategies provide a framework for building ethical foundations and ensuring responsible AI usage in research.

- **Implement Robust Data Encryption:** Data encryption is a fundamental strategy for protecting sensitive information. By utilizing strong encryption algorithms, researchers can ensure that data is rendered unreadable to unauthorized users (Karimian et al., 2022). This process involves converting information into a code that can only be decrypted by individuals who possess the correct keys. Employing encryption not only protects data during transmission but also secures it when stored, mitigating the risk of data breaches (Bakiner, 2023).
- **Conduct Thorough Risk Assessments:** Before initiating research projects that involve sensitive data, conducting comprehensive risk assessments is essential. Researchers should identify potential vulnerabilities and the likelihood of data breaches. This proactive approach allows teams to develop tailored mitigation strategies that address specific risks (Bakiner, 2023). Additionally, regular updates to risk assessments are crucial as new technologies and methodologies emerge, ensuring that ethical considerations remain at the forefront of research practices (Karimian et al., 2022).
- **Establish Clear Data Governance Policies:** Creating clear and comprehensive data governance policies is critical for ensuring ethical research practices. These policies should outline how data is collected, used, stored, and shared throughout the research lifecycle. Researchers should establish guidelines that prioritize transparency, accountability, and ethical compliance. Moreover, engaging stakeholders in the development of these policies fosters a collaborative approach and encourages adherence to ethical standards (Bouhouita-Guermech et al., 2023).
- **Train Researchers in Ethical AI Practices:** Educating researchers on ethical AI practices is vital for fostering a culture of responsibility. Training programs should cover the ethical implications of using AI, focusing on issues such as bias, data privacy, and the potential impact of AI-driven decisions. By instilling a strong ethical framework within research teams, institutions can empower researchers to make informed decisions that prioritize ethical considerations in their work (Karimian et al., 2022).
- **Utilize Anonymization Techniques:** Anonymization techniques are crucial for safeguarding personal data in research involving AI. Researchers should implement methods that remove personally identifiable information (PII) from datasets, ensuring that individual identities cannot be traced back to the data (Ryan et al., 2021). This approach not only protects participants' privacy

but also helps researchers comply with data protection regulations, such as the General Data Protection Regulation (GDPR).

- **Foster an Open Culture of Ethical Reflection:** Encouraging an open culture of ethical reflection within research teams can significantly enhance ethical awareness. Researchers should feel empowered to discuss ethical dilemmas and raise concerns about their methodologies. Regular discussions and workshops focused on ethical challenges can facilitate collaborative problem-solving and reinforce the importance of ethical considerations throughout the research process (Qadhi et al., 2024).
- **Engage with Ethics Committees:** Collaborating with ethics committees is an effective way to ensure that research practices align with ethical standards. Researchers should seek guidance from these committees when designing studies, particularly when sensitive data is involved (Konidena et al., 2024). Ethics committees can provide valuable insights, identify potential ethical breaches, and suggest improvements to research protocols. This engagement fosters a culture of accountability and reinforces the commitment to ethical research practices.
- **Develop Comprehensive Data Sharing Agreements:** When collaborating with external partners or institutions, researchers should establish comprehensive data sharing agreements (Resnik & Hosseini, 2024). These agreements should clearly outline how data will be shared, the purposes for which it will be used, and the measures taken to protect sensitive information. By formalizing these arrangements, researchers can mitigate the risks associated with data sharing and ensure that ethical standards are upheld throughout the collaboration (Jobin et al., 2019).
- **Monitor and Audit Data Practices:** Regular monitoring and auditing of data practices are essential for ensuring compliance with ethical standards. Researchers should implement mechanisms to track data usage, access, and storage throughout the research process. Conducting periodic audits can help identify potential breaches and areas for improvement, fostering a culture of accountability and ethical vigilance within research teams (Resnik & Hosseini, 2024).
- **Stay Informed about Evolving Regulations and Standards:** The landscape of data protection and ethical research practices is continually evolving. Researchers must stay informed about current regulations, industry standards, and best practices related to data security and ethical AI usage (Paraman & Anamalah, 2023). Engaging in ongoing professional development, attending workshops, and participating in relevant forums can help researchers remain aware of emerging trends and adapt their practices accordingly.

THE ROLE OF AI GOVERNANCE FRAMEWORKS IN MAINTAINING ETHICAL STANDARDS FOR RESEARCHERS IN INSTITUTIONS

The emergence of artificial intelligence (AI) has brought about transformative advancements across various fields, including research. However, this rapid evolution also poses significant ethical challenges that necessitate the establishment of robust AI governance frameworks (Paraman & Anamalah, 2023). Such frameworks play a pivotal role in maintaining ethical standards for researchers in academic and research institutions. Below are key points elaborating on the significance of AI governance frameworks in ensuring responsible AI usage by researchers.

- **Establishing Clear Ethical Guidelines and Standards:** AI governance frameworks provide a structured approach to defining ethical guidelines and standards that govern the development and application of AI technologies. These guidelines articulate the moral principles that researchers must adhere to when designing and implementing AI systems (Samuel et al., 2021). By outlining expectations related to transparency, fairness, accountability, and respect for human rights, governance frameworks ensure that researchers are equipped with the necessary ethical frameworks to navigate complex moral dilemmas inherent in AI research (Paraman & Anamalah, 2023). This clarity not only helps in preventing ethical breaches but also fosters a culture of ethical awareness within research institutions.
- **Enhancing Transparency in Research Practices:** Transparency is a cornerstone of ethical research, particularly in the realm of AI, where complex algorithms can often operate as "black boxes." AI governance frameworks promote transparency by requiring researchers to document their methodologies, data sources, and decision-making processes (Paraman & Anamalah, 2023). This documentation enables stakeholders to understand how AI models function and the reasoning behind specific outcomes (Samuel et al., 2021). By enhancing transparency, governance frameworks empower researchers to engage in more responsible practices, as stakeholders—including peers, regulatory bodies, and the public—can scrutinize and evaluate the ethical implications of research findings (Paraman & Anamalah, 2023). This scrutiny not only builds trust in AI systems but also encourages researchers to adopt a more reflective and accountable approach.
- **Mitigating Risks of Bias and Discrimination:** One of the significant ethical challenges associated with AI research is the potential for bias and discrimination in algorithmic decision-making. AI governance frameworks are instrumental in addressing these concerns by mandating rigorous bias as-

sessments and mitigation strategies (Jobin et al., 2019). Researchers are encouraged to evaluate the data sets they use, ensuring they are representative and free from inherent biases. Furthermore, these frameworks advocate for continuous monitoring of AI systems to identify and rectify biases that may emerge post-deployment (Konidena et al., 2024). By embedding such practices within governance frameworks, institutions can significantly reduce the risk of perpetuating discrimination and ensure that AI systems are developed and utilized in a manner that promotes equity and justice.

- **Fostering Collaboration and Interdisciplinary Dialogue:** AI governance frameworks can facilitate collaboration among researchers from diverse fields, including ethics, law, social sciences, and engineering. This interdisciplinary dialogue is crucial for identifying and addressing the multifaceted ethical implications of AI technologies (Konidena et al., 2024). By encouraging collaborative efforts, governance frameworks enable researchers to draw on a wide range of perspectives, fostering a more comprehensive understanding of the ethical landscape surrounding AI research. Such collaboration can lead to the development of innovative solutions to ethical challenges, ensuring that research practices are not only technically sound but also socially responsible and ethically justified (Paraman & Anamalah, 2023).
- **Promoting Responsible Data Management Practices:** The responsible use of data is a fundamental aspect of ethical AI research. Governance frameworks emphasize the importance of data privacy, security, and informed consent. Researchers are guided to implement best practices for data collection, storage, and sharing, ensuring that the rights of individuals whose data is being used are protected (Jobin et al., 2019). By establishing robust data management protocols, AI governance frameworks help mitigate risks associated with data breaches, misuse, and unethical data practices (Paraman & Anamalah, 2023). This commitment to responsible data management enhances the overall integrity of research processes and reinforces the ethical foundations upon which AI technologies are built.
- **Encouraging Continuous Ethical Training and Development:** As AI technologies evolve, so too do the ethical challenges associated with their use. AI governance frameworks play a crucial role in promoting continuous ethical training and development for researchers. By integrating ethical education into research programs, institutions can equip researchers with the knowledge and skills needed to navigate the complexities of AI ethics (Trocin et al., 2023). This ongoing education ensures that researchers remain informed about emerging ethical issues, regulatory changes, and best practices in AI governance. By fostering a culture of ethical vigilance, institutions can bet-

ter prepare their researchers to make informed and responsible decisions throughout the research lifecycle (Konidena et al., 2024).

- **Ensuring Accountability and Oversight Mechanisms:** Accountability is essential for maintaining ethical standards in AI research. Governance frameworks establish mechanisms for oversight and accountability, ensuring that researchers are held responsible for their actions and decisions (Qadhi et al., 2024). This may include the implementation of ethics review boards, regular audits of AI projects, and clear channels for reporting ethical violations. By holding researchers accountable, governance frameworks reinforce the importance of ethical conduct and deter potential misconduct. This accountability not only safeguards the integrity of research practices but also promotes public trust in AI technologies and the institutions that develop them (Trocin et al., 2023).
- **Navigating Legal and Regulatory Compliance:** AI governance frameworks are critical in helping researchers navigate the complex legal and regulatory landscape surrounding AI technologies. By aligning research practices with existing laws and regulations, such frameworks minimize the risk of legal repercussions and ensure compliance with relevant ethical standards. Researchers benefit from guidance on issues such as intellectual property, data protection, and compliance with AI-specific regulations. This alignment fosters a culture of ethical responsibility, as researchers are made aware of the legal implications of their work and the importance of adhering to both ethical and legal standards (Karimian et al., 2022).
- **Contributing to Societal Trust and Acceptance of AI Technologies:** Public trust is vital for the successful integration of AI technologies into society. AI governance frameworks contribute to building this trust by promoting ethical standards and practices within research institutions. When researchers demonstrate a commitment to ethical AI development, transparency, and accountability, it fosters confidence among stakeholders, including the public, industry partners, and policymakers. This trust is essential for ensuring the societal acceptance of AI technologies, which can lead to greater collaboration between researchers and the communities they serve (Trocin et al., 2023). By prioritizing ethical considerations, institutions can play a pivotal role in shaping a positive narrative around AI, ultimately supporting its responsible advancement.

HOW EDUCATION AND TRAINING NEED TO EVOLVE WITH AI ADVANCEMENTS FOR AI ETHICS AMONG RESEARCHERS

The rapid development and integration of artificial intelligence (AI) into various domains have necessitated a shift in how education and training are delivered, especially in the realm of AI ethics. Researchers must be equipped not only with technical expertise but also with a deep understanding of ethical implications to ensure responsible AI use (Karimian et al., 2022). Below is an exploration of how education and training need to evolve to build a strong ethical foundation among researchers.

- **Embedding AI Ethics in the Core Curriculum:** Traditional educational programs often focus on technical aspects of AI, such as machine learning algorithms, data processing, and automation, while ethical considerations are treated as supplementary topics. This needs to change (Kazim & Koshiyama, 2021). AI ethics should be embedded in the core curriculum of any program related to AI development and research. By integrating ethics as a fundamental subject, researchers can gain a well-rounded understanding of the consequences of AI on society, data privacy, human rights, and decision-making processes. Ensuring that ethical principles are not an afterthought but a foundational component of AI education will enable researchers to approach AI development with a balanced perspective, combining innovation with responsibility (Trocin et al., 2023).
- **Incorporating Multidisciplinary Learning Approaches:** AI ethics is not confined to the domain of computer science; it spans law, sociology, psychology, and philosophy. To foster ethical awareness in researchers, educational programs should incorporate multidisciplinary learning approaches. Researchers need to be exposed to diverse fields that influence ethical decision-making in AI development (Karimian et al., 2022). Understanding social impacts, regulatory requirements, and philosophical debates surrounding AI will allow researchers to address complex ethical dilemmas with greater clarity. This approach can also lead to the development of AI systems that are more inclusive and aligned with societal values (Samuel et al., 2021).
- **Training in Ethical Decision-Making Frameworks:** One of the key challenges researchers face in AI development is how to make ethical decisions in ambiguous or unforeseen situations. To prepare for this, training in ethical decision-making frameworks is essential (Nikolinakos, 2023). Educational programs should teach researchers to apply structured frameworks such as utilitarianism, deontological ethics, and virtue ethics when assessing AI technologies. These frameworks offer a foundation for evaluating the moral

implications of AI systems, enabling researchers to critically assess whether their work adheres to ethical standards, minimizes harm, and promotes fairness. Such training equips researchers with the tools to navigate the grey areas where AI development may have unintended or potentially harmful consequences (Kazim & Koshiyama, 2021).

- **Practical Simulations and Case Studies in AI Ethics:** Theoretical knowledge alone is insufficient to foster ethical awareness. Educational programs should incorporate practical simulations and case studies that allow researchers to engage with real-world ethical challenges in AI development (Resnik & Hosseini, 2024). By analysing historical examples where AI has led to ethical breaches—such as biased algorithms, privacy violations, or autonomous systems causing harm—researchers can better understand the gravity of ethical decision-making (Kazim & Koshiyama, 2021). These hands-on learning experiences help researchers anticipate potential ethical pitfalls and build strategies to mitigate risks before they arise in practice. It also cultivates a proactive approach to ethics rather than a reactive one.
- **Ongoing Professional Development in AI Ethics:** I is an ever-evolving field, with advancements occurring at a rapid pace. Ethical considerations in AI development today may be different from those that arise in the future, particularly as new technologies such as quantum computing or brain-machine interfaces emerge. Therefore, educational institutions and research bodies must offer ongoing professional development in AI ethics. This could take the form of workshops, seminars, or certification programs that keep researchers updated on the latest ethical concerns and regulatory changes in AI (Nikolinakos, 2023). Lifelong learning will ensure that researchers remain vigilant in applying ethical principles throughout their careers, adapting their knowledge as technology evolves.
- **Promoting Transparency and Accountability in AI Research:** Education and training for AI researchers must emphasize the importance of transparency and accountability in AI development. Researchers should be trained to document their processes, disclose data sources, and make AI models explainable (Mökander & Axente, 2023). Transparency is crucial for ensuring that AI systems are not only effective but also ethical and fair. Furthermore, accountability mechanisms must be introduced, where researchers are responsible for the outcomes of the AI systems they create (Kazim & Koshiyama, 2021). Training researchers to prioritize these values in their work will lead to greater public trust in AI technologies and reduce the risks associated with opaque or biased AI systems.
- **Developing Ethical Leadership in AI Research:** AI researchers are not just contributors to technological progress; they are often leaders in innovation

within their fields. As such, educational programs must also focus on developing ethical leadership skills among researchers. This includes cultivating a sense of responsibility for setting ethical standards within their teams, organizations, or communities (Telkamp & Anderson, 2022). Researchers must be equipped to lead by example, advocate for ethical practices in AI development, and mentor future generations of researchers on the importance of ethics. Ethical leadership is vital for ensuring that AI advancements are pursued with a commitment to societal well-being and the prevention of harm (Kazim & Koshiyama, 2021).

- **Encouraging Public Engagement and Ethical Dialogue:** AI researchers need to engage with the public and participate in broader ethical dialogues about AI's role in society. Education and training should encourage researchers to communicate their work's ethical implications to non-expert audiences, fostering public awareness and discourse on AI ethics (Cheng et al., 2021). By involving the public in these conversations, researchers can gain diverse perspectives on the societal impacts of AI technologies and ensure that their work aligns with public values and concerns. Public engagement also helps researchers remain accountable to the communities that are affected by their innovations, reinforcing the importance of ethical AI development.
- **Fostering Ethical Collaboration across Global Boundaries:** AI is a global technology, and its ethical challenges are not confined to any single nation. Researchers often collaborate across borders, working with diverse teams that may have different cultural values, regulatory environments, and ethical norms (Cheng et al., 2021). Educational programs must therefore foster a sense of ethical collaboration, preparing researchers to navigate cross-cultural ethical dilemmas and align AI development with international standards. By encouraging global ethical dialogue and cooperation, researchers can contribute to the creation of AI systems that benefit humanity as a whole, rather than prioritizing the interests of specific regions or groups (Mökander & Axente, 2023).
- **Integration of AI Ethics into Research Funding and Publication Processes:** Education and training in AI ethics must extend to the broader research ecosystem, including funding and publication processes. Researchers should be trained to consider ethical implications when applying for funding, ensuring that their projects are aligned with ethical standards from the outset (Cheng et al., 2021). Additionally, ethical considerations should be integral to the peer-review and publication processes, with researchers being held accountable for the ethical dimensions of their work. Training researchers to incorporate ethics into these processes will ensure that ethical AI develop-

ment is rewarded and promoted within the scientific community, encouraging widespread adherence to ethical principles (Cheng et al., 2021).

PREDICTING THE FUTURE ROLE OF AI IN MAINTAINING RESEARCH ETHICS AMONG RESEARCHERS

Artificial Intelligence (AI) is revolutionizing numerous sectors, and research ethics is no exception. As AI becomes more integrated into the research process, its potential to uphold and enhance ethical standards among researchers will be profound. However, this evolution also comes with challenges. The future role of AI in maintaining research ethics involves diverse and complex dimensions that need careful consideration.

- **Automating Ethical Review Processes:** AI systems could be designed to assist in ethical reviews by analysing research proposals for potential ethical risks. Currently, ethics committees manually review each submission, a process prone to delays and inconsistencies due to human bias (Cheng et al., 2021). In the future, AI could ensure faster, more objective evaluations by identifying ethical concerns related to participant consent, confidentiality, or the potential misuse of data. By automating this process, AI can serve as a supplementary tool to human oversight, enhancing the consistency and speed of ethics reviews, while also reducing the burden on institutional review boards.
- **Real-Time Monitoring of Ethical Compliance:** AI can offer real-time monitoring tools that track researchers' adherence to ethical guidelines throughout the research process (Mökander & Axente, 2023). From ensuring data collection is carried out within the confines of informed consent to preventing the misuse of sensitive information, AI can provide continuous oversight. Such systems could flag potential ethical violations as they occur, allowing for immediate corrective actions rather than retrospective enforcement (Telkamp & Anderson, 2022). This could significantly reduce cases of unintentional ethical breaches and improve the overall integrity of research.
- **Bias Detection in Research Methods:** One of the most significant ethical concerns in research is the presence of bias, whether in data collection, analysis, or interpretation. AI has the potential to identify biases in research methods that may go unnoticed by human researchers (Klimova et al., 2023). By analysing large datasets and identifying patterns of skewed or discriminatory practices, AI can ensure that research remains objective and fair. As AI tools become more sophisticated, their ability to detect and correct biases before

they affect research outcomes will be crucial in maintaining ethical standards in scientific inquiry (Resnik & Hosseini, 2024).

- **Ensuring Data Privacy and Confidentiality:** As research increasingly involves large datasets, particularly in sensitive fields such as medicine or social sciences, ensuring data privacy is paramount. AI will play a critical role in protecting the confidentiality of participants by enforcing data privacy regulations through encryption and anonymization techniques (Telkamp & Anderson, 2022). Advanced AI algorithms can help in creating secure data-sharing platforms that allow for collaboration without compromising privacy. By automating compliance with privacy laws like GDPR or HIPAA, AI will help researchers navigate complex data protection landscapes while ensuring ethical research practices are upheld (Klimova et al., 2023).
- **Ethical AI in Autonomous Research Tools:** With AI becoming an integral part of research tools, it is crucial to ensure that the AI itself operates ethically. As AI systems become autonomous, capable of conducting experiments, analysing data, and even formulating hypotheses, the ethical use of these tools must be prioritized. In the future, AI systems will need to be embedded with ethical constraints that align with established research guidelines (Resnik & Hosseini, 2024). Ensuring that AI tools themselves adhere to ethical standards will prevent misuse or the accidental propagation of unethical research practices by autonomous systems.
- **Facilitating Informed Consent:** Informed consent is a cornerstone of ethical research, but ensuring participants fully understand complex research protocols can be challenging. AI could assist in simplifying and automating the process of obtaining informed consent by providing tailored explanations to participants, assessing their understanding, and adapting the information accordingly (Peters et al., 2020). AI systems could generate personalized consent forms based on individual participants' educational levels, cultural backgrounds, or language preferences, ensuring that participants are fully aware of the risks and benefits involved in the research (Cheng et al., 2021).
- **Preventing Plagiarism and Data Fabrication:** AI-based plagiarism detection tools are already widely used, but future advancements could extend their capabilities to prevent data fabrication and manipulation (Cheng et al., 2021). AI systems will be able to track the authenticity of datasets, ensuring that researchers do not engage in unethical practices such as falsifying results. By comparing research outcomes against historical data, AI can detect inconsistencies and anomalies that suggest data manipulation. As AI continues to evolve, these systems could be fully integrated into the peer-review process, making it nearly impossible for fabricated research to pass undetected (Samuel et al., 2021).

- **Promoting Ethical Collaboration in Global Research:** AI-powered platforms will facilitate more transparent and ethical collaboration among researchers from different regions and institutions. By streamlining communication and ensuring equitable contribution to research projects, AI can help prevent issues such as unfair authorship or exploitation of researchers from underprivileged backgrounds (Osasona et al., 2024). These platforms could also provide automated audits of contributions to ensure that credit is distributed fairly, thus reducing ethical conflicts in collaborative research efforts. This could foster a culture of fairness and accountability in global research communities (Peters et al., 2020).
- **Developing Adaptive Ethical Guidelines:** Research ethics is not static; it evolves as new technologies and methodologies emerge. AI can assist in continuously updating ethical guidelines by analysing trends in research practices and identifying new ethical dilemmas. AI-driven systems could help institutional review boards and governing bodies stay ahead of potential ethical issues by suggesting updates to existing codes of conduct based on emerging research trends (Samuel et al., 2021). These adaptive guidelines would ensure that ethical standards keep pace with scientific innovation, helping researchers navigate the ethical complexities of future research landscapes.
- **Enhancing Transparency and Accountability in Research:** AI can be instrumental in improving transparency and accountability in research. By creating digital logs of research activities, AI can offer detailed records of every step taken in a research project, from data collection to publication (Al-kfairy et al., 2024). This will ensure that all actions are traceable, reducing the risk of misconduct such as tampering with results or concealing unethical practices. These digital records could be made available to ethical review boards or even the public, promoting greater transparency in the research process and holding researchers accountable for their actions (Osasona et al., 2024).
- **Balancing Innovation and Ethical Boundaries:** AI's role in research will require a careful balance between fostering innovation and maintaining ethical boundaries. While AI can expedite research and open up new possibilities, it will also need to ensure that these advancements do not come at the expense of ethical considerations. For instance, AI-driven systems could assist in balancing the pressure to publish with the need to uphold ethical standards, ensuring that research is driven by integrity rather than the pursuit of recognition (Al-kfairy et al., 2024). AI's role in this context will be to support a research culture that values ethical rigor as much as scientific advancement.
- **Education and Training on AI Ethics in Research:** AI will play a significant role in educating researchers about ethics in the AI-driven research environment. Future AI systems could be used to create interactive training

modules that immerse researchers in ethical dilemmas and teach them how to navigate complex scenarios (Osasona et al., 2024). By simulating ethical challenges in a controlled environment, AI can help researchers develop a deeper understanding of ethical principles, preparing them to handle real-world situations. This AI-driven approach to ethics education will be critical in ensuring that researchers are equipped to uphold ethical standards in an increasingly AI-dependent research ecosystem.

- **Supporting Equitable Access to Research Opportunities:** Ethical concerns around equitable access to research opportunities, particularly in underserved regions, are likely to be addressed by AI (Osasona et al., 2024). AI platforms can democratize access to research funding, resources, and collaboration opportunities by matching researchers with funding bodies based on merit and need rather than geographic or institutional bias. This equitable distribution of resources will ensure that research opportunities are not limited to privileged groups, promoting a more inclusive research landscape and upholding ethical principles of fairness and justice (Klimova et al., 2023).
- **Ethical Auditing of AI-Driven Research:** As AI becomes more involved in the research process, ensuring that AI-driven research itself adheres to ethical guidelines will become paramount. In the future, AI systems may be developed to audit research that relies heavily on AI, ensuring that AI algorithms do not introduce biases or unethical practices into the research. These systems will help verify that AI-driven studies are conducted responsibly, maintaining the integrity of both the research process and its outcomes (Al-kfairy et al., 2024).

CONCLUSION

The role of AI in maintaining research ethics among researchers is poised to become increasingly pivotal. As AI technologies continue to evolve, they will offer powerful tools to uphold ethical standards, automate ethical reviews, and enhance transparency and accountability across the research process. By enabling real-time monitoring of compliance, detecting biases, and safeguarding data privacy, AI will not only help prevent ethical breaches but also foster a culture of responsibility and fairness in global research communities. Moreover, AI-driven systems will play a critical role in facilitating ethical education, ensuring that future researchers are equipped with the knowledge and tools to navigate complex ethical landscapes. However, as AI becomes more autonomous, the need for adaptive ethical frameworks will grow, ensuring that both AI systems and the research they support remain aligned with societal values and ethical guidelines. In this evolving ecosystem, interdisciplinary collaboration and continuous dialogue will be essential to address emerging challenges and ensure that AI contributes positively to the future of ethical research practices.

REFERENCES

Al-kfairy, M., Mustafa, D., Kshetri, N., Insiew, M., & Alfandi, O. (2024). Ethical Challenges and Solutions of Generative AI: An Interdisciplinary Perspective. *Informatics (MDPI), 11*(3), 3. Advance online publication. DOI: 10.3390/informatics11030058

Bakiner, O. (2023). What do academics say about artificial intelligence ethics? An overview of the scholarship. *AI and Ethics, 3*(2), 513–525. DOI: 10.1007/s43681-022-00182-4

Benefo, E. O., Tingler, A., White, M., Cover, J., Torres, L., Broussard, C., Shirmohammadi, A., Pradhan, A. K., & Patra, D. (2022). Ethical, legal, social, and economic (ELSE) implications of artificial intelligence at a global level: A scientometrics approach. *AI and Ethics, 2*(4), 667–682. DOI: 10.1007/s43681-021-00124-6

Benzinger, L., Ursin, F., Balke, W.-T., Kacprowski, T., & Salloch, S. (2023). Should Artificial Intelligence be used to support clinical ethical decision-making? A systematic review of reasons. *BMC Medical Ethics, 24*(1), 48. DOI: 10.1186/s12910-023-00929-6 PMID: 37415172

Biller-Andorno, N., Ferrario, A., Joebges, S., Krones, T., Massini, F., Barth, P., Arampatzis, G., & Krauthammer, M. (2022). AI support for ethical decision-making around resuscitation: Proceed with care. *Journal of Medical Ethics, 48*(3), 175–183. DOI: 10.1136/medethics-2020-106786 PMID: 33687916

Bouhouita-Guermech, S., Gogognon, P., & Bélisle-Pipon, J.-C. (2023). Specific challenges posed by artificial intelligence in research ethics. *Frontiers in Artificial Intelligence, 6*, 1149082. Advance online publication. DOI: 10.3389/frai.2023.1149082 PMID: 37483869

Cheng, L., Varshney, K. R., & Liu, H. (2021). Socially Responsible AI Algorithms: Issues, Purposes, and Challenges. *Journal of Artificial Intelligence Research, 71*, 1137–1181. DOI: 10.1613/jair.1.12814

D'Antonoli, T. A. (2020). Ethical considerations for artificial intelligence: An overview of the current radiology landscape. *Diagnostic and Interventional Radiology (Ankara, Turkey), 26*(5), 504–511. DOI: 10.5152/dir.2020.19279 PMID: 32755879

Dhirani, L. L., Mukhtiar, N., Chowdhry, B. S., & Newe, T. (2023). Ethical Dilemmas and Privacy Issues in Emerging Technologies: A Review. *Sensors (Basel), 23*(3), 3. Advance online publication. DOI: 10.3390/s23031151 PMID: 36772190

Floridi, L., Cowls, J., Beltrametti, M., Chatila, R., Chazerand, P., Dignum, V., Luetge, C., Madelin, R., Pagallo, U., Rossi, F., Schafer, B., Valcke, P., & Vayena, E. (2018). AI4People—an Ethical Framework for a Good AI Society: Opportunities, Risks, Principles, and Recommendations. *Minds and Machines*, *28*(4), 689–707. DOI: 10.1007/s11023-018-9482-5 PMID: 30930541

Hallamaa, J., & Kalliokoski, T. (2022). AI Ethics as Applied Ethics. *Frontiers of Computer Science*, *4*, 776837. Advance online publication. DOI: 10.3389/fcomp.2022.776837

Huang, C., Samek, T., & Shiri, A. (2021). AI and Ethics: Ethical and Educational Perspectives for LIS. *Journal of Education for Library and Information Science*, *62*(4), 351–365. DOI: 10.3138/jelis-62-4-2020-0106

Jobin, A., Ienca, M., & Vayena, E. (2019). The global landscape of AI ethics guidelines. *Nature Machine Intelligence*, *1*(9), 389–399. DOI: 10.1038/s42256-019-0088-2

Karimian, G., Petelos, E., & Evers, S. M. A. A. (2022). The ethical issues of the application of artificial intelligence in healthcare: A systematic scoping review. *AI and Ethics*, *2*(4), 539–551. DOI: 10.1007/s43681-021-00131-7

Kazim, E., & Koshiyama, A. S. (2021). A high-level overview of AI ethics. *Patterns (New York, N.Y.)*, *2*(9), 100314. Advance online publication. DOI: 10.1016/j.patter.2021.100314 PMID: 34553166

Klimova, B., Pikhart, M., & Kacetl, J. (2023). Ethical issues of the use of AI-driven mobile apps for education. *Frontiers in Public Health*, *10*, 1118116. Advance online publication. DOI: 10.3389/fpubh.2022.1118116 PMID: 36711343

Konidena, B. K., Malaiyappan, J. N. A., & Tadimarri, A. (2024). Ethical Considerations in the Development and Deployment of AI Systems. *European Journal of Technology*, *8*(2), 2. Advance online publication. DOI: 10.47672/ejt.1890

Lund, B. D., Wang, T., Mannuru, N. R., Nie, B., Shimray, S., & Wang, Z. (2023). ChatGPT and a new academic reality: Artificial Intelligence-written research papers and the ethics of the large language models in scholarly publishing. *Journal of the Association for Information Science and Technology*, *74*(5), 570–581. DOI: 10.1002/asi.24750

Mökander, J., & Axente, M. (2023). Ethics-based auditing of automated decision-making systems: Intervention points and policy implications. *AI & Society*, *38*(1), 153–171. DOI: 10.1007/s00146-021-01286-x

Morley, J., Floridi, L., Kinsey, L., & Elhalal, A. (2020). From What to How: An Initial Review of Publicly Available AI Ethics Tools, Methods and Research to Translate Principles into Practices. *Science and Engineering Ethics*, *26*(4), 2141–2168. DOI: 10.1007/s11948-019-00165-5 PMID: 31828533

Ng, D. T. K., Leung, J. K. L., Chu, K. W. S., & Qiao, M. S. (2021). AI Literacy: Definition, Teaching, Evaluation and Ethical Issues. *Proceedings of the Association for Information Science and Technology*, *58*(1), 504–509. DOI: 10.1002/pra2.487

Nikolinakos, N. Th. (2023). Ethical Principles for Trustworthy AI. In Nikolinakos, N. Th. (Ed.), *EU Policy and Legal Framework for Artificial Intelligence, Robotics and Related Technologies—the AI Act* (pp. 101–166). Springer International Publishing., DOI: 10.1007/978-3-031-27953-9_3

Osasona, F., Amoo, O. O., Atadoga, A., Abrahams, T. O., Farayola, O. A., & Ayinla, B. S. (2024). Reviewing the Ethical Implications of AI in Decision Making Processes. *International Journal of Management & Entrepreneurship Research*, *6*(2), 2. Advance online publication. DOI: 10.51594/ijmer.v6i2.773

Ouchchy, L., Coin, A., & Dubljević, V. (2020). AI in the headlines: The portrayal of the ethical issues of artificial intelligence in the media. *AI & Society*, *35*(4), 927–936. DOI: 10.1007/s00146-020-00965-5

Paraman, P., & Anamalah, S. (2023). Ethical artificial intelligence framework for a good AI society: Principles, opportunities and perils. *AI & Society*, *38*(2), 595–611. DOI: 10.1007/s00146-022-01458-3

Peters, D., Vold, K., Robinson, D., & Calvo, R. A. (2020). Responsible AI—Two Frameworks for Ethical Design Practice. *IEEE Transactions on Technology and Society*, *1*(1), 34–47. *IEEE Transactions on Technology and Society*. Advance online publication. DOI: 10.1109/TTS.2020.2974991

Qadhi, S. M., Alduais, A., Chaaban, Y., & Khraisheh, M. (2024). Generative AI, Research Ethics, and Higher Education Research: Insights from a Scientometric Analysis. *Information (Basel)*, *15*(6), 6. Advance online publication. DOI: 10.3390/info15060325

Resnik, D. B., & Hosseini, M. (2024). The ethics of using artificial intelligence in scientific research: New guidance needed for a new tool. *AI and Ethics*. Advance online publication. DOI: 10.1007/s43681-024-00493-8

Ryan, M., Antoniou, J., Brooks, L., Jiya, T., Macnish, K., & Stahl, B. (2021). Research and Practice of AI Ethics: A Case Study Approach Juxtaposing Academic Discourse with Organisational Reality. *Science and Engineering Ethics*, *27*(2), 16. DOI: 10.1007/s11948-021-00293-x PMID: 33686527

Samuel, G., Chubb, J., & Derrick, G. (2021). Boundaries between Research Ethics and Ethical Research Use in Artificial Intelligence Health Research. *Journal of Empirical Research on Human Research Ethics; JERHRE*, *16*(3), 325–337. DOI: 10.1177/15562646211002744 PMID: 33733915

Shneiderman, B. (2020). Bridging the Gap between Ethics and Practice: Guidelines for Reliable, Safe, and Trustworthy Human-centered AI Systems. *ACM Trans. Interact. Intell. Syst.*, *10*(4), 26:1-26:31. DOI: 10.1145/3419764

Stahl, B. C. (2022). From computer ethics and the ethics of AI towards an ethics of digital ecosystems. *AI and Ethics*, *2*(1), 65–77. DOI: 10.1007/s43681-021-00080-1

Stahl, B. C., Antoniou, J., Ryan, M., Macnish, K., & Jiya, T. (2022). Organisational responses to the ethical issues of artificial intelligence. *AI & Society*, *37*(1), 23–37. DOI: 10.1007/s00146-021-01148-6

Telkamp, J. B., & Anderson, M. H. (2022). The Implications of Diverse Human Moral Foundations for Assessing the Ethicality of Artificial Intelligence. *Journal of Business Ethics*, *178*(4), 961–976. DOI: 10.1007/s10551-022-05057-6

Trocin, C., Mikalef, P., Papamitsiou, Z., & Conboy, K. (2023). Responsible AI for Digital Health: A Synthesis and a Research Agenda. *Information Systems Frontiers*, *25*(6), 2139–2157. DOI: 10.1007/s10796-021-10146-4

Vainio-Pekka, H., Agbese, M. O.-O., Jantunen, M., Vakkuri, V., Mikkonen, T., Rousi, R., & Abrahamsson, P. (2023). The Role of Explainable AI in the Research Field of AI Ethics. *ACM Trans. Interact. Intell. Syst.*, *13*(4), 26:1-26:39. DOI: 10.1145/3599974

Chapter 9
Ethics Implications of Artificial Intelligence for K-12 Schools:
Combining Dewey's Pragmatism With Freire's Critical Lens

Adam I. Attwood
https://orcid.org/0000-0002-4718-9272
Austin Peay State University, USA

ABSTRACT

This chapter posits a combined conceptual framework for discussing the ethics of the use of artificial intelligence (AI) in K-12 schools. By combining John Dewey's pragmatism with Paulo Freire's critical pedagogy, the development of ethics guidelines for AI use in schools may be more holistic and human-centered. The potential benefits and risks associated with AI use by teachers, students, and administrators are discussed while emphasizing the need for ethical guidelines that protect the interests of all interested groups involved. Recommendations are presented based on this combined framework for ethical use of generative and non-generative AI that may be of use for teacher educators, K-12 school teachers, and administrators.

INTRODUCTION

As artificial intelligence (AI) rapidly integrates into various facets of society, its influence on education is becoming increasingly pronounced. The adoption of AI in K-12 education presents unprecedented opportunities and challenges, requiring careful ethical considerations to ensure that the technology is used responsibly and

DOI: 10.4018/979-8-3693-9035-1.ch009

equitably. This chapter explores the ethical implications of AI development and implementation in K-12 education, focusing on both generative and non-generative AI. Using a theoretical framework combining John Dewey's pragmatism (Anderson, 2023; Dewey, 1938, 1994; Fesmire, 2003) with Paulo Freire's critical pedagogy (Farag et al., 2022; Freire, 2014; Gibson, 1999), this chapter analyzes the potential benefits and risks associated with AI use by teachers, students, and administrators, emphasizing the need for ethical guidelines that protect the interests of all interested persons involved. A combined framework is posited for establishing ethics guidelines for using AI in schools.

CONCEPTUAL FRAMEWORK

Overview of AI Technology Types

AI technologies in K-12 education can be broadly categorized into generative and non-generative types. Generative AI refers to systems that can create new content, such as text, images, music, or even complex algorithms. Examples include language models like GPT (Generative Pre-trained Transformer), which can produce coherent essays, solve mathematical problems, and generate creative content that might be used in classroom settings. Non-generative AI includes tools designed for data analysis, predictive modeling, and task automation. These AI systems can track student progress, predict academic outcomes, personalize learning experiences, and assist in administrative tasks, such as attendance tracking and resource allocation. Both types of AI offer significant potential to transform education, but their deployment in K-12 schools must be approached with a deep understanding of the ethical concerns they raise (Chiu, 2021; Law, 2024).

Dewey's Pragmatism and Freire's Critical Pedagogy

Dewey's pragmatism is centered on the idea that knowledge and truth are not static or absolute but are instead dynamic and evolving through human experiences. Dewey believed that ideas should be tested through practical application and that their value is determined by their effectiveness in solving problems. An example of this was Dewey's Lab School at the University of Chicago (Rud & Attwood, 2016). This type of approach emphasizes the interplay between thought and action, suggesting that our understanding of the world is shaped by our interactions with it. For Dewey, learning is an active process where individuals engage with their environment, experiment, and adapt based on the outcomes, leading to the continuous refinement of ideas (Anderson, 2023; Dewey, 1938, 1994).

Dewey's (1922, 1938, 1994) pragmatism also extends to his views on democracy and education. He saw democracy not just as a political system but as a way of life that encourages active participation, open communication, and collaborative problem-solving. In education, Dewey advocated for experiential learning, where students learn by doing and reflecting on their experiences. This approach aligns with his belief that education should prepare individuals to be active, thoughtful participants in society, capable of adapting to change and contributing to the common good. Thus, Dewey's pragmatism is deeply rooted in the belief that knowledge is inherently linked to action and that the purpose of thought is to guide effective practice (Dewey, 1938, 1994; Fesmire, 2003).

Freire's (2014) critical pedagogy is an approach that emphasizes the role of education as a tool for social justice and liberation. Freire challenged what he called the traditional "banking model" of education, where students are seen as passive recipients of knowledge. Instead, he advocated for a dialogical, problem-posing method, where education becomes a collaborative process of inquiry between teachers and students. This approach encourages students to critically examine their reality, question oppressive structures, and develop a consciousness that empowers them to enact change. For Freire, education is not just about acquiring knowledge but about fostering critical thinking and awareness that leads to transformative action (Freire, 2014).

Central to Freire's critical pedagogy is the concept of conscientization, which refers to the deepening of awareness that enables individuals to perceive social, political, and economic contradictions and act against oppressive elements of society. Freire believed that education should help learners recognize the ways in which power and oppression shape their lives and inspire them to challenge and transform these conditions. His pedagogy is inherently political, advocating for an education that empowers marginalized communities to reclaim their voices and participate actively in the creation of a more just and equitable society (Farag et al., 2022; Freire, 2014; Gibson, 1999).

As seen in **Table 1**, these two philosophical approaches to education may be summarized into six major aspects for comparison that form the foundation for a combined approach to establishing ethics guidelines. These aspects are drawn from a synthesis of Anderson (2023), Rud and Attwood (2016), Dewey (1994), Farag et al. (2022), Fesmire (2003), Freire (2014), and Gibson (1999). When considering this comparison chart, a combined conceptual framework may be posited that combines these two philosophical approaches to analyzing, evaluating, and conceptualizing education for K-12 contexts. This scholar calls this combination the *critical and pragmatic Dewey-Freire* (*CritPraDF*) framework. This conceptual framework may be used as a guide in educator preparation programs and administrator preparation programs to consider how to write critical pragmatic guidelines for the ethical use of AI technologies in schools.

Table 1. Comparison of John Dewey's pragmatism with Paulo Freire's critical pedagogy

Aspect	John Dewey's Pragmatism	Paulo Freire's Critical Pedagogy
Philosophical Foundation	Knowledge is derived from experience and practical engagement with the world.	Education as a means for social justice and liberation from oppression.
Primary Role of Education	Education is a process of continuous growth, where students learn by doing and reflecting on their experiences.	Education is a practice of freedom, empowering students to critically examine and transform their reality.
Teacher-Student Relationship	Collaborative and interactive; the teacher guides students in active learning and problem-solving.	Dialogical and reciprocal; the teacher and student engage in mutual learning and critical dialogue.
Primary Goal of Education	To prepare students to be adaptable, thoughtful participants in a democratic society.	To raise critical consciousness and empower students to challenge oppression.
View of Knowledge	Knowledge is dynamic, evolving through experimentation and practical application.	Knowledge is constructed through critical reflection on one's social and political context.
Curricular Application	Emphasis on experiential learning, inquiry-based education, and problem-solving.	Emphasis on critical pedagogy, social justice education, and the development of critical consciousness.

Based on the comparison seen in **Table 1**, the combined conceptual framework that is posited here—the critical and pragmatic Dewey-Freire (*CritPraDF*) framework—may be used to approach analysis of AI in schools holistically and position the discussion of ethics from multiple points of view. The perceptions of teachers, students, administrators, and various additional interested persons on the role of AI and what ethics guidelines should be written are important to be included. Implementing this inclusive discussion process will help ensure equitable representation. This combined philosophy in the form of a conceptual framework has five constructs that emphasize the importance of both practical engagement with technology and the critical examination of its social and ethical implications. The five constructions of this framework may be seen as criteria categories for developing guidelines for AI use in schools.

The first construct is "Education as Active and Experiential." The combined philosophy would stress that AI should be used to enhance active learning experiences. AI tools should support students in engaging with real-world problems, encouraging experimentation and critical thinking. In line with Dewey's (1994) pragmatism, AI should facilitate hands-on learning, where students learn by doing, reflecting on outcomes, and adapting their approaches. This practical engagement helps students develop a deeper understanding of concepts and their applications.

The second construct is “Education as a Practice of Freedom.” Drawing on Freire’s (2014) critical pedagogy, this framework would emphasize the role of AI in promoting social justice and empowering students to challenge oppressive structures. AI should not be a passive tool but an instrument that students use to critically analyze and address social, political, and economic issues. This technology should be employed to foster critical consciousness, enabling students to recognize biases in AI and use it as a means to advocate for equity and justice.

The third construct is “Teacher-Student Collaboration.” In this framework, the relationship between teachers and students remains central. AI should be integrated into the classroom in ways that enhance dialogue and collaboration. Teachers would guide students in using AI tools, fostering an environment where knowledge is co-constructed through active participation and critical inquiry. This approach ensures that AI complements, rather than replaces, the human elements of education.

The fourth construct is “Dynamic and Evolving Knowledge.” AI in education should be seen as part of an ongoing process of knowledge construction. This combined philosophy would advocate for the use of AI to explore new ideas, test hypotheses, and refine understanding. Just as knowledge is not static, AI tools should be used in ways that are adaptable and responsive to students' evolving needs and contexts. This dynamic approach aligns with Dewey’s view of knowledge as fluid and ever-changing.

The fifth construct is “Ethical and Responsible Use of AI.” Finally, this framework would highlight the ethical implications of AI use in education. Students and educators should be trained to critically assess AI technologies, understanding their potential biases and limitations. By integrating Freire’s focus on critical consciousness, the framework ensures that AI is used in a way that promotes fairness, equity, and respect for all learners. This combined philosophy provides a flexible, yet defined, approach to evaluating AI in K-12 education. The goal is to help ensure that this technology is used to enrich learning experiences, foster critical thinking, and promote social justice.

AI ETHICS ISSUES FOR K-12 TEACHERS, STUDENTS, AND ADMINISTRATORS

Ethical Considerations in AI Development

AI development for K-12 education must prioritize ethical principles that protect students, teachers, and the broader educational ecosystem. Key ethical concerns include bias and fairness, transparency and accountability, and privacy and data security (Adams et al., 2023; Chauncey & McKenna, 2023; Fiske et al., 2019; Stahl & Eke, 2024; Tlili et al., 2023). AI systems often collect and process large amounts of data to function effectively, raising concerns about how this data is stored, shared, and used. Given that students are minors, special care must be taken to protect their privacy and prevent any misuse of their data. Developers and educators must prioritize data security and ensure that AI systems comply with all relevant laws and regulations, such as the Family Educational Rights and Privacy Act (FERPA) in the United States. This suggests the importance of school personnel including parents and guardians in discussions about how their children's data is used and to obtain informed consent where necessary (Adams et al., 2023).

In addition to safeguarding against bias and ensuring fairness, AI in K-12 education must also be developed with the well-being of students in mind. AI tools can influence how students learn, how they are assessed, and how they interact with peers and teachers. That third item—social interaction—is especially important within the K-12 context because that is where students develop a substantial part of their identities and social behaviors. AI may be changing that social cognitive process (Attwood, 2020). Educators and administrators should be proactive in addressing these potential changes to help support prosocial development. These tools should be designed to support positive educational outcomes and promote the holistic development of students (Tlili et al., 2023). This includes fostering critical thinking, creativity, and social-emotional skills, rather than simply focusing on standardized test scores or other narrow measures of success. Ethical AI development should prioritize creating environments where all students can thrive in schools that support prosocial development (Attwood, 2020). In doing so, the rapid integration of AI technologies in schools may still achieve balance between an increasing reliance on AI and maintenance of the human aspects of interpersonal learning (Yue et al., 2022; Su et al., 2024).

The broader educational ecosystem—including teachers, administrators, and policymakers—should be actively involved in the development and deployment of AI technologies in K-12 education. Teachers should not be reduced to mere facilitators of AI-driven curricula; rather, they should be empowered to use AI as a tool that enhances their professional judgment and expertise. Collaborative approaches that involve educators in the design and implementation of AI systems are essential to ensure that these technologies align with educational goals and values. By fostering a culture of ethical reflection and collaboration, the integration of AI in K-12 education can be guided by principles that protect and uplift all members of the educational community (Adams et al., 2023; Chauncey & McKenna, 2023; Stahl & Eke, 2023; Tlili et al., 2023).

Bias and Fairness

Bias and fairness in the use of artificial intelligence (AI) in K-12 education are complex issues that require careful consideration, especially when examined through the lenses of Dewey's pragmatism and Freire's critical pedagogy. AI systems are only as unbiased as the data they are trained on. In the context of K-12 education, biased AI could reinforce existing inequalities, particularly for marginalized students. For instance, predictive models might inaccurately assess a student's potential based on incomplete or biased data, leading to unfair treatment or misplaced interventions (Stahl & Eke, 2024). Dewey's pragmatism emphasizes experiential learning and the continuous interaction between learners and their environment. He believed that education should be rooted in real-world experiences, promoting critical thinking and problem-solving skills. Freire's critical pedagogy advocates for education as a practice of freedom, where learners critically engage with content to challenge societal structures of oppression. When applied to the context of AI in education, these philosophies offer distinct yet complementary insights into the challenges of bias and fairness.

Generative AI, such as tools that create personalized learning experiences or generate content based on student inputs, can be seen through Dewey's pragmatism to potentially enhance experiential learning. These tools can adapt to students' needs, offering diverse perspectives and problem-solving opportunities that align with Dewey's emphasis on real-world applicability. However, the risk of bias in generative AI lies in the algorithms that drive these tools. If these algorithms are based on biased data or designed without consideration of diverse student backgrounds, they can reinforce existing inequalities, thus undermining the fairness that Dewey's pragmatism seeks to promote.

In contrast, Freire's critical pedagogy may approach generative AI with a critical lens, questioning whose interests are served by these technologies. Freire would likely critique the use of AI that perpetuates a banking model of education, where students passively receive information rather than actively engage with it. Generative AI, if not carefully designed, can replicate this passive learning model by providing pre-packaged knowledge without encouraging critical dialogue or questioning. To align with Freire's ideals, AI must be used to empower students, encouraging them to question and challenge the content they interact with, rather than passively consume it.

Non-generative AI, such as data analytics tools that track student performance or predictive algorithms that suggest interventions, can be aligned with Dewey's pragmatism in terms of providing actionable insights for educators. These tools can help create responsive learning environments that adapt to the needs of students, fostering the kind of experiential learning that Dewey championed. However, the fairness of these tools is contingent on the accuracy and impartiality of the data they use. If the data reflects societal biases, the AI could perpetuate these biases, leading to unfair treatment of students from marginalized groups. Applying Dewey's pragmatism would likely advocate for the continuous evaluation and adjustment of these tools to ensure they serve all students equitably.

In applying Freire's critical lens, however, the approach to non-generative AI may focus on power dynamics in its use and its implications of action by the one in power toward the one with less power. Freire's lens questions who control the data and how it is used, emphasizing the need for transparency and student agency. Freire would argue that AI tools should not be used to impose top-down decisions on students but should instead be tools that facilitate their critical engagement with the learning process. Non-generative AI, when used uncritically, could reinforce power imbalances by marginalizing student voices and prioritizing efficiency over meaningful learning. When viewed through Freire's lens, fairness in AI use may mean ensuring that students have the power to influence how these tools are used in their education.

In applying Dewey's pragmatism and Freire's critical pedagogy to both generative and non-generative AI, the importance of context and agency in addressing bias and fairness may be foundational in writing guidelines and policies for the use of AI technologies. Dewey's approach would focus on the practical applications of AI in creating dynamic and equitable learning environments, while Freire's approach would emphasize the need for critical reflection and resistance to the passive absorption of biased content. Both perspectives call for a thoughtful integration of AI in education, one that considers the diverse needs of students and the potential for AI to either perpetuate or challenge existing inequalities.

Addressing bias and fairness in AI for K-12 education would likely benefit from a synthesis of Dewey's and Freire's philosophies. Educators should strive to create learning environments that are both experiential and critically engaged, using AI as a tool to enhance, rather than diminish, these goals. This means designing AI systems that are transparent, adaptable, and reflective of the diverse experiences and voices of all students. By doing so, educators can harness the potential of AI to support equitable and meaningful learning experiences, in line with both Dewey's vision of education as growth and Freire's vision of education as liberation. The intersection of AI, bias, and fairness in K-12 education presents both challenges and opportunities. These lenses, when combined, offer a potentially valuable conceptual framework for navigating these complexities, emphasizing the need for AI tools that are both practically effective and critically empowering. By applying these philosophical lenses, educators can work towards a more just and equitable use of AI in education, one that truly serves the needs of all students.

Transparency and Accountability

Issues of transparency and accountability should be considered as developers should ensure that AI systems used in education are transparent in their operations. Educators, students, and parents should understand how AI-driven decisions are made, and there should be clear accountability for errors or adverse outcomes. This is particularly critical when AI influences important decisions, such as student assessments or resource allocation (Stahl & Eke, 2024). Transparency and accountability in the use of AI in K-12 education are crucial for ensuring that these technologies serve educational goals in a way that is equitable, ethical, and effective. In this context, Dewey's pragmatism emphasizes the importance of democratic participation and the continuous, reflective interaction between learners and their environment, advocating for educational practices that are open, adaptable, and responsive to the needs of students. Freire's critical pedagogy, however, stresses the need for education to be a liberatory practice, where students critically engage with knowledge to challenge and transform oppressive societal structures. Both perspectives underscore the importance of transparency and accountability in the use of AI in education, though they approach these concepts from different angles.

From a Deweyan perspective, transparency in generative AI is crucial for ensuring that the educational experiences it creates are genuinely beneficial and relevant to students' real-world contexts. Dewey would argue that the processes and algorithms behind generative AI should be open to scrutiny by educators, students, and other contributors. This transparency would allow for continuous reflection and adaptation, ensuring that the AI's outputs are aligned with the educational goals of fostering critical thinking and problem-solving skills. Without transparency, there is a risk that generative AI could create learning experiences that are misaligned with students' needs or, worse, perpetuate existing biases.

Freire's critical pedagogy would further emphasize the need for transparency in generative AI, but from a perspective that focuses on power dynamics and the potential for AI to reinforce existing inequalities. Freire would likely be concerned about the opacity of AI algorithms, which can obscure the ways in which they reproduce societal biases or serve the interests of those in power. When viewed through Freire's lens, transparency in AI is not just about making the technology understandable; it is about ensuring that students and educators have the agency to question, critique, and transform the content generated by AI. This means that AI tools must be designed in a way that allows for critical engagement and resistance, rather than passive consumption of information.

Accountability in the use of generative AI is also a key concern, particularly when considering Dewey's emphasis on democratic participation. Dewey would advocate for systems of accountability that involve all contributors—students, teachers, parents, and the broader community—in the evaluation and oversight of AI tools. This could involve mechanisms for feedback and revision, ensuring that AI systems are held accountable for their impacts on students' learning and development. Dewey would stress the importance of a collaborative approach to accountability, where the educational community works together to ensure that AI is used in ways that promote democratic values and educational equity.

Using Freire's critical lens, accountability in generative AI would need to be closely tied to issues of power and liberation. Freire's lens may suggest that those who create and implement AI systems in education must be accountable to the communities they serve, particularly marginalized groups. This means not only addressing the potential harms of AI but also ensuring that the benefits of AI are equitably distributed. Freire would likely call for participatory processes in which students and educators actively shape the development and use of AI tools, holding those in power accountable for the social and educational impacts of these technologies.

Non-generative AI, such as predictive analytics or adaptive learning systems, also raises important issues of transparency. Dewey's pragmatism would emphasize the need for these AI systems to be transparent in how they collect, analyze, and use data. Transparency is essential for ensuring that the insights provided by these systems are valid and can be trusted by educators and students alike. Dewey would likely advocate for the use of AI tools that are open to inspection and modification, allowing educators to understand and, if necessary, adjust the algorithms to better serve their students' needs.

Freire's critical pedagogy would approach the transparency of non-generative AI with a focus on how these systems might reinforce or challenge existing power structures. Freire would be concerned about the ways in which opaque algorithms could be used to justify decisions that affect students' educational opportunities without giving them a voice in the process. When viewed through Freire's lens, transparency in non-generative AI is about ensuring that students and educators have the knowledge and tools to critically examine and challenge the data-driven decisions that shape their learning environments.

Accountability in non-generative AI, from a Deweyan perspective, may involve creating systems that ensure these tools are used in ways that are fair, just, and aligned with the goals of democratic education. This could involve regular audits of

AI systems to ensure they are not perpetuating bias or inequity, as well as creating avenues for redress if students or educators feel they have been unfairly impacted by AI-driven decisions. Dewey would advocate for a proactive approach to accountability, where educators take an active role in shaping how AI is used in their schools.

For Freire's critical approach, accountability in non-generative AI may be viewed as connected to issues of social justice. Freire's approach would argue that those who design and implement AI systems must be held accountable for the ways in which these tools impact marginalized students. This could involve not only addressing the immediate impacts of AI but also ensuring that these systems are used to actively challenge and dismantle oppressive structures within education. Freire would likely call for a radical rethinking of how accountability is conceptualized in the context of AI, emphasizing the need for participatory, community-based approaches that empower students and educators to shape their educational futures.

Privacy and Security

Recent legislative action prompted by families and school contributors has also addressed issues of privacy and data security (National Conference of State Legislatures, 2024). The collection and analysis of student data by AI systems raises privacy concerns. Developers should implement data protection measures to prevent unauthorized access and misuse of sensitive student information. The use of student data for AI development should be subject to ethical guidelines and policies, ensuring that students' rights are not compromised. What those rights are, however, has been shifting rapidly and does not seem to have consensus yet (Chiu, 2021; Law, 2024). It is possible that there will not be consensus but, instead, a plurality of guidelines and policies that have some overlap.

Applying Dewey's pragmatism to the issue of privacy and security emphasizes the importance of a safe and supportive learning environment that fosters growth, while Freire's critical pedagogy emphasizes the protection of students' rights and the need for education to serve as a means of liberation. Both perspectives offer valuable insights into how privacy and security should be handled in the context of AI in education. Generative AI, which can create personalized learning experiences and generate content for students, poses unique challenges to privacy. From Dewey's pragmatic perspective, maintaining student privacy is essential for creating an environment where students can freely explore and learn without fear of intrusion or misuse of their personal information. Dewey would argue that any AI system used in education must be designed with strong privacy protections, ensuring that students' data is collected and used transparently and responsibly. This is crucial for fostering trust between students, educators, and the technology they interact with, which is a cornerstone of Dewey's vision of a democratic and participatory educational system.

In applying Freire's critical pedagogy lens, the issue of privacy in generative AI with a focus on power dynamics and the potential for exploitation. This lens is focused on concerns about who controls the data collected by AI systems and how that data is used, particularly in relation to marginalized students. Freire's approach may argue that without strong privacy protections, AI could become a tool of oppression, where students' personal information is used to reinforce existing power structures or to manipulate students' learning experiences in ways that serve the interests of those in control. When viewed through Freire's lens, protecting student privacy is not just about preventing data breaches; it is about ensuring that students retain control over their own information and that AI is used to empower rather than exploit them.

Security is another critical concern when it comes to generative AI in education. Dewey's pragmatism would stress the importance of creating AI systems that are secure against external threats, such as hacking or data theft, as well as internal misuse. From a Deweyan lens, the security of AI systems is a fundamental part of ensuring a stable and trustworthy learning environment. He would advocate for the continuous assessment and improvement of AI security measures, emphasizing the need for educational institutions to take an active role in safeguarding students' information and the integrity of the learning process.

From Freire's perspective, the security of AI systems would also be viewed through the lens of social justice. Freire would likely be concerned with how security measures are implemented and whether they serve to protect all students equally, particularly those from marginalized backgrounds. He would argue that security protocols should not only protect students from external threats but also from potential abuses of power within the educational system. When viewed through Freire's lens, true security in AI use may mean ensuring that these technologies do not become tools for surveillance or control but instead contribute to the creation of a safe and liberatory educational environment.

Non-generative AI, such as data analytics tools and predictive algorithms, also raises significant privacy concerns. Dewey's pragmatism would emphasize the need for transparency in how these systems collect and use data, ensuring that students and educators understand what information is being gathered and for what purposes. Dewey would argue that privacy protections in non-generative AI are essential for maintaining the integrity of the educational process, as students must be able to trust that their personal information will not be used in ways that could harm them or limit their opportunities.

Freire's critical pedagogy would take a more critical stance on the use of non-generative AI, particularly in relation to how data is collected and used to make decisions about students. Freire would likely be concerned about the potential for these AI systems to reinforce existing inequalities by using data to track, sort, or categorize students in ways that reflect and perpetuate societal biases. He would argue that privacy protections in non-generative AI must go beyond simply protecting students' personal information; they must also ensure that data is not used to limit students' educational opportunities or to impose oppressive structures within the educational system.

In terms of security, Dewey's pragmatism would advocate for protections to ensure that non-generative AI systems are secure against unauthorized access and manipulation. Dewey would stress the importance of ongoing vigilance and adaptability in the face of evolving security threats, emphasizing that a secure learning environment is essential for students to feel safe and supported in their educational journey. He would likely support a proactive approach to AI security, where educational institutions are continuously improving their practices to keep pace with technological advancements.

Freire's perspective on security in non-generative AI would emphasize the need to protect students not only from external threats but also from the potential misuse of AI within the educational system itself. When viewed through Freire's lens, there is an argument that security measures should be designed to prevent AI from being used to surveil or control students, ensuring that these technologies contribute to a more just and equitable educational environment. For Freire, true security means ensuring that AI serves the needs of students and educators, rather than becoming a tool of oppression or domination.

There is, then, potentially tension between those in power and those who are not. Surveillance may be used as a preventive measure as well as a failsafe to enforce accountability by administrators on teachers and students. Given these complex dynamics, privacy and security in the use of AI in K-12 education are critical issues that must be addressed with careful consideration of both Dewey's and Freire's educational philosophies. Dewey's pragmatism may be used to highlight the importance of creating secure and transparent learning environments that foster trust and democratic participation, while Freire's critical pedagogy may be used to emphasize discussion on the need to protect students' (and teachers') rights and ensure that AI is used to empower rather than exploit. By applying these philosophical frameworks, educators can work towards the responsible and ethical use of AI in education, ensuring that these technologies contribute to a safe, equitable, and liberatory learning environments for all students.

Ethical Implications of AI Use by Teachers

There is a trend in teachers increasingly relying on AI tools to enhance instructional practices, personalize learning, and manage administrative tasks. However, the ethical implications of this reliance are complex. The complexity of the ethical implications of this trend includes issues of pedagogical autonomy, teacher-student relationships, and professional development. While AI can support teachers by providing insights into student performance and suggesting instructional strategies, there is a risk that over-reliance on AI could undermine teachers' professional autonomy. Ethical AI implementation should empower teachers rather than dictate their instructional choices (Yue et al., 2022). AI systems might alter the dynamics of teacher-student relationships. For example, if AI tools are used to monitor student behavior or performance continuously, it could lead to a surveillance culture in classrooms, potentially eroding trust between teachers and students (Attwood, 2020). Teachers need adequate training to understand the ethical implications of using AI in their classrooms. This includes recognizing potential biases in AI tools, understanding the limitations of AI-generated insights, and maintaining an ethical approach to integrating AI into their teaching practices (Su et al., 2023).

AI systems, especially those used for grading, assessment, or personalized learning, often rely on data that may reflect societal biases. If teachers use these tools without understanding or questioning their underlying algorithms, there is a risk that AI could reinforce stereotypes and inequalities, leading to unfair treatment of students. This issue is critical from both a Deweyan and Freirean perspective, as both philosophers emphasize the importance of education in promoting equity and social justice. As such, ethical issues pertaining to the potential for AI to undermine the teacher-student relationship should be proactively addressed in schools as AI may change social learning (Attwood, 2020). AI tools can automate many aspects of teaching, such as grading or providing feedback, which could lead to a more efficient classroom. However, there is a risk that over-reliance on AI could depersonalize education, reducing the opportunity for meaningful interactions between teachers and students. Dewey's pragmatism emphasizes the importance of experiential learning and the social aspects of education, where the teacher's role is central to facilitating students' engagement with real-world problems. Freire's critical pedagogy, similarly, sees education as a dialogical process that requires a human connection to challenge and transform oppressive structures. The ethical implication here is that teachers should carefully balance the use of AI with the need to maintain the relational and humanistic aspects of education.

Privacy is an ethical concern when it comes to teachers using AI. AI systems often require large amounts of data to function effectively, and this data typically includes sensitive student information. Teachers need to be aware of the privacy risks associated with AI, including how student data is collected, stored, and used. From a Deweyan perspective, maintaining student privacy is crucial for creating a safe and trusting learning environment. Freire would argue that protecting students' data is also a matter of power and agency, as misuse of data could lead to surveillance and control rather than empowerment.

Accountability is also a key ethical consideration. When AI tools are used in the classroom, it is essential to determine who is responsible for the decisions made by these systems. If an AI-driven assessment tool provides inaccurate results, who is held accountable—the teacher, the developers of the AI, or the educational institution? Dewey's pragmatism would advocate for a system of shared accountability, where teachers, administrators, and AI developers work together to ensure that AI tools are used ethically and effectively. Freire would emphasize the need for accountability mechanisms that include the voices of students and their communities, ensuring that AI is used in ways that promote liberation and justice rather than oppression.

Given these ethical challenges, it is crucial to provide teachers with appropriate training in the use of AI. One recommendation, grounded in Dewey's pragmatism, is to offer hands-on, experiential training that allows teachers to interact with AI tools in real-world educational contexts. This training should focus on helping teachers understand how AI works, including its potential benefits and limitations, so they can make informed decisions about when and how to use these tools. Teachers should also be encouraged to continuously reflect on their use of AI, adapting their practices based on their experiences and the needs of their students.

A second recommendation, informed by Freire's critical pedagogical lens, is to incorporate critical theories in discussions about the social and ethical implications of AI into teacher training. This training should encourage teachers to question the power dynamics involved in AI use, including who benefits from AI and who might be marginalized by it. Teachers should be empowered to critically engage with AI tools, using them not just as passive recipients but as active participants who can shape how these technologies are used in the classroom. By fostering a critical consciousness about AI, teachers can ensure that these tools are used in ways that promote social justice and the empowerment of all students.

Ethical Implications of AI Use by Students

Students are both the primary beneficiaries and subjects of AI systems in K-12 education. The ethical use of AI by students involves several critical considerations: learning integrity, equity in access, and digital literacy. Generative AI tools, such as essay-writing bots, pose challenges to academic integrity. There is a growing concern that students might use these tools to complete assignments dishonestly, raising questions about how to maintain academic standards in the age of AI. Access to AI tools varies significantly across different schools and communities. Ensuring that all students have equitable access to AI technologies is an ethical imperative. Without careful attention, AI could exacerbate the digital divide, leaving some students at a disadvantage. Students must be equipped with the digital literacy skills needed to critically engage with AI tools. This includes understanding how AI works, recognizing its limitations, and being aware of the ethical considerations surrounding its use. Ethical AI education should be integrated into the curriculum to prepare students for a future where AI plays a central role in society (Chiu, 2021; Law, 2024; Wang, & Lester, 2023).

A major issue of students using generative AI in educational settings is academic integrity, as that definition may be changing with the rapid integration of AI. This is a sort of Industrial Revolution that is changing the way individuals approach work. AI tools, such as those that generate essays, solve complex problems, or even mimic a student's writing style, can be misused to complete assignments without genuine effort. This could potentially disrupt the learning process, as students might not develop the critical thinking and problem-solving skills they need. The ease with which AI can be used poses a challenge for educators in maintaining fairness and ensuring that students' work truly reflects their abilities and understanding.

Another ethical implication is the risk of intellectual dependency. As students increasingly rely on AI to assist with their studies, there is a danger that they may become overly dependent on these tools, losing the ability to think critically and independently. This reliance can stifle creativity and hinder the development of essential cognitive skills. Moreover, if students are not taught to critically assess the outputs of AI, they may accept information uncritically, leading to a superficial understanding of complex topics. This diminishes the value of education as a process of inquiry and exploration, reducing it instead to mere information consumption.

The issue of equity is also notable. Not all students have equal access to AI tools, which can exacerbate existing inequalities in education. Students from disadvantaged backgrounds may be less likely to afford or have access to sophisticated AI resources, putting them at a disadvantage compared to their peers. This digital divide could lead to a widening gap in educational outcomes, where only those with access to AI can fully benefit from its potential, further entrenching social and economic disparities.

From the perspective of Dewey's (1994) pragmatism, the use of AI by students should be evaluated based on its ability to enhance learning through active, experiential engagement. Dewey would likely argue that AI should not be used as a crutch but as a tool that supports deeper understanding and exploration. He would advocate for teaching students how to use AI in ways that encourage inquiry, experimentation, and the practical application of knowledge. This approach would ensure that AI serves as a means to an end, facilitating meaningful learning experiences rather than replacing them.

Freire's (2016) critical pedagogy offers a lens through which to view the ethical use of AI by students that challenges uncritical assumptions. Freire would emphasize the importance of teaching students to approach AI critically, questioning its outputs and understanding the underlying power dynamics. He would advocate for an educational approach that empowers students to use AI as a tool for social justice, enabling them to challenge oppressive structures and engage in transformative action. This perspective encourages students to see AI not just as a tool for personal advancement but as a means to contribute to the collective good.

To address these ethical concerns, two key recommendations are posed. First, educators should incorporate training that aligns with Dewey's pragmatism, teaching students to use AI in ways that promote active learning and critical thinking. This could involve assignments that require students to use AI tools to explore real-world problems, encouraging them to test ideas, analyze outcomes, and reflect on their learning processes. Second, in line with Freire's critical pedagogy, educators should foster a critical consciousness among students, helping them to understand the implications of AI, recognize its potential biases, and use it ethically to challenge injustices. This dual approach would equip students with the skills and mindset needed to navigate the complexities of AI in a responsible and empowered manner.

Ethical Implications of AI Use by Administrators

School administrators play a crucial role in overseeing the implementation of AI systems in K-12 education. Their decisions have significant ethical implications for data-driven decision-making, policy development, and contributor communication. AI can assist administrators in making data-driven decisions regarding student placement, resource allocation, and policy development. However, reliance on AI for such decisions raises ethical concerns about the accuracy and fairness of AI-driven recommendations. Administrators must critically assess AI outputs and ensure that human judgment remains central to decision-making processes. Administrators are responsible for developing policies that govern the use of AI in schools. These policies must address issues such as data privacy, bias mitigation, and the appropriate use of AI tools in classrooms. Ethical policy development should involve input from a diverse range of contributors, including teachers, students, parents, and AI experts. Administrators should also communicate transparently with all contributors about the role of AI in education. This includes explaining how AI tools are used, what data is collected, and how ethical concerns are being addressed. Building trust with contributors is essential for the successful and ethical implementation of AI in schools (Chauncey & McKenna, 2023; Stahl & Eke, 2024; Wang, & Lester, 2023).

A major ethical concern is the potential for AI to influence decisions related to student tracking, discipline, and resource allocation based on biased data. AI systems often rely on historical data, which may contain biases reflecting existing societal inequalities. When administrators use AI to make decisions about students' educational pathways, there is a risk that these biases could be perpetuated, leading to inequitable outcomes. This concern resonates with both Dewey's pragmatism in terms of its emphasis on education as a tool for democracy and social progress, and Freire's critical pedagogy in terms of its advocacy for education as a means of challenging and transforming oppressive structures.

Another ethical issue is the impact of AI on student privacy. School administrators may use AI tools to monitor student behavior, academic performance, and even social interactions. While these tools can provide valuable insights, they also raise significant privacy concerns. The extensive data collection required by AI systems can intrude on students' personal lives and create an environment of surveillance, where students are constantly monitored. Dewey would argue that such practices undermine the democratic and participatory nature of education by creating a climate of distrust. Freire, similarly, would be concerned with how surveillance and data collection could be used to control and suppress student self-efficacy rather than promote their development and liberation.

The ethical implications of AI also extend to the potential for dehumanization in decision-making processes. AI systems, by their nature, are designed to optimize efficiency, often reducing complex human interactions to data points and algorithms. When administrators rely heavily on AI for decisions related to student discipline, admissions, or resource distribution, there is a risk that the human element of education—understanding the individual needs, circumstances, and potential of each student—could be lost. Dewey's pragmatism stresses the importance of education as a lived experience, where individual differences and human relationships are central. Freire's critical pedagogy emphasizes the need for education to be a dialogical process, where students' voices and lived experiences are integral to their learning journey. Overreliance on AI could undermine these core principles, leading to decisions that are more technocratic than educational.

Accountability is another significant ethical concern when it comes to AI use by school administrators. When decisions are influenced or made by AI systems, it becomes important to determine who is responsible for those decisions. If an AI-driven system results in unfair treatment of a student or a flawed resource allocation, who is held accountable—the school administrators, the developers of the AI, or the educational institution as a whole? Dewey would advocate for shared accountability, where all contributors, including administrators, educators, and the broader community, play a role in overseeing the ethical use of AI. Freire would emphasize the importance of ensuring that accountability mechanisms do not solely rest with those in power but also include the voices of students and their communities. Freire's critical approach is especially important here as it emphasizes a "checks and balances" concept necessary for Dewey's (1922) democratic ideals.

In addressing these ethical challenges, it is important to provide K-12 school administrators with training on the responsible use of AI, especially as administrators hold the most power in their school. One recommendation, based on Dewey's pragmatism, is to design training programs that focus on experiential learning and reflective practice. Administrators should be trained not only on how AI systems work but also on how to critically evaluate the data and algorithms that underpin these systems. This training should encourage administrators to reflect on the real-world implications of their decisions, considering how AI impacts students' educational experiences and outcomes. By fostering a reflective and adaptive approach to AI use, administrators can ensure that their decisions align with the broader goals of equity and democracy in education.

Another recommendation, informed by Freire's critical pedagogy, is to incorporate a critical examination of power dynamics into AI training for school administrators. This training should encourage administrators to question the underlying assumptions of AI systems and to critically engage with the ways in which AI may reinforce or challenge existing power structures within the school system. Administrators should approach their leadership role in a proactive way that thoughtfully approaches AI in ways that are informed for the promotion of social justice and the empowerment of all students, rather than merely reinforcing the status quo. While very challenging, this call to humility and self-reflection from Freire's critical lens may help foster the ideals of democracy that is more just. By fostering critical consciousness among administrators, such training may help ensure that AI is used in ways that are ethically sound and aligned with the goals of educational liberation and equity.

CONCLUSION

The critical and pragmatic Dewey-Freire (CritPraDF) framework posited in this study may be used to develop and analyze guidelines for AI usage ethics in educational organizations. This framework has a flexible, yet holistic, approach to critically evaluating pragmatic integration of AI into K-12 education. As such, AI used with this framework offers substantial opportunities to enhance learning, personalize instruction, and streamline administrative tasks. However, these benefits come with substantial ethical challenges that must be carefully navigated.

From the pragmatic lens of Dewey, generative artificial intelligence offers a powerful tool for problem-solving and innovation. It can facilitate experimentation, reflection, and growth in knowledge. However, its potential for automation can also lead to a loss of critical thinking skills and a widening of educational disparities. Freire's critical pedagogy highlights the importance of empowering learners to question and challenge dominant narratives. While AI can provide access to diverse perspectives, it must be used critically to avoid reinforcing existing biases or unjust power structures.

The CritPraDF framework encourages schools to prioritize human-centered design, ensuring that AI tools are developed with people's needs and interests in mind. This involves ongoing evaluation and iteration to ensure that AI is serving its intended purpose and is not creating unintended consequences. The CritPraDF framework encourages critical reflection on the role of technology in society. Schools should foster students' critical awareness of AI by teaching them about the ethical implications of AI development and use. This includes exploring issues such as privacy, data security, and the potential for AI to be used for surveillance or manipulation. By understanding the limitations and potential harms of AI, students can become informed and responsible digital citizens.

To more fully realize the benefits of generative AI, educators should consider how integrating it into learning experiences may be done in ways that promote critical engagement and problem-solving without it causing intellectual atrophy. Each contributor group—teachers, students, administrators, *et cetera*—has a role to play in establishing ethics guidelines and enforcing those guidelines. By combining Dewey's pragmatism with Freire's critical pedagogy, a conceptual process may be used in analyzing the ethical implications of AI development and implementation in K-12 education. This is a rapidly evolving process as AI technologies increase in their integration with school contributors and educational processes as well as throughout society.

REFERENCES

Adams, C., Pente, P., Lemermeyer, G., & Rockwell, G. (2023). Ethical principles for artificial intelligence in K-12 education. *Computers and Education: Artificial Intelligence*, *4*, 100131. Advance online publication. DOI: 10.1016/j.caeai.2023.100131

Anderson, E. (2023). Dewey's moral philosophy. In Zalta, E. N., & Nodelman, U. (Eds.), *The Stanford encyclopedia of philosophy*. Stanford University., https://plato.stanford.edu/archives/spr2023/entries/dewey-moral/

Attwood, A. I. (2020). Changing social learning theory through reliance on the Internet of Things and artificial intelligence. *Journal of Social Change*, *12*(1), 103–111. DOI: 10.5590/JOSC.2020.12.1.08

Chauncey, S. A., & McKenna, H. P. (2023). A framework and exemplars for ethical and responsible use of AI Chatbot technology to support teaching and learning. *Computers and Education: Artificial Intelligence*, *5*, 100182. Advance online publication. DOI: 10.1016/j.caeai.2023.100182

Chiu, T. K. F. (2021). A holistic approach to the design of Artificial Intelligence (AI) education for K-12 schools. *TechTrends*, *65*(5), 796–807. DOI: 10.1007/s11528-021-00637-1

Dewey, J. (1922). *Democracy and education: An introduction to the philosophy of education*. Macmillan.

Dewey, J. (1938). *Experience and education*. Macmillan.

Dewey, J. (1994). The moral writings of John Dewey. In Gouinlock, J. (Ed.), *Prometheus Books* (Revised ed.).

Farag, A., Greeley, L., & Swindell, A. (2022). Freire 2.0: Pedagogy of the digitally oppressed. *Educational Philosophy and Theory*, *54*(13), 2214–2227. DOI: 10.1080/00131857.2021.2010541

Fesmire, S. (2003). *John Dewey and moral imagination: Pragmatism in ethics*. Indiana University Press.

Fiske, A., Henningsen, P., & Buyx, A. (2019). Your robot therapist will see you now: Ethical implications of embodied artificial intelligence in psychiatry, psychology, and psychotherapy. *Journal of Medical Internet Research*, *21*(5), e13216. Advance online publication. DOI: 10.2196/13216 PMID: 31094356

Freire, P. (2014). *Pedagogy of the oppressed* (M. B. Ramos, Trans.; Thirtieth anniversary edition.). Bloomsbury.

Gibson, R. (1999). Paulo Freire and pedagogy for social justice. *Theory and Research in Social Education*, *27*(2), 129–159. DOI: 10.1080/00933104.1999.10505876

Law, L. (2024). Application of generative artificial intelligence (GenAI) in language teaching and learning: A scoping literature review. *Computers and Education Open*, *6*, 100174. DOI: 10.1016/j.caeo.2024.100174

National Conference of State Legislatures. (2024). Social media and children 2024 legislation. https://www.ncsl.org/technology-and-communication/social-media-and-children-2024-legislation

Rud, A. G., & Attwood, A. I. (2016). Dewey Lab School at the University of Chicago. In Peters, M. A. (Ed.), *Encyclopedia of educational philosophy and theory* (pp. 480–484). Springer., DOI: 10.1007/978-981-287-532-7_46-1

Stahl, B. C., & Eke, D. (2024). The ethics of ChatGPT – Exploring the ethical issues of an emerging technology. *International Journal of Information Management*, *74*, 102700. DOI: 10.1016/j.ijinfomgt.2023.102700

Su, J., Guo, K., Chen, X., & Chu, S. K. W. (2024). Teaching artificial intelligence in K–12 classrooms: A scoping review. *Interactive Learning Environments*. *32*(9), 5207-5226. DOI: 10.1080/10494820.2023.2212706

Tlili, A., Shehata, B., Adarkwah, M. A., Bozkurt, A., Hickey, D. T., Huang, R., & Agyemang, B. (2023). What if the devil is my guardian angel: ChatGPT as a case study of using chatbots in education. *Smart Learning Environments*, *10*(1), 15–24. DOI: 10.1186/s40561-023-00237-x

Wang, N., & Lester, J. (2023). K-12 education in the age of AI: A call to action for K-12 AI literacy. *International Journal of Artificial Intelligence in Education*, *33*(2), 228–232. DOI: 10.1007/s40593-023-00358-x

Yue, M., Jong, M. S.-Y., & Dai, Y. (2022). Pedagogical design of K-12 Artificial Intelligence education: A systematic review. *Sustainability (Basel)*, *14*(23), 15620. DOI: 10.3390/su142315620

Chapter 10
Fostering Ethical Awareness in AI Practitioners:
The Role of Interdisciplinary Education in Shaping Future Technologists

Hemlata Parmar
https://orcid.org/0009-0009-1438-5427
Manipal University, India

Utsav Krishan Murari
https://orcid.org/0009-0007-1606-6775
Sharda University, India

ABSTRACT

As artificial intelligence (AI) permeates various sectors, the need for robust ethical guidelines becomes increasingly urgent. This study explores the role of interdisciplinary education in fostering ethical awareness among future technology professionals. By embedding ethics within AI and advanced technology curricula, educational institutions can cultivate AI practitioners who combine technical expertise with a strong sense of social responsibility. The findings emphasize the importance of an ethics-driven framework in AI education, where integrating technology with ethics, philosophy, and social sciences can effectively address AI's ethical challenges. Using case studies and theoretical analysis, this research proposes a model for integrating ethical training at every stage of technological education. Such a model equips AI professionals to tackle issues of bias, justice, accountability, and transparency.

DOI: 10.4018/979-8-3693-9035-1.ch010

Building a strong ethical foundation is essential for creating AI solutions that prioritize human welfare, paving the way toward a more just and responsible future in AI.

1. INTRODUCTION

Artificial Intelligence (AI) has become one of the most transformative forces in modern technology, impacting a broad range of sectors, including healthcare, finance, education, and public administration (Russell & Norvig, 2021). This technological advancement is not limited to operational improvements; rather, it has the potential to reshape society fundamentally. AI has introduced efficiencies and capabilities previously unattainable, but it has also raised significant ethical and moral questions. These questions are central to the ongoing discourse about AI's role in society, addressing issues such as privacy, justice, accountability, and transparency (Binns, 2018). As AI becomes increasingly integrated into critical decision-making processes, the ethical implications of these systems grow more complex and pervasive. Consequently, the need for ethical frameworks and guidelines governing AI deployment has become paramount.

The advent of AI has also brought attention to the limitations of traditional technological education, which often prioritizes technical skills over ethical awareness (Campolo et al., 2017). While the technical training of AI practitioners is undeniably essential, it is becoming clear that technical proficiency alone is insufficient for responsible AI development. The ethical and social consequences of AI applications require a new educational approach that incorporates moral reasoning and ethical decision-making into technical training. This chapter explores the role of interdisciplinary education, specifically the integration of ethics into AI curricula, as a means to cultivate a morally conscious AI workforce.

1.1 AI's Role in Society and the Rise of Ethical Concerns

AI technologies have evolved rapidly over recent years, with significant advancements in machine learning, natural language processing, and computer vision. These advancements have enabled AI to perform complex tasks, from diagnosing diseases and predicting financial market trends to enabling autonomous vehicles and personalizing digital marketing strategies (Goodfellow et al., 2016). However, as these technologies become more powerful and widely adopted, they introduce ethical challenges that traditional governance frameworks struggle to address.

One of the central ethical concerns associated with AI is the issue of bias. Machine learning algorithms, for instance, are often trained on historical data that may contain biases reflective of existing societal inequities (Mehrabi et al., 2021). When these algorithms are deployed in real-world applications—such as hiring, lending, and law enforcement—they can inadvertently reinforce and perpetuate these biases. The infamous COMPAS algorithm, used in the U.S. criminal justice system to assess recidivism risk, demonstrated this issue when it was found to disproportionately classify Black defendants as high risk (Angwin et al., 2016). Such examples underscore the urgent need for AI practitioners who can identify, understand, and address the ethical implications of the technologies they develop.

Another significant ethical issue in AI is the concept of transparency. Many AI systems, particularly those using deep learning techniques, are often described as "black boxes" due to their complex, non-linear nature, which makes it difficult to understand how they reach specific conclusions (Rudin, 2019). In high-stakes areas like healthcare and criminal justice, this lack of transparency can erode public trust and lead to decisions that are difficult to contest or audit. Accountability in AI decision-making is thus another critical issue, as it is challenging to assign responsibility when an opaque algorithm contributes to a decision. For society to benefit equitably from AI advancements, transparency, accountability, and fairness must be embedded in the technology itself—a task that requires more than just technical skill; it requires ethical insight.

1.2 Current Gaps in AI Education and the Need for Ethical Training

Despite the ethical implications of AI, many AI educational programs remain focused on technical competencies, with limited emphasis on ethics. Greene et al. (2019) found that AI curricula at prominent educational institutions frequently lack structured courses in ethics, leaving future practitioners underprepared for the moral challenges they may face in their careers. This emphasis on technical skills is partly a response to industry demand; as the AI field grows, there is a need for a workforce skilled in programming, data analysis, and algorithm development. However, this focus on technical training at the expense of ethical education limits the ability of AI professionals to foresee and mitigate ethical risks.

This gap in education has led scholars to advocate for a more holistic approach that combines technical knowledge with ethical and social sciences (Cath, 2018). Such interdisciplinary education aims to foster critical thinking, moral reasoning, and ethical decision-making skills, equipping AI practitioners to consider the societal impact of their work. Interdisciplinary programs that integrate philosophy, sociology, and law with computer science provide students with the analytical tools needed to identify and address ethical issues in AI deployment. An education that merges technology and ethics is not only more comprehensive but also aligns with the growing recognition that AI development must be guided by values of social responsibility and human well-being.

1.3 The Role of Interdisciplinary Education in Moral Development

Interdisciplinary education, especially when it includes fields like philosophy and social sciences, plays a crucial role in shaping the moral framework of AI practitioners. Moral development is a cognitive process that enables individuals to discern right from wrong and to act upon that discernment in complex situations (Rest et al., 1999). Traditional AI education, which focuses on the acquisition of technical skills, does not address this dimension of professional preparation. However, by incorporating ethics and social sciences into AI curricula, institutions can foster the moral growth necessary for responsible AI practice.

Research has shown that interdisciplinary approaches in education can promote empathy, social awareness, and a sense of accountability (Sjoberg et al., 2020). For instance, case-based learning that combines technical and ethical analysis encourages students to engage deeply with real-world scenarios where AI might be used. This method of instruction can help students develop a nuanced understanding of the ethical dimensions of technology and consider how their work might impact individuals and society as a whole. By nurturing these qualities, interdisciplinary education can help produce AI practitioners who are not only technically skilled but also morally grounded.

1.4 Interdisciplinary Education as a Mitigator of Ethical Risks in AI

Ethics-focused interdisciplinary education can serve as a vital tool in addressing the ethical risks associated with AI. By equipping students with the skills to recognize and mitigate ethical challenges, interdisciplinary education promotes a more responsible approach to AI development. Case studies and theoretical frameworks from disciplines like philosophy and sociology can help students identify potential biases and conflicts of interest, empowering them to develop AI systems that prioritize fairness and equity (Cath, 2018). Such training also encourages students to consider long-term implications, fostering a culture of accountability and transparency.

The practical application of interdisciplinary education in AI is evident in fields like healthcare, where AI's ethical challenges are particularly pronounced. In medical imaging, for instance, machine learning algorithms are increasingly used to assist in diagnosis. However, biases in training data can lead to disparities in diagnostic accuracy, potentially disadvantaging certain demographic groups (Ghassemi et al.,

2020). Interdisciplinary training helps practitioners anticipate and address these biases, ensuring that AI contributes positively to patient outcomes. By teaching future AI practitioners to approach these issues with both technical and ethical rigor, interdisciplinary education can significantly mitigate the ethical risks associated with AI deployment.

1.5 Proposed Model for Integrating Ethics into AI Curricula

To address the ethical challenges in AI, this chapter proposes an interdisciplinary model for AI education that embeds ethical training throughout the curriculum. The proposed model suggests that ethical considerations should be integrated into each stage of AI education, from foundational courses to advanced applications. Introductory courses might cover fundamental ethical principles, while specialized courses could involve in-depth analysis of case studies where ethical dilemmas arise. This progressive approach ensures that ethical thinking is reinforced throughout the educational process, preparing students to navigate the complexities of ethical AI deployment.

Embedding ethics across the curriculum also promotes a culture of ethical awareness and responsibility. As students advance through their education, they encounter increasingly complex scenarios that challenge their moral reasoning and ethical decision-making skills. This cumulative experience not only enhances their technical proficiency but also cultivates an ethical perspective that is essential for responsible AI practice. By fostering an environment where ethical considerations are prioritized, institutions can produce AI practitioners who are capable of balancing innovation with social responsibility.

1.6 The Importance of a Morally Conscious AI Workforce for a Responsible Future

The ultimate goal of interdisciplinary education in AI is to create a workforce that prioritizes the well-being of humanity and the responsible use of AI technology. As AI continues to evolve, the potential consequences of its misuse grow. Ethical AI practitioners play a critical role in ensuring that AI serves society in a manner that respects individual rights, promotes fairness, and avoids harm. A morally conscious AI workforce is essential for navigating the ethical complexities of AI, as it allows practitioners to anticipate and address the potential risks of their work proactively (Floridi & Cowls, 2019).

Furthermore, the societal impact of AI necessitates a new standard for professional responsibility in technology. Just as the medical field holds practitioners to a standard of care that prioritizes patient well-being, the AI field must adopt standards that prioritize ethical considerations and human-centered design. This shift in professional standards requires an educational foundation that instills a commitment to ethical values and social responsibility. Interdisciplinary education, by fostering these values, supports the development of an AI workforce that is prepared to address the challenges of AI in a way that benefits society as a whole.

The integration of interdisciplinary education into AI curricula is essential for cultivating an ethically aware and morally responsible AI workforce. As AI technology continues to advance, the ethical challenges it presents will only intensify, making it imperative for educational institutions to prepare students not only for technical success but also for ethical responsibility. By combining technical and ethical training, interdisciplinary education provides a holistic foundation that enables AI practitioners to navigate the moral complexities of their work, ultimately contributing to a more responsible and equitable future for AI.

2. LITERATURE REVIEW

The integration of ethical considerations into artificial intelligence (AI) education has garnered considerable attention due to the significant implications of AI technologies on society. This literature review explores the role of interdisciplinary education in cultivating ethical AI practitioners, addressing critical themes such as the evolution of AI and its ethical challenges, the integration of ethics into AI education, and the influence of interdisciplinary approaches on moral conscience development.

2.1 The Rise of Artificial Intelligence and Ethical Challenges

The transformative potential of AI has made it an invaluable tool across various sectors, yet it has also raised concerns about unintended social and ethical consequences. According to Bostrom (2014), AI technology has the potential for both immense benefit and significant risk. As AI systems become more autonomous, the likelihood of ethical conflicts, such as privacy violations and discrimination, increases. Research by Campolo et al. (2017) emphasizes that while AI can be used to improve decision-making and increase efficiency, it also presents risks related to fairness and bias, particularly in applications such as facial recognition, predictive policing, and automated hiring systems.

Floridi and Cowls (2019) propose a unified framework of ethical principles for AI, identifying values such as transparency, accountability, and justice as fundamental to responsible AI development. They argue that these principles are essential to mitigate the negative effects of AI on society. Their work highlights the ethical complexities involved in the deployment of AI, particularly concerning the potential for misuse and unintended harm. As these technologies become more widespread, they contend, establishing ethical guidelines becomes crucial.

2.2 The Importance of Ethics in AI Education

The integration of ethics into AI education has been advocated as a means of preparing practitioners to handle the ethical dilemmas inherent in AI technology. As Greene, Hoffmann, and Stark (2019) point out, many current AI curricula prioritize technical skills over ethical training, leading to an unbalanced educational framework that inadequately prepares students for real-world ethical challenges. They argue that a balanced curriculum should incorporate both technical and ethical components to foster a comprehensive understanding of AI's societal impact.

Raji et al. (2020) extend this perspective by discussing how ethical considerations can help prevent harmful outcomes associated with AI applications. They suggest that ethics education should be introduced early in the AI curriculum to ensure that students develop an ethical mindset before acquiring advanced technical skills. This approach, they argue, encourages students to view AI development not merely as a technical challenge but as a socially impactful practice that requires responsibility and care.

2.3 Challenges in Implementing Ethical AI Education

Implementing ethical AI education poses several challenges, including institutional resistance, limited resources, and the need for faculty with expertise in both technology and ethics. Calo (2017) discusses the structural difficulties educational institutions face in integrating ethics into AI programs, citing issues such as curriculum constraints and a lack of qualified instructors. He argues that overcoming these challenges is essential for creating a generation of AI professionals who prioritize ethical considerations.

Moreover, Sloane and Moss (2019) observe that interdisciplinary AI programs often struggle to maintain a cohesive curriculum, as the integration of ethics and technical training requires a careful balance. They recommend developing specialized training programs for educators to bridge the gap between technical expertise and ethical knowledge, ensuring that faculty members are equipped to guide students through both technical and ethical challenges in AI.

2.4 Interdisciplinary Education and Moral Conscience Development

Interdisciplinary education, which integrates fields such as philosophy, social sciences, and ethics with technology, has been proposed as a means of cultivating a deeper understanding of AI's societal impact. Cath (2018) asserts that interdisciplinary education fosters critical thinking and ethical awareness, as it exposes students to a range of perspectives on the implications of AI technology. This approach enables students to consider the broader social and ethical contexts of their work, preparing them to address complex moral issues in their professional practice.

Studies by Sjoberg, Bondesson, and Olsson (2020) further support the value of interdisciplinary education, showing that students who receive training in both ethics and technology demonstrate greater moral awareness and a stronger commitment to responsible AI practices. Their research reveals that interdisciplinary training enhances students' ability to anticipate and address ethical challenges, thus reducing the risk of unintended consequences associated with AI systems.

2.5 Case Studies on Ethical Decision-Making in AI

Several case studies have highlighted the practical benefits of interdisciplinary education in fostering ethical decision-making skills among AI practitioners. For instance, a study by Mittelstadt et al. (2016) analyzed decision-making processes among students trained in both AI and ethics, revealing that they were more likely to identify potential biases and other ethical concerns in AI systems. This suggests that interdisciplinary education can help students develop a proactive approach to ethical challenges, enabling them to design more responsible and fair AI applications.

Another case study by Richards and Smart (2017) examines the ethical challenges encountered by practitioners working in the field of AI. Their findings indicate that individuals with interdisciplinary training were better prepared to handle ethical dilemmas, as they had a more nuanced understanding of AI's societal implications. This research underscores the value of interdisciplinary education in equipping practitioners with the skills needed to navigate complex ethical landscapes.

2.6 Comparative Analysis of Ethical Outcomes in AI Education

Comparative studies have further demonstrated the impact of interdisciplinary education on ethical outcomes among AI practitioners. A study by Jobin, Ienca, and Vayena (2019) compared the ethical awareness of students from traditional AI programs with those from interdisciplinary programs, finding that students in the latter group exhibited significantly higher levels of ethical awareness. They argue that interdisciplinary training helps students develop a broader perspective, enabling them to recognize and address ethical risks in AI development.

Similarly, Lin, Selinger, and Hadley (2020) compared the decision-making processes of practitioners with different educational backgrounds. Their research reveals that individuals with interdisciplinary training are more likely to prioritize ethical values, such as transparency and fairness, over purely technical considerations. This suggests that interdisciplinary education fosters a holistic approach to AI development, encouraging practitioners to consider the ethical implications of their work.

2.7 The Role of Ethics in Shaping Responsible AI Practitioners

Ethics plays a central role in shaping responsible AI practitioners who are equipped to handle the moral complexities of AI technology. Russell, Dewey, and Tegmark (2015) argue that ethics should be integrated into AI education as a core component, rather than an optional add-on. They suggest that a strong foundation in ethics can help practitioners develop the critical thinking skills necessary to address the unique ethical challenges posed by AI, such as bias and accountability.

Yuan et al. (2020) further support this view, emphasizing the importance of ethical training in fostering a sense of responsibility among AI practitioners. They argue that an ethical foundation enables practitioners to consider the potential social consequences of their work, helping to ensure that AI technology is developed in a way that benefits society.

2.8 Ethical Frameworks for AI Education

Developing ethical frameworks for AI education is essential to guide the integration of ethics into AI curricula. Winfield, Michael, and Pitt (2019) propose a framework that emphasizes transparency, accountability, and fairness, arguing that these principles should be incorporated into AI education to promote responsible practices. They recommend that educational institutions adopt these principles as foundational elements in AI programs, ensuring that students are trained to prioritize ethical values alongside technical skills.

Floridi and Taddeo (2018) also advocate for ethical frameworks in AI education, suggesting that principles such as beneficence and non-maleficence should guide AI development. Their work highlights the importance of ethical principles in fostering responsible AI practices, as these frameworks provide practitioners with a moral compass for navigating complex ethical dilemmas.

2.9 Proposed Models for Integrative AI Education

Several scholars have proposed models for integrating ethics into AI education. Dignum (2018) presents a model that combines technical training with ethical reflection, suggesting that this approach helps students develop a comprehensive understanding of AI's societal impact. She argues that by embedding ethics throughout the AI curriculum, educational institutions can prepare students to address ethical challenges at every stage of AI development.

Similarly, Arnold et al. (2019) propose a model for ethical AI education that emphasizes the importance of case studies and real-world examples. They argue that practical applications of ethical principles help students understand the relevance of ethics in AI, making it easier for them to apply these principles in professional contexts.

The literature reviewed here demonstrates the importance of interdisciplinary education in preparing AI practitioners to address ethical challenges. By integrating ethics with technical training, interdisciplinary education fosters a comprehensive understanding of AI's societal implications, enabling practitioners to consider issues such as bias, justice, and accountability. While challenges remain in implementing ethical AI education, studies suggest that interdisciplinary approaches can enhance ethical awareness and decision-making skills among students, ultimately contributing to the development of a more responsible and equitable future for AI technology.

3. METHODOLOGY

This study employed a mixed-methods research design to comprehensively evaluate the impact of interdisciplinary ethics education on the development of ethical awareness among AI students. By integrating quantitative and qualitative approaches, this methodology aimed to capture a broad understanding of how interdisciplinary education can shape ethical perspectives and prepare future AI practitioners to address complex ethical challenges in real-world applications (Creswell & Plano Clark, 2017). The mixed-methods approach allowed for the triangulation of data, enhancing the reliability of the findings by validating quantitative results with qualitative insights (Tashakkori & Teddlie, 2003).

3.1 Research Design

The research design was structured in two phases to systematically address the study objectives. The first phase focused on quantitative data collection via surveys, measuring the extent to which interdisciplinary ethics education influenced students' ethical awareness. The second phase utilized qualitative interviews to gather in-depth insights into students' attitudes and perspectives on ethical issues in AI, thereby providing a more nuanced understanding of their preparedness to handle ethical dilemmas (Yin, 2014).

1. **Phase One: Quantitative Survey**

In the initial phase, a survey was administered to students enrolled in AI programs across multiple institutions, specifically targeting programs that included interdisciplinary ethics courses alongside core technical subjects. The survey included a Likert-scale questionnaire to assess students' ethical awareness, understanding of ethical principles, and perceived readiness to handle ethical challenges. This structured survey enabled the quantitative measurement of ethical awareness, which is essential for establishing baseline trends and patterns (Cohen, Manion, & Morrison, 2018).

The survey items were designed based on validated scales used in previous studies assessing ethics in technology education (Greene et al., 2019). These items aimed to measure the perceived importance of ethics in AI, awareness of ethical issues such as bias and accountability, and self-assessed capability to apply ethical principles in decision-making. The survey was distributed online to ensure ease of access and to capture a larger sample size, thus improving the generalizability of the findings (Bryman, 2012).

- **Data Analysis for Quantitative Survey:**

Responses from the survey were analyzed using descriptive and inferential statistics. Descriptive statistics provided an overview of students' general ethical awareness, while inferential statistics were used to assess differences in ethical awareness between students in interdisciplinary programs and those in technical-only programs. Statistical analyses were performed using SPSS software to identify significant differences and correlations, particularly focusing on how interdisciplinary education may contribute to heightened ethical awareness (Field, 2013).

2. **Phase Two: Qualitative Interviews**

The second phase involved conducting semi-structured interviews with a subset of survey respondents, focusing on students who had undergone interdisciplinary training in ethics and AI. The aim was to gain qualitative insights into how these students perceived and approached ethical challenges in AI. Semi-structured interviews were chosen to allow flexibility, enabling the researcher to probe into specific aspects of the students' experiences with ethics education (Kvale & Brinkmann, 2009).

The interview guide included open-ended questions on topics such as the perceived relevance of ethics in AI, specific instances where ethical awareness was applied, and students' confidence in navigating ethical challenges. This format provided rich, detailed accounts of students' moral reasoning and ethical considerations in the context of AI (Patton, 2015). Interviews were recorded and transcribed verbatim to ensure accuracy in data interpretation and coding.

- **Data Analysis for Qualitative Interviews:**

A thematic analysis was conducted on the interview transcripts, identifying key themes related to students' ethical development and their views on the importance of interdisciplinary education (Braun & Clarke, 2006). NVivo software was used to facilitate coding and to identify recurring patterns, such as enhanced ethical reasoning, awareness of societal implications, and attitudes toward accountability and fairness. This approach allowed for an in-depth understanding of the specific benefits and challenges students perceived in interdisciplinary ethics education (Charmaz, 2006).

3.2 Sampling and Participant Selection

Participants for both the survey and interviews were selected from students enrolled in AI programs at three prominent universities in Greater Noida, a region recognized for its educational institutions that emphasize interdisciplinary approaches to technology and ethics. Universities in Greater Noida were strategically chosen due to their distinctive focus on integrating ethics within AI curricula, providing a fertile ground for assessing the impact of interdisciplinary education on ethical awareness in AI. Greater Noida hosts some of the country's leading academic programs in artificial intelligence, supported by a diverse student body and faculty expertise in both technical and ethical domains.

By selecting institutions from this area, the study ensured a relevant sample, as these universities offer structured interdisciplinary programs that align with the research objective of evaluating the role of ethics in AI education. Additionally, Greater Noida's reputation as an educational hub with a strong emphasis on innovation in technology made it an ideal setting for gathering insights from students with varying levels of prior exposure to ethics education. This selection approach facilitated the inclusion of a broad range of perspectives, as the institutions attract students from diverse backgrounds, enhancing the depth and applicability of the study's findings. A total of 200 students were invited to participate in the survey, with an approximate response rate of 65%, resulting in 130 valid responses. From this group, 20 students were selected for in-depth interviews using purposive sampling, ensuring a representative mix of backgrounds and experiences, as recommended by Flick (2018).

3.3 Validity and Reliability

To ensure validity, the study employed triangulation by using both survey data and interviews to confirm and elaborate on findings from each phase (Denzin, 1978). Content validity of the survey instrument was established through expert review, with scholars in ethics and AI education validating the relevance of survey items (Greene et al., 2019). Reliability was ensured by piloting the survey instrument with a smaller group of students before the main study, allowing for adjustments to the wording of questions to improve clarity and consistency (Bell, Bryman, & Harley, 2018).

3.4 Ethical Considerations

This study adhered to ethical research standards, following institutional guidelines for data collection and informed consent. Participation was voluntary, and respondents were assured of their anonymity and the confidentiality of their responses. Prior to data collection, informed consent was obtained from all participants, and they were informed of their right to withdraw from the study at any point without penalty (Silverman, 2013).

3.5 Limitations

While the mixed-methods approach provided a comprehensive view, there were limitations related to the sample size and generalizability. The study focused on students from institutions with established interdisciplinary programs, which may limit the applicability of findings to institutions without such resources. Future research could expand to include a broader range of educational settings to enhance generalizability (Marshall & Rossman, 2016).

4. DATA INTERPRETATION

4.1 Phase One: Quantitative Survey Data

- **Sample Size:** 130 students
- **Groups:**
- Group A (Interdisciplinary AI Program with Ethics): 65 students

- Group B (Technical AI Program without Ethics): 65 students

Table 1. Survey results on ethical awareness and preparedness (Likert Scale: 1 = Strongly Disagree, 5 = Strongly Agree)

Survey Question	Group A (Mean ± SD)	Group B (Mean ± SD)
1. I feel well-prepared to identify ethical dilemmas in AI applications.	4.2 ± 0.7	2.9 ± 0.9
2. Ethical issues in AI are important to me.	4.5 ± 0.5	3.2 ± 0.8
3. I believe AI practitioners have a responsibility to consider societal impacts.	4.6 ± 0.4	3.3 ± 0.7
4. My program has provided me with sufficient ethical training.	4.3 ± 0.6	2.6 ± 0.8
5. I feel confident that I can apply ethical principles in real-world AI scenarios.	4.0 ± 0.8	2.8 ± 0.9
6. I am aware of biases that can be embedded in AI systems.	4.4 ± 0.6	3.0 ± 0.9

Data Analysis

- **Descriptive Statistics:**
  - Group A students showed consistently high mean scores across all survey items, suggesting a strong sense of ethical awareness and preparedness.
  - Group B students had lower mean scores, especially in questions assessing preparedness and awareness of bias.
- **Inferential Statistics:**
  - **Independent Samples t-test:** Conducted to compare the mean scores between Group A and Group B.
    - **Results:** Statistically significant differences were found in all items ($p < 0.05$), indicating that students in interdisciplinary programs with ethics training had a significantly higher level of ethical awareness and preparedness than those in technical-only programs.

4.2 Phase Two: Qualitative Interview Data

- **Sample Size for Interviews:** 20 students
- **Groups:** 10 from Group A (Interdisciplinary) and 10 from Group B (Technical)

4.2.1 Interview Themes and Sample Responses

Theme 1: Importance of Ethics in AI

- **Group A (Interdisciplinary):**
  - "Our ethics courses made me realize how critical it is to question the data sources we use, especially since AI applications affect real lives."
  - "Learning about historical cases of AI bias has prepared me to think twice about fairness when designing algorithms."
- **Group B (Technical):**
  - "I haven't had much training in ethics. I feel that the main focus has been on coding and the technical aspects of AI."
  - "I'm aware that AI can have social impacts, but I don't feel well-equipped to address those issues."

Theme 2: Ethical Preparedness and Awareness of Bias

- **Group A (Interdisciplinary):**
  - "I feel that my interdisciplinary education helps me understand the potential biases and social implications better. For example, I'm more cautious about using biased datasets."
  - "We discussed a lot about accountability, which I think is crucial, as AI can often operate in 'black box' ways that make it hard to understand."
- **Group B (Technical):**
  - "Bias in AI is something I've heard about, but we haven't covered it in detail. I'm more focused on algorithm efficiency."
  - "I would like to know more about transparency issues, but it's not something my program emphasizes."

Theme 3: Confidence in Applying Ethical Principles

- **Group A (Interdisciplinary):**
  - "I feel confident because our program emphasized real-world applications, so we practiced addressing ethical issues hands-on."
  - "Having a background in ethics allows me to foresee potential problems before they arise, which is essential for responsible AI design."
- **Group B (Technical):**

- o “I think I would benefit from ethical training, as I sometimes find it hard to see the broader societal impacts of my work.”
- o “I feel more prepared to tackle technical issues than ethical ones, which can be challenging in real-world settings.”

5. RESULTS

The results of this study reveal significant differences between students enrolled in interdisciplinary AI programs with ethics training (Group A) and those in purely technical AI programs without such training (Group B). Findings suggest that interdisciplinary education enhances ethical awareness, preparedness, and a commitment to responsible AI practice, aligning with prior research highlighting the value of ethics in technology education (Greene et al., 2019; Floridi & Cowls, 2019).

5.1 Quantitative Survey Findings

The quantitative survey assessed several dimensions of ethical awareness and preparedness, focusing on students' ability to recognize ethical issues, their perceived importance of ethics, and their self-assessed readiness to apply ethical principles in real-world AI contexts.

1. **Ethical Awareness and Preparedness:**
   - **Group A (Interdisciplinary Program with Ethics):** Students from interdisciplinary programs with ethics training scored consistently higher on measures of ethical awareness and preparedness. Specifically, Group A students reported a mean score of **4.2** (SD = 0.7) on their readiness to identify ethical dilemmas, compared to **2.9** (SD = 0.9) for Group B. This difference highlights that ethics-integrated programs effectively equip students to recognize ethical issues in AI applications, consistent with findings from studies that advocate for ethics as a core component of technology education (Cath, 2018; Russell et al., 2015).
   - **Group B (Technical Program without Ethics):** In contrast, students in technical-only programs scored lower on all survey items related to ethical awareness. For instance, Group B’s mean score of **3.0** (SD = 0.9) on awareness of biases in AI systems was significantly lower than Group A's score of **4.4** (SD = 0.6). These findings align with previous research indicating that technical-only programs may fall short in preparing students for the ethical complexities of AI (Calo, 2017; Sjoberg et al., 2020).

Supporting Studies: The disparity in awareness and preparedness between Group A and Group B aligns with Floridi and Cowls' (2019) assertion that ethics training is crucial for responsible AI practice. Floridi and Cowls emphasize that technical knowledge alone is insufficient for AI practitioners to navigate ethical challenges, underscoring the need for interdisciplinary education to foster comprehensive ethical awareness. Furthermore, Greene et al. (2019) argue that without ethics training, AI practitioners may overlook critical issues like transparency, fairness, and accountability, leading to AI applications that can perpetuate societal biases.

2. **Perceived Importance of Ethics in AI Practice:**
    - The survey also highlighted a significant gap between the two groups in their perception of ethics' importance in AI practice. Group A students rated the importance of ethics at **4.5** (SD = 0.5), while Group B students rated it at **3.2** (SD = 0.8). This difference suggests that interdisciplinary education instills a stronger sense of responsibility among AI students, reinforcing the findings of Cath (2018), who argues that ethics training in technology programs cultivates a moral framework necessary for responsible AI design and deployment.

Supporting Studies: This result is consistent with research by Russell et al. (2015), who argue that ethics should be embedded in AI education to prepare practitioners for the societal impacts of AI. Russell and colleagues emphasize that an understanding of ethics enhances practitioners' commitment to accountability and public good, which are essential in fields where technologies directly impact individuals and communities.

3. **Inferential Statistics:**
    - **Independent samples t-tests** confirmed that differences between Group A and Group B were statistically significant across all survey items ($p < 0.05$), indicating that interdisciplinary programs with ethics training have a substantial effect on students' ethical awareness and preparedness. These statistical findings provide quantitative support for the view that ethics education is indispensable in cultivating ethically aware AI practitioners, in line with previous research underscoring the critical role of ethics in AI education (Bostrom, 2014; Campolo et al., 2017).

5.2 Qualitative Interview Findings

The qualitative interviews further illustrated how interdisciplinary ethics education influences students' understanding of AI's societal impact and their approach to ethical decision-making. Three main themes emerged from the interviews: the perceived importance of ethics in AI, awareness of biases and ethical risks, and confidence in applying ethical principles.

1. **Importance of Ethics in AI:**
    - **Group A (Interdisciplinary):** Students from interdisciplinary programs consistently highlighted the importance of ethics in AI. They recognized the societal implications of their work, expressing a commitment to ethical standards. For example, one Group A student stated, "Learning about historical cases of AI bias has prepared me to think twice about fairness when designing algorithms." This aligns with Cath (2018), who posits that interdisciplinary education fosters a greater appreciation for the ethical dimensions of AI, encouraging students to prioritize values like fairness and social responsibility.
    - **Group B (Technical):** While Group B students acknowledged the relevance of ethics, they often perceived it as secondary to technical skills. One student noted, "I'm aware that AI can have social impacts, but I don't feel well-equipped to address those issues." This response reflects findings by Greene et al. (2019), who argue that purely technical programs may leave students underprepared for addressing ethical challenges, potentially limiting their ability to foresee and mitigate ethical risks in AI applications.
2. **Awareness of Bias and Ethical Risks:**
    - **Group A (Interdisciplinary):** Students from interdisciplinary programs demonstrated a nuanced understanding of biases and ethical risks in AI, frequently mentioning the importance of accountability and transparency. One participant shared, "Our program taught us to be cautious about using biased datasets, and I now know how essential it is to consider fairness in every stage of AI development." This perspective supports findings by Floridi and Cowls (2019), who emphasize that interdisciplinary ethics education raises awareness of the social impact of AI, fostering a commitment to mitigating bias and ensuring equitable outcomes.
    - **Group B (Technical):** In contrast, Group B students displayed limited awareness of biases in AI. Many focused primarily on technical performance and efficiency, often overlooking ethical implications. As one

respondent commented, "Bias in AI is something I've heard about, but we haven't covered it in detail." This response underscores the need for ethics education, as highlighted by Sjoberg et al. (2020), who argue that a lack of ethics training may result in practitioners inadvertently perpetuating biases and inequalities through AI systems.

3. **Confidence in Applying Ethical Principles:**
   - **Group A (Interdisciplinary):** Interdisciplinary-trained students reported a high level of confidence in applying ethical principles to real-world scenarios, crediting hands-on case studies and ethics-focused discussions in their programs for fostering confidence. One student shared, "I feel confident because our program emphasized real-world applications, so we practiced addressing ethical issues hands-on." This finding is supported by Kalluri (2020), who suggests that experiential learning in ethics enables AI students to apply theoretical knowledge practically, enhancing their ability to handle ethical challenges effectively.
   - **Group B (Technical):** Conversely, students without ethics training felt less prepared to handle ethical dilemmas. A Group B student remarked, "I would benefit from ethical training, as I sometimes find it hard to see the broader societal impacts of my work." This aligns with Calo (2017), who argues that purely technical training can limit students' understanding of AI's societal implications, underscoring the importance of integrating ethics in technical education.

5.3 Summary of Findings

Overall, the results indicate that interdisciplinary ethics education in AI programs significantly enhances students' ethical awareness, preparedness, and understanding of AI's societal impact. Group A students demonstrated a higher ethical consciousness, greater confidence in addressing ethical challenges, and a more profound commitment to fairness and accountability than Group B students. These findings align with prior research on the benefits of ethics-integrated education in cultivating responsible AI practitioners, suggesting that interdisciplinary ethics education is essential for preparing a new generation of AI professionals who are attuned to the ethical complexities of their work (Floridi & Cowls, 2019; Cath, 2018; Greene et al., 2019).

6. DISCUSSION

This study investigated the role of interdisciplinary ethics education in shaping ethical awareness, preparedness, and decision-making skills among AI students. The results provide compelling evidence that students in interdisciplinary AI programs with integrated ethics training (Group A) demonstrate significantly higher ethical awareness, a greater understanding of AI's societal implications, and enhanced confidence in applying ethical principles, compared to students in technical-only AI programs without such training (Group B). These findings contribute to a growing body of literature that underscores the importance of ethics in AI education (Floridi & Cowls, 2019; Cath, 2018).

6.1 Importance of Ethics in AI Education

The elevated ethical awareness and preparedness observed in Group A aligns with previous research emphasizing that interdisciplinary education plays a crucial role in fostering a moral framework for AI practitioners (Russell et al., 2015; Kalluri, 2020). This study's findings reflect the assertions of Floridi and Cowls (2019), who argue that AI education should not be limited to technical training but must also include an ethical component to address the societal and moral complexities of AI deployment. Group A students' higher ratings in the perceived importance of ethics demonstrate that interdisciplinary education instills a sense of responsibility, encouraging students to consider the broader impact of their work on society.

Conversely, Group B's comparatively lower scores on ethical awareness and preparedness indicate that technical-only programs may inadvertently foster a narrow focus on technical skills, which can limit practitioners' understanding of AI's societal risks and ethical challenges (Calo, 2017). The interviews with Group B students further reinforced this point, with many expressing a desire for more ethics training to help them navigate complex ethical dilemmas, suggesting a gap in the traditional technical curriculum. As Greene et al. (2019) argue, without structured ethics education, AI practitioners may lack the foundational knowledge to address critical issues such as accountability, fairness, and bias in their work.

6.2 Awareness of Bias and Ethical Risks

This study also highlights the critical role of interdisciplinary education in fostering awareness of bias and ethical risks associated with AI systems. Group A students demonstrated a heightened awareness of these issues, frequently mentioning the importance of accountability, fairness, and transparency. These insights align with prior studies, such as Cath (2018), who emphasizes that interdisciplinary programs encourage students to think critically about the ethical implications of technology. The higher awareness of biases among Group A students is consistent with the findings of Kalluri (2020), who suggests that ethics education prepares AI practitioners to identify and mitigate biases that might otherwise lead to unfair outcomes. By understanding and addressing such risks, interdisciplinary-trained AI practitioners are better positioned to contribute to the responsible and equitable development of AI technologies.

In contrast, Group B students had limited knowledge of bias and ethical risks, often focusing more on technical efficiency than on fairness and accountability. This outcome aligns with Russell et al. (2015), who caution that purely technical training may result in practitioners overlooking the ethical dimensions of their work. This study's findings thus support the view that ethics-integrated programs can help AI students cultivate a broader understanding of their work's potential impact, which is essential for mitigating ethical risks in AI deployment.

6.3 Confidence in Applying Ethical Principles

The study found that students in interdisciplinary programs (Group A) expressed significantly more confidence in applying ethical principles to real-world scenarios than those in technical-only programs (Group B). Group A students attributed their confidence to practical experiences with case studies, ethics-based projects, and discussions, suggesting that hands-on ethical training enhances students' ability to handle complex moral questions. This finding aligns with Kalluri (2020) and Sjoberg et al. (2020), who argue that experiential learning is key to fostering ethical decision-making skills in AI practitioners. Through real-world applications, students learn to navigate ethical issues effectively, increasing their confidence in resolving dilemmas that may arise in AI practice.

In contrast, Group B students reported lower confidence in addressing ethical challenges, reflecting a lack of training in this area. This observation is consistent with Calo's (2017) findings that technical-only training may hinder students' ability to address the ethical dimensions of AI effectively, underscoring the need for interdisciplinary approaches to AI education. Integrating ethical training within AI curricula appears to play a pivotal role in preparing students to apply ethical considerations in practice, equipping them to handle AI's complex societal implications.

7. CONCLUSION

This study provides substantial evidence supporting the integration of ethics into AI education through interdisciplinary programs. The findings underscore that interdisciplinary ethics education fosters a heightened ethical awareness, preparedness, and confidence among AI students, equipping them to address ethical dilemmas that may arise in their professional careers. In line with existing literature, this study reaffirms that purely technical training may be insufficient to prepare AI practitioners for the ethical challenges posed by AI technologies (Floridi & Cowls, 2019; Greene et al., 2019).

The higher ethical awareness observed among interdisciplinary-trained students suggests that a combined focus on technical and ethical education enables AI practitioners to foresee and mitigate potential societal risks, such as biases embedded in algorithms and lack of accountability in AI-driven decision-making processes. These findings support the argument that interdisciplinary programs cultivate a more socially responsible AI workforce, as suggested by Cath (2018) and Sjoberg et al. (2020).

7.1 Implications for AI Education

This study's results hold significant implications for AI education policy and curriculum design. Educational institutions and policymakers should consider embedding ethics across AI curricula, from introductory courses to advanced topics. Such a structure ensures that ethical considerations are consistently reinforced throughout students' educational journey, fostering a deeper and more enduring sense of responsibility. By equipping AI students with the tools to address ethical challenges, interdisciplinary programs contribute to developing a workforce capable of creating AI systems that prioritize social well-being and fairness.

7.2 Limitations and Future Research

Despite its contributions, this study has limitations. The sample focused on students from institutions with established interdisciplinary programs, which may limit the generalizability of the findings. Future research could expand to include a broader range of educational settings, exploring how varying levels of ethics integration affect ethical awareness across different demographics and regions. Additionally, longitudinal studies could provide insights into how ethics training impacts AI practitioners' decision-making over time, examining the long-term effects of interdisciplinary education on ethical AI practice.

In conclusion, this study underscores the critical role of interdisciplinary ethics education in preparing responsible AI practitioners. By instilling ethical awareness, readiness, and confidence in AI students, interdisciplinary programs contribute to a more socially responsible and equitable future in AI technology.

REFERENCES

Angwin, J., Larson, J., Mattu, S., & Kirchner, L. (2016). *Machine Bias: There's Software Used Across the Country to Predict Future Criminals*. And It's Biased Against Blacks. ProPublica.

Arnold, T., Lee, M., & Yampolskiy, R. (2019). The Ethics of AI: Challenges and Opportunities. *Ethics and Information Technology*.

Bell, E., Bryman, A., & Harley, B. (2018). *Business Research Methods*. Oxford University Press.

Binns, R. (2018). "Fairness in Machine Learning: Lessons from Political Philosophy." *Proceedings of the 2018 Conference on Fairness, Accountability, and Transparency*.

Bostrom, N. (2014). *Superintelligence: Paths, dangers, strategies*. Oxford University Press.

Braun, V., & Clarke, V. (2006). Using Thematic Analysis in Psychology. *Qualitative Research in Psychology*, *3*(2), 77–101. DOI: 10.1191/1478088706qp063oa

Bryman, A. (2012). *Social Research Methods*. Oxford University Press.

Calo, R. (2017). "Artificial Intelligence Policy: A Primer and Roadmap." University of California, Davis, Law Review.

Campolo, A., Sanfilippo, M., Whittaker, M., & Crawford, K. (2017). *AI Now Report 2017*. AI Now Institute.

Cath, C. (2018). Governing Artificial Intelligence: Ethical, Legal, and Technical Opportunities and Challenges. *Philosophical Transactions. Series A, Mathematical, Physical, and Engineering Sciences*, *376*(2133), 20180080. DOI: 10.1098/rsta.2018.0080 PMID: 30322996

Charmaz, K. (2006). Constructing Grounded Theory: A Practical Guide through Qualitative Analysis. *Sage (Atlanta, Ga.)*.

Cohen, L., Manion, L., & Morrison, K. (2018). *Research Methods in Education*. Routledge.

Creswell, J. W., & Plano Clark, V. L. (2017). Designing and Conducting Mixed Methods Research. *Sage (Atlanta, Ga.)*.

Denzin, N. K. (1978). *The Research Act: A Theoretical Introduction to Sociological Methods*. McGraw-Hill.

Dignum, V. (2018). *Responsible Artificial Intelligence: How to Develop and Use AI in a Responsible Way*. Springer.

Field, A. (2013). Discovering Statistics Using IBM SPSS Statistics. *Sage (Atlanta, Ga.)*.

Flick, U. (2018). An Introduction to Qualitative Research. *Sage (Atlanta, Ga.)*.

Floridi, L., & Cowls, J. (2019). A Unified Framework of Five Principles for AI in Society. *Harvard Data Science Review*. Advance online publication. DOI: 10.1162/99608f92.8cd550d1

Floridi, L., & Taddeo, M. (2018). What is data ethics? *Philosophical Transactions of the Royal Society A*. PMID: 30322997

Ghassemi, M., Oakden-Rayner, L., & Beam, A. L. (2020). The False Hope of Current Approaches to Explainable Artificial Intelligence in Health Care. *The Lancet. Digital Health*. PMID: 34711379

Goodfellow, I., Bengio, Y., & Courville, A. (2016). *Deep Learning*. MIT Press.

Greene, D., Hoffmann, A. L., & Stark, L. (2019). "Better, Nicer, Clearer, Fairer: A Critical Assessment of the Movement for Ethical Artificial Intelligence and Machine Learning." *Proceedings of the 52nd Hawaii International Conference on System Sciences*. DOI: 10.24251/HICSS.2019.258

Jobin, A., Ienca, M., & Vayena, E. (2019). The global landscape of AI ethics guidelines. *Nature Machine Intelligence*, *1*(9), 389–399. DOI: 10.1038/s42256-019-0088-2

Kalluri, P. (2020). Don't ask if artificial intelligence is good or fair, ask how it shifts power. *Nature*, *583*(7815), 169. DOI: 10.1038/d41586-020-02003-2 PMID: 32636520

Kalluri, P. (2020). Don't ask if artificial intelligence is good or fair, ask how it shifts power. *Nature*, *583*(7815), 169. DOI: 10.1038/d41586-020-02003-2 PMID: 32636520

Kvale, S., & Brinkmann, S. (2009). InterViews: Learning the Craft of Qualitative Research Interviewing. *Sage (Atlanta, Ga.)*.

Lin, P., Selinger, E., & Hadley, P. (2020). "Ethics and AI: The Role of Interdisciplinary Education." Journal of Ethical AI.

Marshall, C., & Rossman, G. B. (2016). Designing Qualitative Research. *Sage (Atlanta, Ga.)*.

Mehrabi, N., Morstatter, F., Saxena, N., Lerman, K., & Galstyan, A. (2021). A Survey on Bias and Fairness in Machine Learning. *ACM Computing Surveys*.

Mittelstadt, B. D., Allo, P., Taddeo, M., Wachter, S., & Floridi, L. (2016). The ethics of algorithms: Mapping the debate. *Big Data & Society*, *3*(2), 2053951716679679. DOI: 10.1177/2053951716679679

Patton, M. Q. (2015). Qualitative Research and Evaluation Methods. *Sage (Atlanta, Ga.)*.

Raji, I., Bender, E., Smith, T., & Boyd, K. (2020). "Out of One, Many: The Rise of AI Ethics." Journal of AI Research.

Rest, J. R., Narvaez, D., Bebeau, M. J., & Thoma, S. J. (1999). *Postconventional Moral Thinking: A Neo-Kohlbergian Approach*. Psychology Press. DOI: 10.4324/9781410603913

Richards, N. M., & Smart, W. D. (2017). *How Should the Law Think About Robots?* The Law and AI Journal.

Rudin, C. (2019). Stop Explaining Black Box Machine Learning Models for High Stakes Decisions and Use Interpretable Models Instead. *Nature Machine Intelligence*, *1*(5), 206–215. DOI: 10.1038/s42256-019-0048-x PMID: 35603010

Russell, S., Dewey, D., & Tegmark, M. (2015). Research Priorities for Robust and Beneficial Artificial Intelligence. *AI Magazine*, *36*(4), 105–114. DOI: 10.1609/aimag.v36i4.2577

Russell, S., & Norvig, P. (2021). *Artificial Intelligence: A Modern Approach* (4th ed.). Pearson.

Silverman, D. (2013). Doing Qualitative Research: A Practical Handbook. *Sage (Atlanta, Ga.)*.

Sjoberg, L., Bondesson, M., & Olsson, K. (2020). "Interdisciplinary Education for Ethical AI: A Study on Morality in AI Curriculums." Journal of Ethical AI.

Sloane, M., & Moss, E. (2019). "AI Ethics and Education: A Framework for a Responsible Future." *Proceedings of the AAAI Conference on AI Ethics and Society*.

Tashakkori, A., & Teddlie, C. (2003). *Handbook of Mixed Methods in Social & Behavioral Research*. Sage.

Winfield, A., Michael, K., & Pitt, J. (2019). AI and Ethics: A Framework for Responsible AI. *AI & Society*.

Yin, R. K. (2014). Case Study Research: Design and Methods. *Sage (Atlanta, Ga.)*.

Chapter 11
Navigating AI and Children's Rights in Corporate Practices:
Ethical AI Integration in Corporate Responsibility

Christopher Bohlens
https://orcid.org/0009-0001-9653-2743
Leuphana University of Lüneburg, Germany

ABSTRACT

This part investigates the convergence of AI and children's rights in corporate contexts, stressing the necessity for ethical AI in corporate social responsibility. It assesses AI's potential to bolster corporate accountability, specifically in the eradication of child labor within supply chains, and to enhance children's access to education and healthcare. The discussion encompasses the ethical and legal dilemmas associated with AI, including algorithmic biases and data privacy issues, while proposing ethical marketing strategies that protect children's digital privacy. It emphasizes the significance of collaborative AI governance, advocating for multi-stakeholder partnerships to weave children's rights into corporate frameworks. The study concludes with strategic guidance for corporations, policymakers, and civil society to embrace ethical AI frameworks, advocating for a proactive, child-centered methodology. Ultimately, it urges corporations to responsibly leverage AI, ensuring that technological progress supports and promotes children's rights in a changing environment.

DOI: 10.4018/979-8-3693-9035-1.ch011

INTRODUCTION AND BACKGROUND

Overview of AI's Integration in Corporate Operations and Its Potential Impact on Children's Rights

The swift progression of artificial intelligence (AI) has markedly redefined corporate operations across a multitude of sectors. Enterprises are harnessing AI technologies to enhance supply chain efficiencies, improve customer interactions, and refine decision-making methodologies (Ransbotham et al., 2018). Although AI confers significant advantages in terms of operational efficiency and financial performance, it concurrently introduces intricate challenges, particularly with respect to the rights of children. As firms assimilate AI into their operational frameworks, the prospective ramifications for children, encompassing both advantageous and detrimental effects—become increasingly salient.

Children constitute a susceptible demographic that can be deeply influenced by AI-driven corporate methodologies. Matters such as data privacy, exposure to unsuitable content, and the reinforcement of societal prejudices through algorithmic decision-making are of critical concern (UNICEF, 2021). For example, marketing algorithms employing AI may unintentionally direct inappropriate advertisements towards children, while automated systems in educational or recruitment settings could exacerbate pre-existing inequalities, thereby impacting children's future prospects (Eubanks, 2018).

AI's Role in Corporate Operations

AI technologies have emerged as fundamental components of contemporary business strategies. Machine learning algorithms scrutinize extensive datasets to discern patterns and insights that guide strategic decision-making (Davenport, 2018; Davenport & Harris, 2007). Within supply chain management, AI bolsters transparency and efficiency through demand forecasting, route optimization, and compliance monitoring with labor standards (Min, 2010). The domain of customer service has been transformed by AI-enhanced chatbots and personalized recommendations, which facilitate improved engagement and satisfaction (Huang & Rust, 2018).

Nevertheless, the implementation of AI is accompanied by notable risks. An overreliance on data-driven paradigms may yield unintended repercussions if the underlying data reflects historical biases or lacks inclusivity (Brauner et al., 2023). In relation to children's rights, such biases can culminate in discriminatory practices that disproportionately impact young individuals from marginalized backgrounds.

Potential Impact on Children's Rights

Data Privacy and Protection

The engagement of children with digital platforms results in the generation of data that is subject to collection, analysis, and potential exploitation. Artificial Intelligence systems that process this data may not provide adequate safeguards for children's privacy, consequently leading to breaches that could have enduring ramifications (Livingstone et al., 2022). The commercialization of data pertaining to children invokes ethical dilemmas regarding consent and the possible misappropriation of personal information.

Algorithmic Bias and Discrimination

Artificial Intelligence algorithms that are developed using biased datasets have the capacity to reinforce and exacerbate pre-existing social disparities. For instance, predictive policing algorithms have faced criticism for disproportionately targeting marginalized communities, a phenomenon that also pertains to youth encounters with law enforcement (Richardson et al., 2019). In the educational sphere, AI-driven evaluative tools may place students who do not conform to the normative profiles delineated by the data at a disadvantage, thereby impacting educational outcomes and future opportunities (Raja et al., 2024; Williamson et al., 2023; Williamson & Eynon, 2020).

Exposure to Inappropriate Content

Artificial Intelligence utilized in content curation and recommendation systems can unintentionally expose children to detrimental or age-inappropriate material. Platforms that depend on AI for the filtration and suggestion of content may not consistently possess the capability to effectively evaluate suitability for younger demographics, consequently leading to potential psychological and emotional detriment (Hassan, 2022).

Importance of Addressing Children's Rights Within AI Governance Frameworks

The integration of children's rights within artificial intelligence governance frameworks is imperative to mitigate associated risks and responsibly leverage the advantages of AI technology. It is essential for corporations to implement a child-centric paradigm that prioritizes the safety, privacy, and overall welfare of minors who are impacted by their AI systems (Kaesling, 2023).

Regulatory Compliance and Ethical Standards

It is of paramount importance to comply with international legal frameworks such as the General Data Protection Regulation (GDPR) and the Children's Online Privacy Protection Act (COPPA). These legislative measures establish rigorous standards for data protection and privacy, particularly in relation to individuals under the age of majority (FTC, 2020). In addition to fulfilling legal obligations, organizations are urged to conform to ethical principles that promote fairness, transparency, and accountability within AI systems (Jobin et al., 2019).

Stakeholder Engagement and Inclusive Governance

Collaboration with diverse stakeholders, including children, parents, educators, and organizations advocating for child rights—can yield critical insights regarding the potential ramifications of AI technologies. Inclusive governance practices ensure that a multitude of perspectives informs the design and execution of AI initiatives, ultimately resulting in more equitable and effective outcomes (Bryson & Winfield, 2017).

Corporate Social Responsibility and Sustainable Practices

The incorporation of children's rights into corporate social responsibility (CSR) strategies strengthens an organization's dedication to ethical practices and sustainable development objectives (SDGs) (UN Global Compact Netzwerk Deutschland e. V., 2015). This methodology not only safeguards the welfare of children but also bolsters corporate reputation, fosters consumer trust, and contributes to enduring commercial success (M. E. Porter & Kramer, 2019; S. R. Porter, 2006).

Wrap-up

The assimilation of AI into corporate frameworks presents both opportunities and challenges related to the rights of children. As AI continues to redefine the business environment, it is essential for corporations to proactively consider the potential effects on young individuals. By embedding children's rights within AI governance frameworks, organizations can ensure that technological progress yields positive societal contributions while protecting the interests of the youngest and most vulnerable populations.

The pursuit of responsible AI necessitates a collaborative approach that reconciles technological advancement with ethical imperatives. Corporations occupy a crucial position in spearheading this initiative, establishing benchmarks for the ethical integration of AI that upholds and promotes the rights of children in an increasingly digitized society.

AI AS A TOOL FOR ENHANCING CORPORATE ACCOUNTABILITY

Using Artificial Intelligence to Monitor and Eradicate Child Labor within Supply Chains

The phenomenon of globalization in supply chains has engendered intricate challenges for enterprises endeavoring to maintain ethical labor standards, particularly in relation to child labor. As supply chains proliferate across diverse nations and encompass numerous suppliers, ensuring adherence to labor regulations becomes increasingly onerous (Gold et al., 2015). The application of Artificial Intelligence (AI) possesses the potential to revolutionize corporate accountability by furnishing advanced instruments for the monitoring, detection, and eradication of child labor practices within supply chains.

AI-Driven Supply Chain Surveillance

AI innovations, including machine learning and data analytics, empower corporations to scrutinize extensive volumes of data sourced from various components within the supply chain. These technological advancements are capable of identifying patterns and discrepancies that may signify unethical practices. For example, AI algorithms are adept at processing supplier information to uncover inconsistencies in labor documentation, production metrics, and workforce demographics that may indicate the existence of child labor (Bai & Sarkis, 2020).

The integration of blockchain technology with AI further augments transparency by establishing immutable records of transactions and processes throughout the supply chain (Saberi et al., 2019). AI can assess blockchain data to authenticate the legitimacy of supplier certifications and compliance documentation, thereby ensuring that each segment of the supply chain conforms to ethical standards.

Predictive Analytics for Risk Assessment

Predictive analytics employs statistical algorithms and machine learning methodologies to evaluate historical and real-time data, thereby forecasting potential risks within the supply chain. By incorporating socioeconomic indicators and regional labor practices into AI models, corporations are able to identify suppliers or geographic areas with an elevated probability of engaging in child labor (Chowdhury et al., 2016). This anticipatory methodology allows organizations to concentrate their auditing and compliance initiatives on high-risk zones, thereby optimizing resource allocation.

Automated Compliance Validation

Natural Language Processing (NLP), a specialized branch of AI, facilitates the automation of compliance verification by scrutinizing legal documents, contracts, and communication records for conformity with labor regulations and corporate policies (Hoek, 2020). AI systems possess the capability to flag deviations from contractual commitments pertaining to labor practices, thereby prompting immediate remedial measures. Furthermore, AI-enhanced image recognition technologies can monitor work environments through video surveillance, identifying potential instances of child labor while upholding privacy considerations (Sodhi & Tang, 2018).

Case Studies on AI-Powered Compliance and Risk Management

Case Study 1: IBM's Trust Your Supplier Platform

The "Trust Your Supplier" blockchain network developed by IBM incorporates artificial intelligence to augment supplier management and compliance verification (Chapman, 2020). The platform employs sophisticated AI algorithms to authenticate supplier credentials, evaluate risk factors, and oversee adherence to labor standards, notably the prohibition of child labor. Through the automation of these processes, IBM significantly minimizes the time and expenditure related to supplier onboarding and the continuous monitoring of compliance.

Case Study 2: Unilever's Responsible Sourcing Policy

Unilever utilizes artificial intelligence and data analytics to oversee its suppliers' conformity with its Responsible Sourcing Policy (Arya, 2024; Unilever, 2017, 2021, 2024b, 2024a). The corporation harnesses AI to scrutinize supplier data, pinpointing risks associated with labor practices, environmental repercussions, and human rights infringements. This methodology empowers Unilever to collaborate with suppliers in a proactive manner to rectify issues and ensure compliance with ethical standards.

Case Study 3: Coca-Cola's Satellite Monitoring

The Coca-Cola Company employs AI-driven satellite imagery to oversee agricultural suppliers for compliance with labor and environmental regulations (Marr, 2019; Quincey, 2022). Artificial intelligence processes satellite data to identify unauthorized alterations in land use and indicators of unlawful practices, such as child labor within agricultural operations. This technological advancement permits Coca-Cola to enforce its supplier standards in a remote and efficient manner.

Case Study 4: Mars Incorporated's Supply Chain Transparency

Mars Incorporated has formed partnerships with artificial intelligence and blockchain enterprises to bolster transparency within its cocoa supply chain (Mars, 2020; Rashid & Kausik, 2024; Sainani, 2024; Tsai, 2017). By leveraging artificial intelligence, Mars examines data concerning supplier practices, identifying risks linked to child labor and implementing corrective measures. This initiative enhances traceability and accountability, thereby ensuring the ethical sourcing of raw materials.

Benefits of AI in Enhancing Accountability

Enhanced Transparency

Artificial Intelligence technologies facilitate the acquisition of real-time insights pertaining to supply chain operations, thereby augmenting transparency for both corporations and their stakeholders (Kamble et al., 2019). This augmented visibility empowers organizations to swiftly identify and rectify unethical practices, consequently cultivating trust among consumers and investors.

Efficient Resource Allocation

Through the identification of high-risk suppliers and geographical areas, AI enables corporations to optimize the allocation of resources dedicated to audits and compliance initiatives (Chae, 2015). This strategic methodology significantly enhances the efficacy of monitoring programs while concurrently minimizing superfluous expenditures.

Proactive Risk Management

The predictive capabilities of AI empower corporations to foresee and mitigate risks prior to the occurrence of violations (Papadopoulos et al., 2017). The early detection of potential child labor practices not only prevents harm but also diminishes legal and reputational repercussions.

Improved Reporting and Compliance

The automation of data collection and analysis facilitates the streamlining of reporting processes, thereby ensuring adherence to legal mandates and corporate social responsibility obligations (G. Wang et al., 2016). Reports generated by AI deliver precise and timely information to regulatory bodies and stakeholders.

Challenges and Considerations

Data Quality and Accessibility

The efficacy of artificial intelligence systems is contingent upon the accessibility and integrity of data sourced from suppliers. Irregular or erroneous data may obstruct the capacity of AI to identify unethical practices (Papadopoulos et al., 2017). It is imperative for corporations to collaborate with suppliers to enhance protocols for data collection and dissemination.

Privacy and Ethical Concerns

The deployment of AI and surveillance technologies engenders apprehensions regarding privacy and the potential for data misuse (Martin & Freeman, 2004). Corporations are obligated to reconcile the necessity for oversight with respect for individual rights, thereby ensuring adherence to data protection legislation.

Algorithmic Bias and Fairness

Artificial intelligence algorithms may embody biases inherent in the training datasets, potentially resulting in the inequitable targeting of specific suppliers or geographical regions (Mehrabi et al., 2021). It is essential to construct AI systems that exhibit transparency and undergo regular audits to mitigate bias.

Technical and Financial Barriers

The integration of AI technologies necessitates substantial financial investment and specialized technical knowledge (Bag et al., 2020). Small and medium-sized enterprises may encounter significant obstacles in the adoption of these technologies without external assistance or cooperative efforts.

Synthesis

Artificial Intelligence (AI) presents robust mechanisms for augmenting corporate responsibility, particularly in the oversight and eradication of child labor within supply chains. Through the utilization of AI technologies, enterprises can attain profound insights, enhance transparency, and adopt proactive strategies to uphold ethical labor standards. The illustrative case studies of IBM, Unilever, Coca-Cola, and Mars exemplify the pragmatic applications of AI in advocating ethical practices and underscore the potential for extensive integration across diverse industries.

Nevertheless, the deployment of AI necessitates a meticulous evaluation of challenges associated with data integrity, privacy concerns, and algorithmic biases. Corporations are obliged to engage in ethical AI development, ensuring that such technologies are implemented responsibly and effectively to safeguard the rights of children. Collaboration with various stakeholders, including suppliers, regulatory bodies, and civil society organizations, is imperative to optimize the advantages of AI and cultivate a culture of accountability.

LEVERAGING AI FOR POSITIVE SOCIETAL IMPACT

AI-Driven Initiatives to Improve Children's Access to Education and Healthcare

Artificial Intelligence (AI) possesses considerable potential to mitigate social disparities, particularly in augmenting access to educational and healthcare services for children, especially within marginalized communities. Through the strategic application of AI technologies, organizations can play a pivotal role in ameliorating discrepancies in resource allocation, service quality, and individualized support, thus cultivating a beneficial societal outcome.

AI in Education

Personalized Learning Experiences

AI-empowered educational platforms facilitate personalized learning by adjusting content to align with the distinct needs, learning modalities, and paces of individual students (Raja et al., 2024; Zawacki-Richter et al., 2019). Machine learning algorithms scrutinize students' interactions, performance metrics, and feedback to customize educational resources, thereby establishing a bespoke learning trajectory for each learner.

For instance, intelligent tutoring systems (ITS) employ AI to replicate personalized tutoring experiences, providing explanations, hints, and assessments that dynamically adjust in response to the learner's advancement (Vanlehn, 2011). Such systems have demonstrated efficacy in enhancing educational outcomes by specifically addressing knowledge deficiencies and reinforcing concepts necessitating further focus.

Accessibility in Remote and Underserved Areas

AI technologies enhance the availability of quality education in geographically isolated regions where resources are limited. Educational chatbots and AI-driven mobile applications can disseminate curriculum content independent of physical infrastructure or continuous internet access (Altmiller & Pepe, 2022; Chatterjee & Chakraborty, 2021; Elkins, 2024; Stoilova et al., 2021). These instruments empower children in underprivileged communities to obtain education that would otherwise remain beyond their reach.

Language Translation and Inclusivity

Natural Language Processing (NLP) facilitates the instantaneous translation of educational materials into a plethora of languages, thereby addressing linguistic heterogeneity and fostering inclusivity (Floridi et al., 2018; Fritsch, 2024; Kadam et al., 2022; *Multilingual AI Chatbot Frameworks | Restackio*, n.d.; T. Wang et al., 2023). Artificial Intelligence has the potential to mitigate linguistic barriers that frequently obstruct educational prospects for children residing in multilingual communities.

AI in Healthcare

Predictive Analytics for Health Risk Assessment

The capacity of artificial intelligence to process and analyze extensive datasets significantly contributes to the domain of predictive analytics within healthcare, facilitating the identification of health risks and enabling timely interventions (Esteva et al., 2019). Specifically for pediatric populations, this capability implies that potential health concerns may be detected at an earlier stage, thereby allowing for the prompt implementation of preventative measures.

For example, artificial intelligence algorithms possess the capability to analyze medical records, genetic information, and environmental variables to forecast the probability of illnesses such as asthma, diabetes, or malnutrition within particular demographic groups (Obermeyer & Emanuel, 2016). This predictive capability empowers healthcare practitioners to allocate resources efficiently and prioritize healthcare delivery for children who are deemed to be at higher risk.

Telemedicine and Remote Diagnostics

Artificial intelligence enhances the efficacy of telemedicine by augmenting remote diagnostics and patient monitoring, which is particularly critical for pediatric patients situated in regions with inadequate healthcare infrastructure (Topol, 2019). AI-driven diagnostic tools are capable of interpreting medical imagery, analyzing clinical symptoms, and offering preliminary evaluations, thereby facilitating timely medical consultations.

Supporting Mental Health

The applications of artificial intelligence are progressively being utilized to bolster mental health services directed at children and adolescents. Chatbots and virtual assistants serve to provide initial counseling, monitor emotional health, and offer coping mechanisms (Gaggioli et al., 2019). Although these technological interventions do not serve as substitutes for professional mental health care, they can enhance support services, particularly in contexts where access to such care is limited.

Corporate Partnerships with NGOs and Governments to Deploy AI in Underserved Communities

Collaborative Initiatives

Corporate entities can enhance the societal ramifications of artificial intelligence by forming strategic alliances with non-governmental organizations (NGOs) and governmental bodies. Such synergistic collaborations amalgamate technological proficiency with empirical insights from the field, thereby ensuring that AI interventions are culturally congruent and adeptly address the specific requirements of communities (Sloane et al., 2022).

Case Study: Microsoft's AI for Health Initiative

The AI for Health initiative by Microsoft engages in partnerships with international health organizations to expedite medical research and optimize healthcare delivery mechanisms (Microsoft, 2024). By furnishing AI tools and resources, Microsoft endorses initiatives aimed at mitigating maternal and infant mortality, a significant issue that critically impacts the wellbeing of children.

Case Study: IBM's Partnership with UNICEF

IBM has established a collaborative relationship with UNICEF through the "Magic Box" initiative, which employs artificial intelligence and big data analytics to tackle humanitarian issues confronting children (UNICEF, 2020). This platform synthesizes data from diverse sources to inform strategic responses to epidemics, natural disasters, and educational requirements.

Ethical Considerations in Partnerships

In the deployment of AI solutions, it is imperative for corporations and their partners to adhere to rigorous ethical standards, particularly in relation to data privacy and informed consent. Engaging with local communities to gain insights into their apprehensions and expectations is vital for the successful execution of these initiatives (Floridi et al., 2018).

Capacity Building and Sustainability

Collaborative efforts should prioritize the enhancement of local capabilities to sustain and expand AI initiatives. This encompasses the training of community members, educators, and healthcare professionals in the proficient use of AI tools, thereby ensuring long-term viability beyond the initial implementation phase (Haenssgen, 2018; Marginson, 2016; Musa, 2024).

Benefits of Leveraging AI for Societal Impact

Reducing Inequality

The enhancement of accessibility to educational and healthcare services through artificial intelligence can significantly mitigate socioeconomic inequalities that influence the developmental trajectories and future prospects of children (UN, 2020). Technological advancements function as a great equalizer, furnishing resources that might have been previously inaccessible owing to geographic or economic impediments.

Empowering Communities

AI-driven initiatives that incorporate community engagement empower individuals by integrating them into the decision-making framework. This cultivates a sense of ownership and amplifies the probability of success and sustainability of the initiatives (Heeks, 2020).

Innovation and Economic Growth

Allocating resources towards AI-enhanced social programs acts as a catalyst for innovation and has the potential to foster economic advancement. As children attain improved educational and healthcare outcomes, they are more inclined to make positive contributions to society in the future (World Bank, 2018).

CHALLENGES AND MITIGATION STRATEGIES

Digital Divide and Infrastructure Limitations

A notable challenge confronting society is the digital divide, which delineates the disparity between individuals possessing access to technological resources and those lacking such access (Dijk, 2020; Livingstone et al., 2022; Stoilova et al., 2021). Limitations in infrastructure, exemplified by inconsistent internet connectivity and insufficient availability of devices, impede the deployment of artificial intelligence solutions in certain communities.

Mitigation Strategy: Corporations are encouraged to allocate resources towards infrastructure development by providing affordable devices or endorsing initiatives aimed at enhancing internet connectivity. Alternative approaches, such as the development of offline AI applications, warrant consideration as well.

Cultural Sensitivity and Appropriateness

Artificial intelligence solutions must be designed with cultural sensitivity and relevance to the communities they are intended to serve. Neglecting to account for local customs, languages, and educational methodologies may yield ineffective or potentially detrimental interventions (M. E. Porter & Kramer, 2019).

Mitigation Strategy: It is imperative to engage local stakeholders throughout the design and implementation stages. By integrating local knowledge and preferences, AI tools can be customized to align with the specific needs of the community.

Data Privacy and Security

The process of collecting and analyzing data from children engenders significant concerns regarding privacy and security. Adhering to legal frameworks and ethical standards is of paramount importance (ICO, 2018, 2023).

Mitigation Strategy: The establishment of comprehensive data protection protocols, securing informed consent, and complying with regulations such as the General Data Protection Regulation (GDPR) and other pertinent laws is essential.

Key Takeaways

Harnessing artificial intelligence for the advancement of societal welfare presents a transformative opportunity for corporations to enhance the well-being of children on a global scale. By augmenting access to educational resources and healthcare services, AI possesses the capability to address entrenched systemic inequalities and empower marginalized communities. Collaborative initiatives with non-governmental organizations and governmental bodies serve to magnify these impacts, amalgamating resources and expertise to develop sustainable solutions.

Nonetheless, it is imperative to navigate the complexities associated with the implementation of artificial intelligence with careful consideration. Tackling challenges such as the digital divide, cultural sensitivity, and data privacy is crucial to ensuring that AI initiatives are ethical, effective, and inclusive. Through a commitment to responsible AI practices, corporations have the potential to play a decisive role in fostering a more equitable and prosperous future for all children.

NAVIGATING ETHICAL AND LEGAL CHALLENGES

Addressing Biases and Ethical Dilemmas in AI Systems Affecting Children

The incorporation of Artificial Intelligence (AI) within corporate frameworks has precipitated profound ethical quandaries, particularly in relation to the rights of children. AI systems, despite providing considerable advantages, are vulnerable to biases that may detrimentally affect children, particularly those hailing from marginalized demographics (Mehrabi et al., 2021). It is imperative for corporations to take proactive measures to confront these biases and ethical dilemmas to ensure that their AI technologies advance fairness and uphold the rights of children.

Algorithmic Bias and Its Impact on Children

Algorithmic bias emerges when AI systems yield biased outcomes due to flawed assumptions inherent in the machine learning methodology (Friedman & Nissenbaum, 1996). Such biases frequently arise from imbalanced training data that fail to encapsulate the diversity of the user population. For children, the ramifications of algorithmic bias can be profoundly significant:

- **Educational Inequities:** AI-enhanced educational instruments may preferentially support specific linguistic or cultural demographics, thereby disadvantaging children who do not conform to the predominant profiles (Holmes et al., 2019).
- **Discriminatory Practices:** Predictive algorithms employed within social services may perpetuate stereotypes, resulting in disparate treatment of children based on race, gender, or socioeconomic status (O'Neil, 2016).

- **Restricted Opportunities:** Biased AI systems utilized in recruitment or admissions processes may constrain opportunities for certain cohorts of children as they transition into higher education or the workforce (Ajunwa et al., 2016).

Ethical Dilemmas in AI Deployment

In addition to biases, the deployment of AI raises ethical considerations:

- **Transparency and Explainability:** Numerous AI systems function as "black boxes," creating challenges in comprehending the decision-making processes involved (Doshi-Velez & Kim, 2017). The absence of transparency obstructs accountability, particularly when AI decisions have adverse effects on children.
- **Consent and Autonomy:** Children may lack a comprehensive understanding of the ramifications associated with data collection and AI processing, prompting inquiries regarding informed consent (Milkaite & Lievens, 2019).
- **Surveillance and Privacy:** AI technologies implemented for monitoring purposes can encroach upon children's privacy rights, especially if executed without appropriate safeguards (Mantelero, 2018).

Strategies for Mitigating Biases and Ethical Risks

Diverse and Inclusive Data Sets

To mitigate algorithmic bias, corporations should ensure that AI systems are developed utilizing diverse and representative data sets (Gebru et al., 2021). This encompasses data that accurately reflects various genders, ethnicities, socioeconomic statuses, and abilities.

Algorithmic Fairness and Auditing

The application of fairness-conscious machine learning methodologies can assist in alleviating biases (Zemel et al., 2013). Routine algorithmic evaluations can detect and rectify discriminatory trends within AI systems (Raji et al., 2020).

Ethical Design Principles

The integration of ethical tenets such as fairness, accountability, and transparency into the design of AI systems is of paramount importance (Floridi et al., 2018; Floridi & Cowls, 2019). Approaches such as explainable AI (XAI) enhance comprehension of AI decision-making mechanisms (Samek et al., 2019).

Stakeholder Engagement

Engagement with stakeholders, including experts in child rights, educators, parents, and children themselves, can yield invaluable insights into potential ethical challenges (Livingstone et al., 2022). Inclusive design methodologies ensure that AI technologies address the requirements of all users.

Legal Frameworks and Corporate Policies for Protecting Children's Data Privacy

The safeguarding of children's data privacy constitutes both a legal and ethical obligation for corporations that implement artificial intelligence technologies. Numerous international and domestic regulations establish benchmarks for data protection, incorporating distinct provisions specifically addressing minors.

International Legal Frameworks

General Data Protection Regulation (GDPR)

The GDPR (Regulation - 2016/679 - EN - GDPR - EUR-Lex, 2016) enacted by the European Union imposes rigorous stipulations on the processing of personal data, particularly enhancing protections for the data of children (Voigt & Von Dem Bussche, 2017). Noteworthy provisions encompass:

- **Parental Consent:** Mandatory for the processing of personal data pertaining to children younger than 16 years (Article 8).
- **Right to Erasure:** Children possess the entitlement to request the deletion of their personal data (Article 17).
- **Clear Communication:** Information disseminated to children must be articulated in straightforward and comprehensible language suitable for their age group (Article 58).

United Nations Convention on the Rights of the Child (UNCRC)

The United Nations Convention on the Rights of the Child (UNCRC) underscores the significance of the right to privacy and the necessity for protection against unlawful interference (Convention on the Rights of the Child, 1990). It advocates for the safeguarding of children's personal information across all contexts, inclusive of digital environments.

National Regulations

Children's Online Privacy Protection Act (COPPA)

In the United States, the Children's Online Privacy Protection Act (COPPA) governs the collection of personal information from children under the age of 13 by online services (FTC, 2013, 2020). It stipulates:

- **Verifiable Parental Consent:** Required prior to the collection of data.
- **Privacy Policies:** Obligation for clear disclosure regarding information practices.
- **Data Security:** Implementation of reasonable measures to ensure data confidentiality. China's Personal Information Protection Law (PIPL)

China's Personal Information Protection Law (PIPL), which came into effect in November 2021, includes explicit provisions for the management of children's personal information (China Briefing, 2021). It necessitates:

- **Parental Consent:** Required for the processing of data concerning minors under the age of 14.
- **Special Protections:** The enforcement of enhanced protocols for data security and regulatory compliance.

Corporate Policy Development

Corporations are mandated to formulate comprehensive policies to ensure compliance with legal obligations and to safeguard children's data privacy:

- **Data Minimization:** Collect solely the essential data and restrict processing activities.

- **Privacy by Design:** Integrate data protection measures from the initial stages of AI system development (European Union Agency for Cybersecurity (ENISA), 2015).
- **Transparency and Consent:** Furnish clear information regarding data practices and secure informed consent where applicable (Wikipedia, 2024).
- **Third-Party Compliance:** Guarantee that partners and service providers conform to equivalent privacy standards.

Challenges in Implementation

Cross-Border Data Transfers

Global corporations encounter complexities when transferring data across jurisdictions that impose varying legal standards (G. Greenleaf, 2019; R. K. Greenleaf, 2002). Mechanisms such as Standard Contractual Clauses (SCCs) and Binding Corporate Rules (BCRs) are vital yet necessitate meticulous implementation.

Evolving Legal Landscape

The rapid pace of technological development frequently surpasses legislative processes, resulting in regulatory voids and ambiguities (Pagallo et al., 2017, 2017). Corporations must remain vigilant regarding emerging laws and adapt their policies in response.

Balancing Innovation and Compliance

Achieving equilibrium between the utilization of AI capabilities and adherence to stringent data protection regulations presents significant challenges (Taddeo & Floridi, 2018). Excessively restrictive policies may stifle innovation, whereas lenient measures could lead to legal repercussions and damage to reputation.

Best Practices for Corporations

Regular Privacy Impact Assessments

Implement Privacy Impact Assessments (PIAs) to systematically identify and address potential risks associated with data processing activities that involve minors (ICO, 2018, 2023).

Employee Training and Awareness

Facilitate comprehensive education for employees regarding data protection legislation, organizational policies, and ethical considerations pertinent to the deployment of artificial intelligence (Burns, 1978). Foster a culture that prioritizes compliance and accountability.

Appointment of Data Protection Officers (DPOs)

In entities that handle significant volumes of personal data, the designation of a Data Protection Officer ensures specialized oversight and proficiency in matters of data protection (GDPR Article 37).

Engagement with Regulatory Bodies

Sustain proactive communication with data protection authorities to remain informed about compliance obligations and exemplary practices (Gellert, 2015).

Ending

Effectively addressing the ethical and legal complexities associated with AI deployment necessitates a proactive and holistic strategy on the part of corporations. Tackling algorithmic biases and ethical conundrums is imperative to safeguard the welfare of children and uphold their rights. Through the implementation of rigorous legal compliance frameworks and the cultivation of an ethical organizational culture, corporations can responsibly harness AI technologies.

The dedication to safeguarding children's rights within the realm of AI governance not only mitigates potential risks but also enhances the organization's reputation and fosters trust among various stakeholders. As artificial intelligence continues to progress, sustained vigilance, adaptation, and collaboration with legal specialists and advocates for children's rights are essential to ensure that technological innovations serve the interests of all societal members, particularly the most vulnerable populations.

AI IN MARKETING AND CHILDREN'S DATA PRIVACY

Risks Associated with AI in Marketing and Data Collection

The emergence of Artificial Intelligence (AI) has fundamentally transformed marketing methodologies, empowering organizations to scrutinize consumer behavior, tailor content, and enhance campaigns with an unparalleled degree of accuracy (Jarek & Mazurek, 2019). Although these innovations present considerable advantages for enterprises, they introduce distinct risks concerning children, who represent an especially vulnerable population within the digital economy.

Targeted Advertising and Children

AI-driven algorithms possess the capability to analyze extensive datasets in order to construct intricate user profiles, inclusive of children, predicated on their online behaviors (Milkaite & Lievens, 2019, 2020). This functionality facilitates exceptionally personalized advertising, which, in relation to children, prompts several critical issues:

- **Exposure to Inappropriate Content:** Children may encounter advertisements that are misaligned with their developmental stage, encompassing those that endorse unhealthy dietary choices, violent gaming experiences, or products intended for adult audiences (Kelly et al., 2015).
- **Manipulative Marketing Techniques:** AI can capitalize on the cognitive susceptibilities of children by disseminating persuasive content that they may lack the ability to scrutinize critically (Nairn & Fine, 2008).
- **Data Collection Without Informed Consent:** Children may unintentionally divulge personal information without fully comprehending the ramifications, resulting in unauthorized data gathering and profiling (Livingstone et al., 2022).

Behavioral Tracking and Profiling

AI systems frequently depend on behavioral tracking to enhance marketing initiatives. This encompasses the aggregation of data regarding browsing history, application usage, geographic location, and even biometric details (Montgomery et al., 2017). For children, such methodologies can infringe upon their privacy rights and lead to:

- **Long-Term Data Accumulation:** Information amassed during childhood has the potential to create a lasting digital identity, thereby influencing future opportunities and privacy (Lievens & Verdoodt, 2018).
- **Security Risks:** The retention of children's data heightens the likelihood of security breaches, potentially exposing sensitive information to nefarious entities (O'Brolcháin et al., 2016).

Lack of Transparency

AI algorithms function with a level of intricacy that is frequently inscrutable to users, complicating the ability of parents and guardians to comprehend how children's data is utilized in marketing endeavors (Pasquale, 2015). This deficiency in transparency obstructs informed decision-making and the provision of consent.

Strategies for Ethical Marketing Practices and Safeguarding Children's Digital Privacy

To alleviate the perils associated with artificial intelligence in marketing directed towards children, corporations must implement ethical practices that prioritize the rights and welfare of minors.

Implementing Age-Appropriate Design

The Age-Appropriate Design Code, established by the United Kingdom's Information Commissioner's Office (ICO), delineates guidelines for online services that are likely to be utilized by children (ICO, 2018, 2023). Fundamental principles encompass:

- **Data Minimization:** Only the data that is essential for the service's functionality should be collected, thereby diminishing the potential for misuse.
- **High Privacy by Default:** Initial settings ought to guarantee optimal privacy, necessitating active opt-in for any sharing of data.
- **Transparency:** Offer clear and accessible information regarding data practices that are tailored to the comprehension levels of children.

Parental Controls and Consent Mechanisms

The integration of comprehensive parental control features empowers guardians to regulate the content and data accessible to their children (Jordan, 2025). Efficient consent mechanisms should:

- **Verify Age Appropriately:** Employ non-intrusive methods to ascertain a user's age without the accumulation of excessive data.
- **Facilitate Parental Oversight:** Allow parents to consent to data collection and oversee their children's interactions with the service.

Ethical AI Design in Marketing

The development of artificial intelligence systems with ethical considerations as a priority guarantees that marketing practices do not exploit or endanger children (Dignum, 2019).

- **Avoiding Manipulative Techniques:** Artificial intelligence should not be employed to generate content that manipulates the emotions or decisions of children.
- **Bias Mitigation:** It is imperative to ensure that algorithms do not perpetuate harmful stereotypes or discriminate against any demographic of children (Weidinger et al., 2022).
- **Explainable AI:** Create artificial intelligence models that offer comprehensible explanations regarding decision-making processes, thereby enhancing transparency (Samek et al., 2019).

Compliance with Legal Frameworks

Compliance with statutes and regulations that safeguard the digital privacy of children is paramount.

- **Children's Online Privacy Protection Act (COPPA):** In the United States, COPPA mandates verifiable parental consent prior to the collection of personal information from children under the age of 13.
- **General Data Protection Regulation (GDPR):** The GDPR within the European Union enforces specific protections for children's data, including the right to be forgotten and age-appropriate consent (Albrecht, 2016; Hof, 2017).
- **California Consumer Privacy Act (CCPA):** The CCPA grants additional rights to residents of California, including children, concerning personal data collected by businesses (State of California, 2018).

Corporate Policies and Best Practices

Corporations ought to formulate internal policies that surpass legal mandates to cultivate trust and exhibit a commitment to ethical standards (Gillespie, 2020).

- **Regular Audits and Assessments:** Undertake periodic evaluations of marketing practices and artificial intelligence systems to ensure adherence and identify opportunities for enhancement.
- **Employee Training:** Provide education for staff regarding ethical considerations and legal responsibilities associated with children's data privacy.
- **Stakeholder Engagement:** Engage child development specialists, ethicists, and community representatives in the formulation of policies.

Case Studies of Ethical Marketing Initiatives

LEGO Group's Responsible Marketing

The LEGO Group has instituted rigorous protocols to ensure ethical marketing practices directed towards children, which include:

- **No Third-Party Data Sharing:** Prohibiting the dissemination of children's data to external entities.
- **Transparent Communication:** Delivering unequivocal information to both parents and children regarding the utilization of data (LEGO, 2024a, 2024b).

SuperAwesome's Kid-Safe Advertising Platform

SuperAwesome provides a technological framework specifically developed for secure digital interactions with children, thereby ensuring adherence to COPPA and GDPR-K regulations (EU-Startups, 2024; SuperAwesome, 2024).

- **Parental Consent Management:** Enabling processes for the verification of parental consent.
- **Contextual Advertising:** Delivering advertisements predicated on the contextual content rather than on personal data.

Emerging Technologies and Future Considerations

As artificial intelligence and marketing technologies progress, novel challenges and opportunities are expected to emerge.

Artificial Intelligence Act Proposal

The proposed Artificial Intelligence Act by the European Commission seeks to regulate artificial intelligence systems through a risk-based framework, incorporating provisions pertinent to the rights of children (Proposal for a Regulation of the European Parliament and of the Council Laying down Harmonised Rules on Artificial Intelligence (Artificial Intelligence Act) and Amending Certain Union Legislative Acts, 2021).

- **High-Risk Systems:** Artificial intelligence systems that engage with children may be categorized as high-risk, necessitating enhanced compliance protocols.

Privacy-Enhancing Technologies (PETs)

Recent advancements in privacy-enhancing technologies, such as differential privacy and federated learning, can safeguard children's data while facilitating beneficial artificial intelligence applications (Yang et al., 2019).

- **Anonymization Techniques:** Implementing measures to ensure that data cannot be traced back to specific children.
- **On-Device Processing:** Conducting data analysis directly on user devices to reduce data transmission and storage requirements.

Closing Remarks

Artificial intelligence has undeniably revolutionized marketing, providing formidable instruments for engagement and expansion. Nevertheless, in the context of children, corporations bear an elevated responsibility to ensure that these technologies do not violate their rights or well-being. By embracing ethical marketing methodologies, instituting comprehensive data privacy protections, and adhering to legal frameworks, businesses can responsibly navigate the intricate landscape of artificial intelligence in marketing.

Safeguarding children's digital privacy represents not only a legal duty but also a moral obligation that fosters trust with consumers and the broader society. As technological advancements persist, sustained vigilance, ethical dedication, and collaboration with regulators and stakeholders will be crucial in protecting the interests of children in the digital era.

COLLABORATIVE APPROACHES TO AI GOVERNANCE

The Role of Corporations in Multi-Stakeholder Collaborations

The swift evolution of Artificial Intelligence (AI) has engendered intricate ethical, legal, and societal dilemmas that cannot be resolved by any singular entity independently. It is imperative to adopt collaborative strategies for AI governance that encompass corporations, governmental entities, non-governmental organizations (NGOs), academic institutions, and civil society to guarantee that AI technologies are developed and implemented in a responsible manner, especially in relation to the rights of children (Coeckelbergh, 2020).

Multi-Stakeholder Collaborations

Corporations and Governments

Corporations assume a pivotal role in influencing AI policy formulation via partnerships with governmental agencies. Such cooperative endeavors can facilitate the establishment of regulatory frameworks that harmonize innovation with the safeguarding of children's rights. An illustrative example is the Partnership on AI, which encompasses leading technology firms such as Apple, Amazon, Google, and Microsoft, collaborating with governmental entities to confront ethical challenges associated with AI.

Engagement with NGOs and Civil Society

Non-governmental organizations dedicated to children's rights offer invaluable insights for AI governance. Collaborations between corporations and entities like UNICEF foster the creation of AI systems that address the distinctive needs and vulnerabilities of children (UNICEF, 2011, 2014a, 2014b, 2021). For instance, UNICEF's "AI for Children" initiative provides a framework for policymakers and businesses to devise child-centric AI strategies.

Academic Partnerships

Academic institutions provide essential research expertise and ethical scrutiny vital for the responsible advancement of AI technologies. Corporations frequently engage in partnerships with universities to investigate the societal implications of AI and to devise instruments that alleviate risks to children. The collaboration between IBM and the Massachusetts Institute of Technology (MIT) on the MIT-IBM Watson AI Lab serves as a paradigm of such synergistic endeavors (MIT, 2017).

Global Initiatives

The Global Partnership on Artificial Intelligence (GPAI)

The Global Partnership on Artificial Intelligence (GPAI) represents a transnational initiative that unites specialists from various sectors, including industry, government, academia, and civil society, with the aim of advancing the responsible development of artificial intelligence. Corporations engaged in GPAI actively contribute to specialized working groups that concentrate on data governance, ethical AI practices, and the evolving nature of work, all of which bear significant ramifications for the rights of children (GPAI, 2024).

IEEE Global Initiative on Ethics of Autonomous and Intelligent Systems

The Institute of Electrical and Electronics Engineers (IEEE) has instituted a framework of standards and guidelines aimed at fostering ethical practices in the development of artificial intelligence. Corporations involved in this initiative play a pivotal role in formulating policies that prioritize transparency, accountability, and the welfare of users, encompassing the interests of children (IEEE, 2019; Shahriari & Shahriari, 2017; Stahl & Wright, 2018).

Best Practices for Integrating Children's Rights into Corporate Social Responsibility (CSR) Programs Through AI

The incorporation of children's rights into Corporate Social Responsibility (CSR) initiatives necessitate intentional methodologies that prioritize the welfare and safeguarding of minors within the scope of artificial intelligence (AI) technologies.

Developing Child-Centered AI Policies

Assessment of AI Impact on Children

It is imperative for corporations to execute comprehensive impact assessments to ascertain the effects of their AI technologies on children. This process entails a thorough evaluation of potential hazards and advantages, taking into account variables such as data privacy, exposure to detrimental content, and the presence of algorithmic biases (OECD, 2019, 2021; Perset et al., 2024).

Adoption of Ethical Frameworks

The establishment of ethical AI frameworks that explicitly recognize children's rights guarantees that the development of AI is congruent with overarching social responsibilities. The UNICEF Policy Guidance on AI for Children delineates principles that corporations may adopt, including the promotion of equity, the assurance of explainability, and the prioritization of the best interests of children (UNICEF, 2011, 2011, 2014a, 2021, 2022).

Collaborating with Child Rights Organizations

Partnerships for Program Development

Collaboration with child rights organizations enables corporations to devise AI applications that effectively address the distinct challenges encountered by children. For example, such partnerships can facilitate the development of educational AI tools customized to meet varied learning requirements or health monitoring systems that protect children's well-being (Save the Children Deutschland e.V., 2023)

Stakeholder Engagement and Consultation

The involvement of children, families, educators, and community members yields valuable insights into the practical ramifications of AI technologies. Engaging in inclusive dialogue equips corporations with a deeper understanding of cultural contexts and ethical considerations, resulting in more efficacious and widely accepted AI solutions (Livingstone et al., 2022).

Case Studies

Microsoft's AI for Accessibility

The AI for Accessibility initiative spearheaded by Microsoft allocates resources towards projects that harness artificial intelligence to empower individuals with disabilities, including minors. Through strategic partnerships with non-governmental organizations and educational institutions, Microsoft endorses programs that cultivate inclusive educational tools, thereby augmenting learning opportunities for children with special needs (Microsoft, 2024).

Google's "Be Internet Awesome Program"

In collaboration with organizations dedicated to online safety, Google has developed the "Be Internet Awesome" initiative, which serves to educate children on the principles of digital citizenship and safety. This program leverages artificial intelligence to deliver interactive experiences that instruct children on matters of privacy, security, and responsible conduct in the digital realm (Google, 2024).

Summary

The pursuit of collaborative frameworks for the governance of artificial intelligence is imperative in confronting the intricate challenges that arise from AI technologies, particularly in relation to the rights of children. Corporations play a crucial role in fostering multi-stakeholder collaborations, offering resources, expertise, and influence that are essential for the formulation of ethical AI practices. By weaving children's rights into corporate social responsibility initiatives through alliances, ethical policymaking, and stakeholder engagement, corporations can guarantee that AI technologies yield beneficial outcomes for the welfare of children globally.

The adoption of collaborative governance not only alleviates potential risks but also promotes innovation that is aligned with social responsibility and sustainability. As artificial intelligence continues to progress, sustained cooperation among all stakeholders will be critical in addressing ethical dilemmas and in harnessing AI's capabilities for the advancement of future generations.

RECOMMENDATIONS AND FUTURE DIRECTIONS

Frameworks for Ethical AI Utilization in Business with Emphasis on Children's Rights

The incorporation of artificial intelligence (AI) into business practices necessitates a thorough framework that prioritizes the rights of children. Corporations are compelled to implement ethical guidelines, transparent governance frameworks, and proactive strategies to guarantee that AI technologies yield positive contributions to the welfare of minors.

Strategic Recommendations for Corporations

1. **Adopt Ethical AI Principles:** Enact universal ethical guidelines such as the OECD Principles on AI, which underscore human-centered values, equity, and accountability (OECD, 2019, 2021, 2024).
2. **Establish Robust Governance Mechanisms:** Formulate internal policies and oversight committees that are dedicated to monitoring AI applications that impact children, ensuring adherence to ethical standards and legal frameworks (Floridi et al., 2018; Floridi & Cowls, 2019).
3. **Engage in Stakeholder Collaboration:** Foster collaboration with governmental bodies, non-governmental organizations, child rights advocacy groups, and academic institutions to acquire diverse perspectives and jointly develop solutions that safeguard the interests of children (Sloane et al., 2022).
4. **Invest in Ethical AI Training:** Facilitate ongoing educational programs for employees regarding ethical AI practices, children's rights, and data privacy to cultivate a culture of accountability (Jobin et al., 2019).
5. **Implement Transparency and Accountability Measures:** Employ explainable AI methodologies to enhance the transparency of AI systems, thereby enabling audits and accountability in decisions that affect children (Samek et al., 2019).

Recommendations for Policymakers and Civil Society

- **Develop Comprehensive Regulations:** It is imperative for governments to devise legislative frameworks that specifically address the implications of AI on children, including rigorous data protection protocols and ethical AI usage guidelines (Gasser & Almeida, 2017).
- **Promote International Standards:** Global organizations can facilitate the establishment of international standards for ethical AI, ensuring the consistent safeguarding of children's rights across various jurisdictions (UNESCO, 2021).
- **Support Research and Innovation:** Advocate for and finance research pertaining to the effects of AI on children to inform evidence-based policymaking and innovative solutions (A. N. Institute, 2019).

Future Directions

The forthcoming terrain of AI in corporate environments presents both challenges and opportunities for the advancement of children's rights:

- **Embracing Privacy-Enhancing Technologies:** The adoption of technologies such as differential privacy and federated learning can safeguard children's data while enabling enterprises to extract valuable insights (Yang et al., 2019)
- **Focus on Inclusive AI Development:** Prioritize diversity within AI development teams to mitigate biases and create more equitable AI systems (A. I. N. Institute, 2019).
- **Strengthening Global Collaboration:** Enhanced international cooperation can foster unified strategies in regulating AI and protecting children's rights on a global scale (ITU, 2024).
- **Empowering Children and Youth:** Integrate AI literacy initiatives into educational curricula to empower the younger generation to proficiently navigate and influence the AI technologies that impact their lives (Long & Magerko, 2020).

CONCLUSION

The Significance of a Proactive, Child-Centered Framework for AI in Corporate Strategy

The convergence of artificial intelligence (AI) and the rights of children presents both a significant challenge and a considerable opportunity for global corporations. As AI technology becomes progressively embedded within business practices, its impact on the lives of children—through sectors such as education, healthcare, marketing, and digital interaction—expands dramatically. This situation mandates a proactive, child-centered framework within corporate strategies to ensure that AI technologies are conceived and implemented in manners that prioritize the welfare and rights of children.

A child-centered framework necessitates the acknowledgment of children as pivotal stakeholders, taking into account their distinct needs and vulnerabilities in all decisions pertaining to AI. It compels corporations to transcend mere adherence to legal mandates, infusing ethical considerations into the core of their AI initiatives. By adopting such an approach, businesses can alleviate risks including breaches of data privacy, algorithmic biases, and exposure to inappropriate material, which disproportionately impact children.

Furthermore, the integration of children's rights into AI strategies aligns with broader corporate social responsibility (CSR) objectives while enhancing a corporation's reputation and credibility. It signifies a dedication to ethical practices and social welfare, which can distinguish a corporation within a competitive marketplace increasingly attuned to ethical concerns.

A Call to Action for the Inclusion of Children's Rights in the Advancing AI Landscape

As custodians of technological advancement, corporations possess a crucial responsibility in shaping an AI-driven future that protects and advocates for children's rights. The following actions represent fundamental steps toward achieving this objective:

- **Incorporate Ethical Principles in AI Development:** Embed frameworks that prioritize fairness, transparency, and accountability. Ensure that AI systems are constructed with safeguards that protect children's data privacy and mitigate biases.
- **Foster Multi-Stakeholder Collaboration:** Collaborate with governments, NGOs, educators, parents, and children to inform AI policies and practices. Collective efforts enrich perspectives and cultivate solutions that are socially responsible and efficacious.
- **Invest in Education and Awareness:** Champion initiatives that enhance digital literacy among children and relevant stakeholders. Educate employees regarding the ethical ramifications of AI and the significance of children's rights.
- **Establish Comprehensive Governance Structures:** Create oversight committees and implement regular auditing processes to monitor AI applications that impact children. Remain informed about evolving legal regulations and ethical standards to ensure continuous compliance and enhancement.
- **Encourage Innovation for Social Welfare:** Utilize AI's potential to tackle societal challenges encountered by children, such as access to quality education and healthcare. Invest in AI initiatives that contribute positively to the development and welfare of children.

The transforming AI landscape offers an unprecedented opportunity for corporations to exemplify responsible leadership, illustrating how technology can be applied judiciously to benefit society's most vulnerable populations. By integrating children's rights into AI strategies, businesses not only safeguard children but also contribute to the establishment of a sustainable and equitable future.

In summation, the ethical incorporation of artificial intelligence within corporate frameworks transcends a mere strategic benefit and embodies a fundamental moral obligation. The decisions made in the present will profoundly influence the futures of children and the nature of the world they will come to inhabit. It is imperative that corporations capitalize on this critical juncture to ensure that artificial intelligence functions as a catalyst for constructive transformation, safeguarding the dignity and rights of children universally.

REFERENCES

Ajunwa, I., Friedler, S. A., Scheidegger, C., & Venkatasubramanian, S. (2016, March 10). *Hiring by Algorithm: Predicting and Preventing Disparate Impact*. https://www.semanticscholar.org/paper/Hiring-by-Algorithm%3A-Predicting-and-Preventing-Ajunwa-Friedler/bd31ad5e998629998f35db9a10d858b36e603248

Albrecht, J. P. (2016). How the GDPR Will Change the World. *European Data Protection Law Review*, *2*(3), 287–289. DOI: 10.21552/EDPL/2016/3/4

Altmiller, G., & Pepe, L. H. (2022). Influence of Technology in Supporting Quality and Safety in Nursing Education. *The Nursing Clinics of North America*, *57*(4), 551–562. DOI: 10.1016/j.cnur.2022.06.005 PMID: 36280294

Arya, M. (2024, February 22). Case Study: Unilever's Sustainable Living Plan. *Medium*. https://medium.com/@thisismayank/case-study-unilevers-sustainable-living-plan-2226b2e5d5f5

Bag, S., Wood, L. C., Xu, L., Dhamija, P., & Kayikci, Y. (2020). Big data analytics as an operational excellence approach to enhance sustainable supply chain performance. *Resources, Conservation and Recycling*, *153*, 104559. DOI: 10.1016/j.resconrec.2019.104559

Bai, C., & Sarkis, J. (2020). A supply chain transparency and sustainability technology appraisal model for blockchain technology. *International Journal of Production Research*, *58*(7), 2142–2162. DOI: 10.1080/00207543.2019.1708989

Brauner, P., Hick, A., Philipsen, R., & Ziefle, M. (2023). What does the public think about artificial intelligence?—A criticality map to understand bias in the public perception of AI. *Frontiers of Computer Science*, *5*, 1113903. Advance online publication. DOI: 10.3389/fcomp.2023.1113903

Bryson, J., & Winfield, A. (2017). Standardizing Ethical Design for Artificial Intelligence and Autonomous Systems. *Computer, 50*(5), 116–119. *Computer*. Advance online publication. DOI: 10.1109/MC.2017.154

Burns, J. M. (1978). *Leadership*. HarperCollins.

Chae, B. (2015). Insights from hashtag #supplychain and Twitter Analytics: Considering Twitter and Twitter data for supply chain practice and research. *International Journal of Production Economics*, *165*, 247–259. DOI: 10.1016/j.ijpe.2014.12.037

Chapman, S. (2020, May 17). *IBM launches its blockchain produce tracker, IBM Food Trust.* https://supplychaindigital.com/technology/ibm-launches-its-blockchain-produce-tracker-ibm-food-trust

Chatterjee, I., & Chakraborty, P. (2021). Use of Information Communication Technology by Medical Educators Amid COVID-19 Pandemic and Beyond. *Journal of Educational Technology Systems, 49*(3), 310–324. DOI: 10.1177/0047239520966996

China Briefing. (2021, August 24). *The PRC Personal Information Protection Law (Final): A Full Translation.* China Briefing News. https://www.china-briefing.com/news/the-prc-personal-information-protection-law-final-a-full-translation/

Chowdhury, P., Lau, C., & Pittayachawan, S. (2016, July 3). *Supply risk mitigation of small and medium enterprises: A social capital approach.* DOI: 10.13140/RG.2.2.23117.03045

Coeckelbergh, M. (2020). *AI Ethics.* The MIT Press. DOI: 10.7551/mitpress/12549.001.0001

Convention on the Rights of the Child. (1990). https://www.ohchr.org/en/instruments-mechanisms/instruments/convention-rights-child

Davenport, T. H. (2018, January 30). Artificial Intelligence for the Real World. *Harvard Business Review.* https://hbr.org/webinar/2018/02/artificial-intelligence-for-the-real-world

Davenport, T. H., & Harris, J. G. (2007). *Competing on Analytics: The New Science of Winning.* Harvard Business Review Press.

Dignum, V. (2019). *Responsible Artificial Intelligence: How to Develop and Use AI in a Responsible Way.* DOI: 10.1007/978-3-030-30371-6

Doshi-Velez, F., & Kim, B. (2017). *Towards A Rigorous Science of Interpretable Machine Learning* (arXiv:1702.08608). arXiv. https://doi.org//arXiv.1702.08608DOI: 10.48550

Elkins, D. (2024, June 12). *Impact of Technology on Nursing Education: Revolutionizing Learning.* Davis & Elkins College. https://www.dewv.edu/impact-of-technology-on-nursing-education-revolutionizing-learning/

Esteva, A., Robicquet, A., Ramsundar, B., Kuleshov, V., DePristo, M., Chou, K., Cui, C., Corrado, G., Thrun, S., & Dean, J. (2019). A guide to deep learning in healthcare. *Nature Medicine, 25*(1), 24–29. DOI: 10.1038/s41591-018-0316-z PMID: 30617335

EU-Startups. (2024, October 1). *SuperAwesome.* EU-Startups. https://www.eu-startups.com/directory/superawesome/

Eubanks, V. (2018). *Automating Inequality: How High-tech Tools Profile, Police, and Punish the Poor* (Illustrated Edition). St Martin's Press.

European Union Agency for Cybersecurity (ENISA). (2015, January 12). *Privacy and Data Protection by Design* [Report/Study]. ENISA. https://www.enisa.europa.eu/publications/privacy-and-data-protection-by-design

Floridi, L., & Cowls, J. (2019). A Unified Framework of Five Principles for AI in Society. *Harvard Data Science Review*. Advance online publication. DOI: 10.1162/99608f92.8cd550d1

Floridi, L., Cowls, J., Beltrametti, M., Chatila, R., Chazerand, P., Dignum, V., Luetge, C., Madelin, R., Pagallo, U., Rossi, F., Schafer, B., Valcke, P., & Vayena, E. (2018). AI4People—An Ethical Framework for a Good AI Society: Opportunities, Risks, Principles, and Recommendations. *Minds and Machines*, *28*(4), 689–707. DOI: 10.1007/s11023-018-9482-5 PMID: 30930541

Friedman, B., & Nissenbaum, H. (1996). Bias in computer systems. *ACM Transactions on Information Systems*, *14*(3), 330–347. DOI: 10.1145/230538.230561

Fritsch, T. (2024). Chatbots: An Overview of Current Issues and Challenges. In Arai, K. (Ed.), *Advances in Information and Communication* (pp. 84–104). Springer Nature Switzerland., DOI: 10.1007/978-3-031-53960-2_7

FTC. (2013, July 25). *Children's Online Privacy Protection Rule ("COPPA")*. Federal Trade Commission. https://www.ftc.gov/legal-library/browse/rules/childrens-online-privacy-protection-rule-coppa

FTC. (2020, July 20). *Complying with COPPA: Frequently Asked Questions*. Federal Trade Commission. https://www.ftc.gov/business-guidance/resources/complying-coppa-frequently-asked-questions

Gaggioli, A., Villani, D., Serino, S., Banos, R., & Botella, C. (2019). Editorial: Positive Technology: Designing E-experiences for Positive Change. *Frontiers in Psychology*, *10*, 1571. Advance online publication. DOI: 10.3389/fpsyg.2019.01571 PMID: 31338053

Gasser, U., & Almeida, V. A. F. (2017). A Layered Model for AI Governance. *IEEE Internet Computing, 21*(6), 58–62. *IEEE Internet Computing*. Advance online publication. DOI: 10.1109/MIC.2017.4180835

Gebru, T., Morgenstern, J., Vecchione, B., Vaughan, J. W., & Wallach, H., III. H. D., & Crawford, K. (2021). *Datasheets for Datasets* (arXiv:1803.09010). arXiv. https://doi.org//arXiv.1803.09010DOI: 10.48550

Gellert, R. (2015). Data protection: A risk regulation? Between the risk management of everything and the precautionary alternative. *International Data Privacy Law*, *5*(1), 3–19. DOI: 10.1093/idpl/ipu035

Gillespie, T. (2020). Content moderation, AI, and the question of scale. *Big Data & Society*, *7*(2), 2053951720943234. DOI: 10.1177/2053951720943234

Gold, S., Trautrims, A., & Trodd, Z. (2015). Modern slavery challenges to supply chain management. *Supply Chain Management*, *20*(5), 485–494. DOI: 10.1108/SCM-02-2015-0046

Google. (2024, October 1). *Be Internet Awesome*. Be Internet Awesome. https://beinternetawesome.withgoogle.com/en-us

GPAI. (2024, October 1). *Global Partnership on Artificial Intelligence—GPAI*. https://gpai.ai/

Greenleaf, G. (2019). *Global Data Privacy Laws 2019: 132 National Laws & Many Bills* (SSRN Scholarly Paper 3381593). https://papers.ssrn.com/abstract=3381593

Greenleaf, R. K. (2002). *Servant Leadership: A Journey Into the Nature of Legitimate Power and Greatness*. Paulist Press.

Haenssgen, M. J. (2018). The struggle for digital inclusion: Phones, healthcare, and marginalisation in rural India. *World Development*, *104*, 358–374. DOI: 10.1016/j.worlddev.2017.12.023

Hassan, A. (2022). The Influence of Artificial Intelligence on Smart Learning: An Overview. In Hamdan, A., Hassanien, A. E., Mescon, T., & Alareeni, B. (Eds.), *Technologies, Artificial Intelligence and the Future of Learning Post-COVID-19: The Crucial Role of International Accreditation* (pp. 591–615). Springer International Publishing., DOI: 10.1007/978-3-030-93921-2_31

Heeks, R. (2020). ICT4D 3.0? Part 1—The components of an emerging "digital-for-development" paradigm. *The Electronic Journal on Information Systems in Developing Countries*, *86*(3), e12124. DOI: 10.1002/isd2.12124

Hoek, R. (2020). Research opportunities for a more resilient post-COVID-19 supply chain – closing the gap between research findings and industry practice. *International Journal of Operations & Production Management, ahead-of-print*. DOI: 10.1108/IJOPM-03-2020-0165

Holmes, W., Bialik, M., & Fadel, C. (2019). *Artificial Intelligence in Education*. Promise and Implications for Teaching and Learning.

Huang, M.-H., & Rust, R. (2018). Artificial Intelligence in Service. *Journal of Service Research, 21*(2), 109467051775245. DOI: 10.1177/1094670517752459

ICO. (2018). *Applications—Children and the GDPR*. https://ico.org.uk/media/for-organisations/guide-to-the-general-data-protection-regulation-gdpr/children-and-the-gdpr-1-0.pdf

ICO. (2023, September 14). *Children and the UK GDPR*. ICO. https://ico.org.uk/for-organisations/uk-gdpr-guidance-and-resources/childrens-information/children-and-the-uk-gdpr/

IEEE. (2019, April 4). *Now Available! Ethically Aligned Design, First Edition! - IEEE Robotics and Automation Society*. https://www.ieee-ras.org/about-ras/latest-news/now-available-ethically-aligned-design-first-edition

Institute, A. I. N. (2019). *Discriminating Systems: Gender, Race, and Power in Ai - Report*. Ai Now Institute.

Institute, A. N. (2019, December 12). AI Now 2019 Report. *AI Now Institute*. https://ainowinstitute.org/publication/ai-now-2019-report-2

ITU. (2024, October 1). *Artificial Intelligence*. https://www.itu.int/en/action/ai/Pages/default.aspx

Jarek, K., & Mazurek, G. (2019). Marketing and Artificial Intelligence. *Central European Business Review, 8*(2), 46–55. DOI: 10.18267/j.cebr.213

Jobin, A., Ienca, M., & Vayena, E. (2019). The global landscape of AI ethics guidelines. *Nature Machine Intelligence, 1*(9), 389–399. DOI: 10.1038/s42256-019-0088-2

Jordan, T. (2025). *Parental Control: A Guide to Raising Balanced Kids in the Digital Era*. Wiley.

Kadam, K., Godbole, S., Joijode, D., Karoshi, S., Jadhav, P., & Shilaskar, S. (2022). Multilingual Information Retrieval Chatbot. In Gunjan, V. K., & Zurada, J. M. (Eds.), *Modern Approaches in Machine Learning & Cognitive Science: A Walkthrough* (pp. 107–121). Springer International Publishing., DOI: 10.1007/978-3-030-96634-8_10

Kaesling, K. (2023). A Rights-Based Approach to Children's Digital Participation in the Multi-Level System of the European Union. In Dethloff, N., Kaesling, K., & Specht-Riemenschneider, L. (Eds.), *Families and New Media: Comparative Perspectives on Digital Transformations in Law and Society* (pp. 73–110). Springer Fachmedien., DOI: 10.1007/978-3-658-39664-0_4

Kamble, S., Gunasekaran, A., & Arha, H. (2019). Understanding the Blockchain technology adoption in supply chains-Indian context. *International Journal of Production Research*, *57*(7), 2009–2033. DOI: 10.1080/00207543.2018.1518610

Kelly, B., King, M., Psy, L., Chapman Mnd, K., Boyland, E., Bauman, A. E., & Baur, L. A. (2015). A hierarchy of unhealthy food promotion effects: Identifying methodological approaches and knowledge gaps. *American Journal of Public Health*, *105*(4), e86–e95. DOI: 10.2105/AJPH.2014.302476 PMID: 25713968

LEGO. (2024a, October 1). *Datenschutzrichtlinie*. https://kids.lego.com/de-de/legal/privacy-policy

LEGO. (2024b, October 1). *Privacy—Sustainability—LEGO.com*. https://www.lego.com/de-de/sustainability/digital-child-safety/privacy?locale=de-de

Lievens, E., & Verdoodt, V. (2018). Looking for needles in a haystack: Key issues affecting children's rights in the General Data Protection Regulation. *Computer Law & Security Report*, *34*(2), 269–278. DOI: 10.1016/j.clsr.2017.09.007

Livingstone, S., Third, A., Burton, P., Witting, S., Lievens, E., Steinberg, S., Gilutz, S., & Souter, D. (2022). The UN committee on the rights of the child's general comment on the digital environment. *AoIR Selected Papers of Internet Research*. DOI: 10.5210/spir.v2022i0.12960

Long, D., & Magerko, B. (2020). What is AI Literacy? Competencies and Design Considerations. *Proceedings of the 2020 CHI Conference on Human Factors in Computing Systems*, 1–16. DOI: 10.1145/3313831.3376727

Mantelero, A. (2018). AI and Big Data: A blueprint for a human rights, social and ethical impact assessment. *Computer Law & Security Report*, *34*(4), 754–772. DOI: 10.1016/j.clsr.2018.05.017

Marginson, S. (2016). The worldwide trend to high participation higher education: Dynamics of social stratification in inclusive systems. *Higher Education*, *72*(4), 413–434. DOI: 10.1007/s10734-016-0016-x

Marr, B. (2019, March 11). *The Amazing Ways Coca Cola Uses Artificial Intelligence And Big Data To Drive Success*. Forbes. https://www.forbes.com/sites/bernardmarr/2017/09/18/the-amazing-ways-coca-cola-uses-artificial-intelligence-ai-and-big-data-to-drive-success/

Mars. (2020, September 8). *Mars Builds on Cocoa Supply Chain Transparency | Mars Global*. https://www.mars.com/news-and-stories/articles/cocoa-supply-chain-transparency

Martin, K. E., & Freeman, R. E. (2004). The Separation of Technology and Ethics in Business Ethics. *Journal of Business Ethics, 53*(4), 353–364. DOI: 10.1023/B:BUSI.0000043492.42150.b6

Mehrabi, N., Morstatter, F., Saxena, N., Lerman, K., & Galstyan, A. (2021). A Survey on Bias and Fairness in Machine Learning. *ACM Comput. Surv., 54*(6), 115:1-115:35. DOI: 10.1145/3457607

Microsoft. (2024, October 1). AI for Health. *Microsoft Research.* https://www.microsoft.com/en-us/research/project/ai-for-health/

Milkaite, I., & Lievens, E. (2019). Children's Rights to Privacy and Data Protection Around the World: Challenges in the Digital Realm. *European Journal of Law and Technology, 10*(1). https://ejlt.org/index.php/ejlt/article/view/674

Milkaite, I., & Lievens, E. (2020). Child-friendly transparency of data processing in the EU: From legal requirements to platform policies. *Journal of Children and Media, 14*(1), 5–21. DOI: 10.1080/17482798.2019.1701055

Min, H. (2010). Artificial intelligence in supply chain management: Theory and applications. *International Journal of Logistics, 13*(1), 13–39. DOI: 10.1080/13675560902736537

MIT. (2017, September 7). *IBM and MIT to pursue joint research in artificial intelligence, establish new MIT-IBM Watson AI Lab.* MIT News | Massachusetts Institute of Technology. https://news.mit.edu/2017/ibm-mit-joint-research-watson-artificial-intelligence-lab-0907

Montgomery, K. C., Chester, J., & Milosevic, T. (2017). Children's Privacy in the Big Data Era: Research Opportunities. *Pediatrics, 140*(Suppl 2), S117–S121. DOI: 10.1542/peds.2016-1758O PMID: 29093045

Multilingual AI Chatbot Frameworks | Restackio. (n.d.). Retrieved October 17, 2024, from https://www.restack.io/p/ai-for-language-translation-answer-multilingual-ai-chatbot-frameworks-cat-ai

Musa, A. Elsadig. (2024). *Developing Digital Inclusion Through Globalization and Digitalization.* IGI Global.

Nairn, A., & Fine, C. (2008). Who's messing with my mind?: The implications of dual-process models for the ethics of advertising to children. *International Journal of Advertising, 27*(3), 447–470. DOI: 10.2501/S0265048708080062

O'Brolcháin, F., Jacquemard, T., Monaghan, D., O'Connor, N., Novitzky, P., & Gordijn, B. (2016). The Convergence of Virtual Reality and Social Networks: Threats to Privacy and Autonomy. *Science and Engineering Ethics*, *22*(1), 1–29. DOI: 10.1007/s11948-014-9621-1 PMID: 25552240

O'Neil, C. (2016). *Weapons of Math Destruction: How Big Data Increases Inequality and Threatens Democracy*. Crown Publishing Group.

Obermeyer, Z., & Emanuel, E. J. (2016). Predicting the Future—Big Data, Machine Learning, and Clinical Medicine. *The New England Journal of Medicine*, *375*(13), 1216–1219. DOI: 10.1056/NEJMp1606181 PMID: 27682033

OECD. (2019). *Child labour risks in global supply chains*. Ending Child Labour, Forced Labour and Human Trafficking in Global Supply Chains. https://mneguidelines.oecd.org/child-labour-risks-in-global-supply-chain.htm

OECD. (2021, January 7). *Children in the digital environment*. OECD. https://www.oecd.org/en/publications/children-in-the-digital-environment_9b8f222e-en.html

OECD. (2024). *MNE Guidelines—Organisation for Economic Co-operation and Development*. OECD Guidelines for Multinational Enterprises on Responsible Business Conduct. https://mneguidelines.oecd.org/mneguidelines/

Pagallo, U., Durante, M., & Monteleone, S. (2017). What Is New with the Internet of Things in Privacy and Data Protection? Four Legal Challenges on Sharing and Control in IoT. In Leenes, R., van Brakel, R., Gutwirth, S., & De Hert, P. (Eds.), *Data Protection and Privacy: (In)visibilities and Infrastructures* (pp. 59–78). Springer International Publishing., DOI: 10.1007/978-3-319-50796-5_3

Papadopoulos, T., Gunasekaran, A., Dubey, R., & Fosso Wamba, S. (2017). Big data and analytics in operations and supply chain management: Managerial aspects and practical challenges. *Production Planning and Control*, *28*(11-12), 873–876. DOI: 10.1080/09537287.2017.1336795

Pasquale, F. (2015). *The Black Box Society: The Secret Algorithms That Control Money and Information*. Harvard University Press. https://www.jstor.org/stable/j.ctt13x0hch

Perset, K., Russo, L., & Oder, N. (2024, June 11). *OECD AI Review of Germany: A new phase for Germany's AI strategy*. OECD.AI Policy Observatory. https://oecd.ai/en/wonk/review-germany-2024

Porter, M. E., & Kramer, M. R. (2019). Creating Shared Value. In Lenssen, G. G., & Smith, N. C. (Eds.), *Managing Sustainable Business: An Executive Education Case and Textbook* (pp. 323–346). Springer Netherlands., DOI: 10.1007/978-94-024-1144-7_16

Porter, S. R. (2006). Institutional Structures and Student Engagement. *Research in Higher Education*, *47*(5), 521–558. DOI: 10.1007/s11162-005-9006-z

Proposal for a Regulation of the European Parliament and of the Council Laying down Harmonised Rules on Artificial Intelligence (Artificial Intelligence Act) and Amending Certain Union Legislative Acts (2021). https://eur-lex.europa.eu/legal-content/EN/TXT/?uri=celex%3A52021PC0206

Quincey, J. (2022). *The Coca-Cola Company 2022 Business & Sustainability Report*. https://www.coca-colacompany.com/content/dam/company/us/en/reports/coca-cola-business-sustainability-report-2022.pdf

Raja, S., Jebadurai, D. J., Ivan, L., Mykola, R. V., Ruslan, K., & Nadiia, P. R. (2024). Impact of Artificial Intelligence in Students' Learning Life. In Khamis, R., & Buallay, A. (Eds.), *AI in Business: Opportunities and Limitations* (Vol. 2, pp. 3–17). Springer Nature Switzerland., DOI: 10.1007/978-3-031-49544-1_1

Raji, I. D., Smart, A., White, R. N., Mitchell, M., Gebru, T., Hutchinson, B., Smith-Loud, J., Theron, D., & Barnes, P. (2020). Closing the AI accountability gap: Defining an end-to-end framework for internal algorithmic auditing. *Proceedings of the 2020 Conference on Fairness, Accountability, and Transparency*, 33–44. DOI: 10.1145/3351095.3372873

Ransbotham, S., Gerbert, P., Reeves, M., Kiron, D., & Spira, M. (2018). Artificial Intelligence in Business Gets Real. *MIT Sloan Management Review*. https://sloanreview.mit.edu/projects/artificial-intelligence-in-business-gets-real/

Rashid, A. B., & Kausik, M. A. K. (2024). AI revolutionizing industries worldwide: A comprehensive overview of its diverse applications. *Hybrid Advances, 7*, 100277. Regulation - 2016/679 - EN - GDPR - EUR-Lex (2016). https://eur-lex.europa.eu/eli/reg/2016/679/ojDOI: 10.1016/j.hybadv.2024.100277

Richardson, R., Schultz, J., & Crawford, K. (2019). *Dirty Data, Bad Predictions: How Civil Rights Violations Impact Police Data, Predictive Policing Systems, and Justice* (SSRN Scholarly Paper 3333423). https://papers.ssrn.com/abstract=3333423

Saberi, S., Kouhizadeh, M., Sarkis, J., & Shen, L. (2019). Blockchain technology and its relationships to sustainable supply chain management. *International Journal of Production Research*, *57*(7), 2117–2135. DOI: 10.1080/00207543.2018.1533261

Sainani, D. (2024, September 24). *Innovative Partnership for Transparent Logistics and Transportation Missions: Mars Launches Pioneering Project With Rigterink, Shippeo and Elain.* AFP.Com. https://www.afp.com/en/news/1313/innovative-partnership-transparent-logistics-and-transportation-missions-mars-launches-pioneering-project-rigterink-shippeo-and-elain-202409243139011

Samek, W., Montavon, G., Vedaldi, A., Hansen, L. K., & Müller, K.-R. (Eds.). (2019). *Explainable AI: Interpreting, Explaining and Visualizing Deep Learning* (1st ed. 2019 Edition). Springer.

Save the Children Deutschland e. V. (2023, March 28). *Child Protection in the Digital World: Why is it needed?* Save the Children International. https://www.savethechildren.net/blog/child-protection-digital-world-why-it-needed

Shahriari, K., & Shahriari, M. (2017). IEEE standard review — Ethically aligned design: A vision for prioritizing human wellbeing with artificial intelligence and autonomous systems. *2017 IEEE Canada International Humanitarian Technology Conference (IHTC)*, 197–201. DOI: 10.1109/IHTC.2017.8058187

Sloane, M., Moss, E., Awomolo, O., & Forlano, L. (2022). Participation Is not a Design Fix for Machine Learning. *Proceedings of the 2nd ACM Conference on Equity and Access in Algorithms, Mechanisms, and Optimization*, 1–6. DOI: 10.1145/3551624.3555285

Sodhi, M. S., & Tang, C. S. (2018). *Research Opportunities in Supply Chain Transparency* (SSRN Scholarly Paper 3189990). https://papers.ssrn.com/abstract=3189990

Stahl, B., & Wright, D. (2018). Ethics and Privacy in AI and Big Data: Implementing Responsible Research and Innovation. *IEEE Security and Privacy*, *16*(3), 26–33. DOI: 10.1109/MSP.2018.2701164

State of California. (2018). *California Consumer Privacy Act (CCPA) | State of California—Department of Justice—Office of the Attorney General.* https://oag.ca.gov/privacy/ccpa

Stoilova, M., Livingstone, S., & Khazbak, R. (2021). *Investigating Risks and Opportunities for Children in a Digital World: A Rapid Review of the Evidence on Children's Internet Use and Outcomes.* Innocenti Discussion Papers. [Innocenti Discussion Papers], DOI: 10.18356/52f3f4b1-en

SuperAwesome. (2024, October 1). *Youth Marketing Solutions.* SuperAwesome. https://www.superawesome.com/

Taddeo, M., & Floridi, L. (2018). How AI can be a force for good. *Science*, *361*(6404), 751–752. DOI: 10.1126/science.aat5991 PMID: 30139858

Topol, E. J. (2019). High-performance medicine: The convergence of human and artificial intelligence. *Nature Medicine*, *25*(1), 44–56. DOI: 10.1038/s41591-018-0300-7 PMID: 30617339

Tsai, K. (2017, June 13). Transparency-One and Mars Food launch revolutionary supply chain solution. *Transparency-One*. https://www.transparency-one.com/de/mars-food-revolutionary-supply-chain-solution/

UN. U. N. S. D. D. D. and O. B. (2020). *The Sustainable Development Goals Report 2020*. https://unstats.un.org/sdgs/report/2020/

UN Global Compact Netzwerk Deutschland e. V. (2015). *Guide to Corporate Sustainability | UN Global Compact*. https://unglobalcompact.org/library/1151

UNESCO. (2021). *AI and education: Guidance for policy-makers*. UNESCO., DOI: 10.54675/PCSP7350

UNICEF. (2011). *Children's Rights and Business Principles | UNICEF*. https://www.unicef.org/documents/childrens-rights-and-business-principles

UNICEF. (2014a, April). *Children's Rights in Sustainability Reporting—A guide for integrating children's rights into the GRI reporting framework*. UNICEF. https://www.business-humanrights.org/en/latest-news/pdf-childrens-rights-in-sustainability-reporting-a-guide-for-integrating-childrens-rights-into-the-gri-reporting-framework/

UNICEF. (2014b, August). *Children Are Everyone's Business Workbook 2.0—A guide for integrating children's rights into policies, impact assessments and sustainability reporting*. UNICEF. https://www.business-humanrights.org/en/latest-news/pdf-children-are-everyones-business-workbook-20-a-guide-for-integrating-childrens-rights-into-policies-impact-assessments-and-sustainability-reporting/

UNICEF. (2020, December 10). *Magic Box Inspires Child Innovators | UNICEF USA*. https://www.unicefusa.org/stories/unicef-magic-box-commended-inspiring-kids-be-social-innovators

UNICEF. (2021, November 19). *Policy guidance on AI for children | Innocenti Global Office of Research and Foresight*. https://www.unicef.org/innocenti/reports/policy-guidance-ai-children

UNICEF. (2022, October 27). *Charting the course | UNICEF*. https://www.unicef.org/reports/charting-course

Unilever. (2017). *Unilever Responsible Sourcing Policy*. https://www.unilever.com/files/92ui5egz/production/f51492642f57b314b05466b6194792e02d075d76.pdf

Unilever. (2021, March). *Unilever Sustainable Living Plan 2010 to 2020*. https://www.unilever.com/files/92ui5egz/production/16cb778e4d31b81509dc5937001559f1f5c863ab.pdf

Unilever. (2024a). *Our History | Unilever global company website—2010—2020*. Unilever. https://www.unilever.com/our-company/our-history-and-archives/2010-2020/

Unilever. (2024b). *The importance of sustainable farming & sourcing*. Unilever. https://www.unilever.com/sustainability/nature/sustainable-sourcing/

van der Hof, S. (2017). I Agree... Or Do I?: A Rights-Based Analysis of the Law on Children's Consent in the Digital World. *Wisconsin International Law Journal*, *34*(2), 409–445.

van Dijk, J. (2020). *The Digital Divide*. John Wiley & Sons.

Vanlehn, K. (2011). The Relative Effectiveness of Human Tutoring, Intelligent Tutoring Systems, and Other Tutoring Systems. *Educational Psychologist*, *46*(4), 197–221. DOI: 10.1080/00461520.2011.611369

Voigt, P., & Von Dem Bussche, A. (2017). *The EU General Data Protection Regulation (GDPR)*. Springer International Publishing., DOI: 10.1007/978-3-319-57959-7

Wang, G., Gunasekaran, A., Ngai, E. W. T., & Papadopoulos, T. (2016). Big data analytics in logistics and supply chain management: Certain investigations for research and applications. *International Journal of Production Economics*, *176*, 98–110. DOI: 10.1016/j.ijpe.2016.03.014

Wang, T., Lund, B. D., Marengo, A., Pagano, A., Mannuru, N. R., Teel, Z. A., & Pange, J. (2023). Exploring the Potential Impact of Artificial Intelligence (AI) on International Students in Higher Education: Generative AI, Chatbots, Analytics, and International Student Success. *Applied Sciences (Basel, Switzerland)*, *13*(11), 11. Advance online publication. DOI: 10.3390/app13116716

Weidinger, L., Uesato, J., Rauh, M., Griffin, C., Huang, P.-S., Mellor, J., Glaese, A., Cheng, M., Balle, B., Kasirzadeh, A., Biles, C., Brown, S., Kenton, Z., Hawkins, W., Stepleton, T., Birhane, A., Hendricks, L. A., Rimell, L., Isaac, W., & Gabriel, I. (2022). Taxonomy of Risks posed by Language Models. *2022 ACM Conference on Fairness, Accountability, and Transparency*, 214–229. DOI: 10.1145/3531146.3533088

Wikipedia. (2024). Article 29 Data Protection Working Party. In *Wikipedia*. https://en.wikipedia.org/w/index.php?title=Article_29_Data_Protection_Working_Party

Williamson, B., & Eynon, R. (2020). Historical threads, missing links, and future directions in AI in education. *Learning, Media and Technology*, *45*(3), 1–13. DOI: 10.1080/17439884.2020.1798995

Williamson, B., Eynon, R., Knox, J., & Davies, H. (2023). *Critical perspectives on AI in education: Political economy, discrimination, commercialization, governance and ethics* (pp. 553–570). DOI: 10.4337/9781800375413.00037

World Bank. (2018). *World Development Report 2018: Learning to Realize Education's Promise* [Text/HTML]. World Bank. https://www.worldbank.org/en/publication/wdr2018

Yang, Q., Liu, Y., Chen, T., & Tong, Y. (2019). *Federated Machine Learning: Concept and Applications* (arXiv:1902.04885). arXiv. https://doi.org//arXiv.1902.04885DOI: 10.48550

Zawacki-Richter, O., Marín, V. I., Bond, M., & Gouverneur, F. (2019). Systematic review of research on artificial intelligence applications in higher education – where are the educators? *International Journal of Educational Technology in Higher Education*, *16*(1), 39. DOI: 10.1186/s41239-019-0171-0

Zemel, R., Wu, Y., Swersky, K., Pitassi, T., & Dwork, C. (2013). Learning Fair Representations. *Proceedings of the 30th International Conference on Machine Learning*, 325–333. https://proceedings.mlr.press/v28/zemel13.html

Chapter 12

AI, Refugee Status Determination, and Human Dignity:

Ethical Challenges in Refugee Migration Management With Comparative Insights

Abhinash Mohanta
https://orcid.org/0000-0002-8732-3202
KIIT University, India

Kyvalya Garikapati
https://orcid.org/0000-0003-0990-6514
KIIT University, India

Narayan Chandra Sarangi
https://orcid.org/0000-0002-8303-2942
XIM University, India

Shivani Samanta
https://orcid.org/0009-0003-6857-6145
South Asian University, India

Sanskruti Mohanty
https://orcid.org/0009-0009-9530-0412
Utkal University, India

DOI: 10.4018/979-8-3693-9035-1.ch012

ABSTRACT

This paper explores the ethical implications and challenges of using Artificial Intelligence (AI) in Refugee Status Determination (RSD), examining its potential to improve efficiency while emphasizing human dignity, fairness, and justice. The study analyzes current ethical frameworks, referencing the G20 AI Principles, SDG 16, OECD AI Principles and UNESCO Recommendation on the Ethics of AI. Through comparative insights and case studies based on interviews, core ethical issues—such as algorithmic bias, transparency, and accountability are assessed in AI's RSD applications. Human rights concerns, including privacy and equitable access, are critically discussed. Recommendations emphasize data inclusivity, transparency, human oversight, and stakeholder engagement to ensure ethical AI in RSD. Finally, future considerations address ongoing advancements, urging continued ethical scrutiny and collaborative dialogue to balance technological progress with the preservation of refugee rights.

INTRODUCTION

Immigration has become a unfolding phenomenon in the immigration crisis half of humanity is on the move by force of war, climate change, or search for a better life. This surge has put a lot of pressure in existing system with regard to asylum and refugee cases. Of these, Refugee Status Determination or RSD still holds importance in determining whether they be given refugees status due to a well-founded fear of being persecuted and harmed (Alexander, 1999). So far RSD has been a rather manual and knowledge-intensive process, where human discretion and professional judgement prevail when considering specific cases against the context of international and national norms. The conventional approaches, however, are sometimes limited by lack of economic capital, organisational hitches, and human element which may lead to or result in time wasting, irregularities, and even biasness.

To address these limitations, artificial intelligence (AI) has emerged as a promising tool to enhance RSD processes (Ling et al., 2020). As widely defined as the capacity of the machines to imitate human intelligence AI can contribute to the automation and increased efficiency of the components of RSD, including data manipulation and assessment of risks and documents. Applications involve the use of preferential algorithms, NLP, as well as biometric recognition systems, all of which raise the capacity to tackle backlog situations and processes cases much faster. These technologies are in pilot in the countries with relatively high influx of refugees in an effort to enhance efficiency and equity in management of migration, with AI taking over light and repetitive tasks, and bringing to the human decision-making systems' table analysis of large data sets (Automating Immigration and Asylum, 2023).

Nevertheless, the application of AI in RSD is also followed by a set of ethical issues that need much consideration. The expected capability of using AI in facilitating and enhancing the RSD processes raises questions on how such systems may affect and respect human rights, and human dignity (Artificial Intelligence & Human Rights: Opportunities & Risks, n.d). Thus, as automated systems assume functions previously performed by human agents, a much more comprehensive assessment of the ethical foundations of AI utilisation in RSD is required (Weber, 2019). This is especially so given that most of the decisions made entail the future of the refugees' individuals who are seeking asylum.

Human dignity as a principle of international human rights law is the cornerstone of the RSD process (Smrkolj, 2009). It entails that equal worth be given to each person and that every human being should be valued and honoured with consideration for his or her history and weakness. As RSD processes are concerned, it is about each individual who is in a desperate search for protection or security from fatal hazards and therefore decision-making relating to each claimant should uphold the dignity of the individual.

The implication of AI in RSD has far-reaching consequences on human dignity especially if AI Chase oriented systems neglect the dignity of the subject by encouraging mechanical model dependent optimization (Inverardi, 2021). This might result in generalising such situations where AI deals with the data as algorithms only reducing every claimant to be just a risk profile among other more elaborate data inputs instead of hearing their individual, context-specific narratives of their misfortune. To successfully and properly implement AI into RSD, this respect for human dignity has to be incorporated into the AI as a form of safeguarding against rampant technological advancement diminishing the ability to evaluate people compassionately on cases to case bases.

Also, it important to note that for Refugee Status Determination, ethical application of AI requires a focus on the concept of fairness (Cameron et al., 2021). As AI systems improve the standard of reducing some types of human bias, they also bring novel threats, including algorithm bias. These biases can be in-built from the datasets that are fed into the AI models, the type and nature of models developed or in the unexpected results obtained such that they reinforce social prejudices. For example, if datasets used in refugee applications include biases, then the outcomes generated by the AI models will also be biassed (Memon et al., 2024). Sustaining such risks suggests a need for careful analysis, review and validation of AI systems and their impacts including the systemic issue of bias across different cultural, legal and ethereal frameworks.

Transparency and accountability are also a difficult part of the ethical use of AI in a Refugee Status Determination (RSD)(Molnar, 2019). Due to many AI models' intricacies, particularly when using machine learning, the end user is frequently not privy to how particular decisions in their life are made. Thus, in RSD the amount of decision making that can have large scale effects, decision-making with insufficient transparency erodes credibility and does not allow for the possibility of challenging the decision. The openness of AI processes and especially XAI models, as well as the clear documentation of criteria for decisions, will be helpful to enable the chances of scrutinization and potential further questioning of claimants and similar stakeholders in case of an AI-based decision (Janssen et al., 2020). Such openness is critical not only for the non-biasness of the RSD systems but also for users to put their trust in the underlying AI technologies.

The ethical implications of AI in RSD are further complicated by the diverse cultural and legal contexts in which these systems operate. A universal approach to fairness and human dignity may not be feasible, as perceptions of fairness and the right to asylum can differ significantly across regions and communities. This makes it necessary that AI systems be developed to accommodate changeable legal frameworks as well as changing social culture so that they meet the needs of different refugees. (Automating Immigration and Asylum, 2023)

In the Indian context, the situation surrounding refugees is particularly unique and complex. India itself remains a country having no commitment under the 1951 Refugee Convention and its 1967 protocol and therefore has no formal policy of granting asylum to refugees (Kaur, 2013). Refugees in India are classified into two distinct groups: There are those that are recognised by the government of India commonly called the Indian recognised groups and those that are recognised by the United Nations High Commission for refugees (Bhattacharjee, 2008). The latter category mainly consists of people who have left conflict or persecution in countries like Afghanistan, Myanmar, and other countries. India seldom grants refugee status to any asylum seeker and the UNHCR, which is responsible for the RSD in India, mainly determines the status of refugees in the region. The UNHCR's processes are self-generated from the funds provided by the organisation and enable the organisation to offer legal protection and assistance to those considered needing asylum (Alexander, 1999). However, this system gives the mechanism for the recognition of the refugee status and reveals the oriented usage of the international formations as well as the resources in dealing with this issue based on the principle that domestic rights for refugees are not stable.

In the recent years, the Government of India has awakened to the reality of AI and its possibility in various sectors, which has been captured in its various policy framework and strategy papers (Chatterjee, 2020). Recently, NITI Aayog, a government-supported 'policy advisory body', laid out a National Strategy on AI to outline how India should approach AI. This strategy is reserved to that of AI as "the capacity of machines to accomplish intellectual activities" and stresses on the prospective development impact of AI (IndiaAI 2023: Expert Group Report, 2023). It aims at trying to guide the development of Artificial Intelligence to meet these needs and this is because in a Welfare State like Kenya there are sectors such as health, transportation and education among others that the government has regulated most.

As the present findings have shown, AI can significantly affect the RSD process due to which it is crucial to provide a detailed analysis of the ethical challenges associated with its application. This paper shows that incorporation of AI in RSD has the potential for improvement on its efficiency and its fairness, however without adequate ethical intervention it undermines the rights and dignity of refugees. With AI progressing and its roles in migration management developing it is of utmost importance that these technologies are developed along with the principles of human rights, equal treatment and openness (Nalbandian, 2022).

In particular, the ethical considerations discussed in this paper - human dignity, fairness, and transparency are foundational to any responsible deployment of AI in RSD (Shneiderman, 2020). Through analysing these principles, this paper seeks to contribute understanding in relation to using AI to improve the RSD process, with regard to the humanitarian principles in protection of refugees.

The next sections of this paper will therefore discuss various case studies and empirical research findings on the use of AI in RSD and show the advantages and constraints arising from the migration management systems underpinned by AI. These insights will assist in making recommendations about how to create ethical principles for the use of AI within Purpose of RSD. With this understanding, the paper aims to help expand the understanding of how AI can be applied to migration management and how its application can be in compliance with human dignity and human rights.

THE ETHICAL FRAMEWORK OF AI IN RSD

The integration of AI for RSD has the potential to revolutionise the refugee migration management and yet brings with it some darker ethical issues that requires proper ethical consideration (Refugee protection in the artificial intelligence era, 2022). This framework is necessary not only to inform the design and implementation of AI technologies, but also to provide core principles by which these technologies could be aligned. Since AI structures are being introduced into domains that have direct impacts on human lives, or as they augment these roles, the ethicist component to their design, development and deployment becomes the most critical factor so that it does not further erode the dignity and rights of refugees (Molnar, 2019).

Ethical issues in this context refer to the application of paradigmatic approaches to ethical reasoning in organisational research with a focus on RSD in the age of Artificial Intelligence.

Importance of an Ethical Framework in AI-RSD Implementation

To integrate AI properly in RSD processes, an ethical code has to envelop every aspects of the development, deployment and used of AI (Ibiricu & Made, 2020). Information processing power of AI technologies; analysis and identification of patterns; ability to generate recommendations; these aspects of AI have potential to improve refugee processing. Nevertheless, if not regulated by ethical principles, these efficiencies may undermine rather than enhance the principles of equity, privacy, and openness, which are a key to humanitarian practise in the refugee status determination (Cameron et al., 2021).

Ethical guidelines offer an organised procedure to avoid precluding unfavourable impacts on human dignity in the execution of artificial intelligence. It has been suggested that such a framework is very important especially in RSD because decisions made have far reaching implications for the lives of persons who are, by and large, marginal (Talhouk et al., 2019). For example, persecuted refugees rely on the results of RSD to have protection, stability, and the opportunity to reconstruct existence. The roles and responsibilities that AI performs in RSD can be ethically addressed through frameworks that establish guidelines for addressing contradictions between organisational needs and bureaucratic compliance on the one hand and human rights and human dignity on the other (Fukuda-Parr & Gibbons, 2021). This structured approach also helps to keep the part of AI more supportive and less interposing to the overall process in RSD.

Traditional Ethical Frameworks and Their Relevance to AI in RSD

The use of AI in the humanitarian field requires the reconsideration of the classical ethical theories which may well provide timeless well-established knowledge of the general well-being of mankind, their equity and justice. In the case of RSD the two major ethical theories: Utilitarian and Deontological has been proven to bring highly useful philosophies (Freeman, 1994).

Utilitarianism Ethics: This ethical framework emphasizes maximizing overall well-being or happiness and minimizing harm (Robertson et al., 2007). When applied to AI in RSD, this theory would mean that even if AI is not absolutely fair in decision making, its utilitarian advantage of faster decision-making could lead to faster resettlement of refugees, and so overall welfare would be increased (Wolff, 2006).

For instance, when an AI system can process multiple variables, the identification of refugees who require protection may be made faster, thereby realising the greatest value of timely action to refugees most at risk. However, there are drawback to utilitarianism as well: while mere focus on outcomes is interesting, it might be missed rights of people (Gerver, 2020). The fast throughput of many cases may result in routine or bureaucratic evaluations which may overlook individual clients who do not fall within the algorithmic profile. Therefore, although the strategies defined by the principles of utilitarianism may help AI developers identify actions that will bring the highest benefits to the largest number of people, it is necessary to follow the processes and make efforts to protect the personal peculiarities of each applicant from optional contempt (Lippert-Rasmussen & Vitikainen, 2020).

Deontological Ethics: This framework, based on principles and duties, focuses specifically on rights and the inherent duties that are attached to their implementation (Vorster, 2009). In RSD, a deontological approach though can take longer ensures that such rights of each refugee are respected without referring to the efficiency (Boswell, 2017). For example, the obligation not to use human beings simply as a means – one of the fundamental Immanuel Kant's principles – explains why reliance on AI also must not turn refugees into mere figures on a chart. Unlike performance and utilitarian goals, deontological approaches suggest that the AI systems should remember that transparency, fairness and accountability should be there to each individual. For RSD, this concept implies that AI processes should merely complement human actions so that the refugee has to endure assessment by an empathic human being even if an AI system does not know every aspect of the claimant's storey (Algorithmic Fairness for Asylum Seekers and Refugees (AFAR), 2023). Deontological ethics therefore calls for an AI structure that strongly defends individual worth notwithstanding the thousands of citizens involved (Barale, 2023).

Adapting Ethical Frameworks to Address Unique AI Dilemmas in RSD

Traditional ethical principles offer foundational guidance, yet the novel challenges presented by AI demand adaptations to these frameworks to address dilemmas specific to RSD (Rupp, 2016). These adaptations should consider AI's potential for bias, its opacity in decision-making, and the risk of depersonalizing refugee assessments.

Balancing Efficiency with Human Dignity: Both utilitarianism and deontological ethics can be adapted to emphasize a balanced approach that respects efficiency gains while upholding human dignity (Freund et al., 2023). For instance, when AI models are used to screen potential refugees in terms of the initial signs that make them eligible to get asylum, an ethical framework should require that an interpreter always evaluates instances where context or fine print make a potential refugee ineligible. The division of work where AI operate at the backend to manage large volumes of data while human agents focus on complex interactions with claimants, may provide the link between the dual objectives of effective operations and the recognition of the dignity of every person (Young et al., 2021).

Mitigating Algorithmic Bias and Ensuring Fairness: The challenge of prejudice in algorithms and especially in the RSD model is equally wanting in fairness and equity if the AI applications apply previous templates which worse could be laden with prejudice (Ferrara, 2023). Another consideration when developing an ethical framework is that procedures need to be formulated to cheque frequently the AI systems to see that the data fed into them for learning are not bigoted and the manner in which the systems operate must be open. While the utilitarian approach to the understanding of fairness can entail improvement of the algorithms which reveal and reduce the possibilities of bias, the deontological viewpoint will underline the importance of preserving compliance with individual claimants' rights by means of various procedural measures. For example, if data-driven biases are observed, an ethical framework must force a rebalancing of the model and confirm that the outputs of the system are favourable to the fair treatment of the minority groups (Chohlas-Wood et al., 2023).

Maintaining Transparency and Accountability: There are also ways that traditional ethical frameworks can be extended in order to provide greater definitional clarity about the openness of AI decision making especially where it concerns life crucial decisions such as RSD (Dehghani et al., 2024). Due to the nature of many AI developments, worrying on ethical principles should require methods where the involved stakeholders must comprehend and where necessary contest AI outcomes. Modifying deontological principles to call for explainability and justifiability of every AI decision guarantees that refugees may understand the rationale of decisions made on their lives (Aizenberg & Hoven, 2020). Additionally, the framework should allow human agents to override the suggested decisions in selected situations due to normative or situational requirements. This not only puts contingent on AI the resulting effects stemming from the refugees' lack of rights but also reconfirms refugees of their inherent rights within the RSD process.

Implementing Privacy Safeguards for Sensitive Data: Analysing refugee data presents deserves extra protection measures since the data itself is confidential and can be easily misused (Policy on the Protection of Personal Data of Persons of Concern to UNHCR, 2023). The principles of the traditional ethical theories have to be enriched with the respect of personal data as personal right. It should be marked by clear policies and procedures as to handling data for analysed use, it should follow both the utilitarian approach where one would not want to harm others, and deontological rationality that aims at respecting people's rights (Mühlhoff, 2023). Among such measures, one can name the restricted use of data, including its collection in limited volumes, storage with the observance of all possible measures to ensure that nobody else gets access to it, and its usage in cases when the refugee has given his/her consent to such utilisation. These practises extend the principle of personal rights for each person over data efficiency as a value.

Empathy and the Human Element in AI-Assisted RSD: The requirement for empathy when scoring in RSD assessments MSP addresses both the rules of utilitarian and deontological theories since such interactions allow getting to know each applicant better (Gîrju, 2021). An ethical development plan for AI-RSD should extend the "human in the loop" models, which means ensuring that all AI outputs are in a condition to be reviewed by a human counterpart capable of empathy and cultural sensitivity. These two are aligned in a way that honours the realism aim of organisational purpose and satisfies the deontological virtue of recognising the individuality of any claimant to dignity. In practise, this could mean using AI systems that identify cases that require attention, so that human agents can provide tailored advice where more nuanced planning is needed.

CORE ETHICAL CHALLENGES IN AI IMPLEMENTATION

Implementing AI into Refugee Status Determination (RSD); Holds certain benefits and certain ethical risks that can only be managed within the reasonable and rational consideration of technology and politics to serve the enhancement of fairness, transparency, and human dignity. Since the goals of refugee status determination are critically important, the ethical foundations of AI implementation cannot be limited to technical issues only; they also have to consider refugees' real-life experiences and their basic rights as refugees (Cameron et al., 2021). In the next part of the paper, we focus on the main ethical issues arising from bias and discrimination, transparency and accountability, and privacy violations related to AI-oriented RSD, and we present recommendations for their management.

Bias and Discrimination

AI algorithms have an inherent risk of bias, which can significantly impact refugee populations, who are already in vulnerable situations. Bias in AI can emerge from various sources, including historical data that reflects societal inequalities, biases introduced during data collection, or those embedded in algorithmic design choices (Feuerriegel et al., 2020). It is for this reason when bias algorithms are incorporated in the process of RSD, some individuals or some groups of people end up being discriminated during the process of determination of refugee status hence faces serious ethical and even humanitarian ramifications.

The implications of bias in RSD are far-reaching. Biased AI could lead to discriminatory outcomes where particular ethnic, national, or religious groups are either overrepresented or underrepresented in decision-making outcomes (Ferrara, 2023). For example, if an AI system is trained on data where certain nationalities historically have lower acceptance rates, it may inadvertently prioritize cases in a way that reflects and perpetuates this imbalance. To refugees, such biases mean the difference between having their status recognised and receiving protection or being endangered, or even persecuted. This is particularly concerning as refugees often have limited recourse to challenge the outputs of AI systems, further exacerbating their vulnerability.

To mitigate bias, several strategies can be implemented. One approach is to ensure that training datasets are diverse and representative, reflecting various backgrounds and experiences to minimize skewed outcomes. Furthermore, continuous auditing and testing of algorithms are essential to identify biases and address them through regular model adjustments and improvements. Integrating a "human-in-the-loop" approach can also help ensure that machine-generated decisions are reviewed and, if necessary, overridden by trained personnel with a deeper understanding of the refugee context (Bansak et al., 2018). This allowed for some human intervention which also cheques on bias that the algorithm may not easily detect and this therefore made its mechanisms slightly more general when it comes to RSD evaluations. Furthermore, a set of ethical principles of ad hoc industries in the frame of creation of RSD based on AI should be formulated to exclude biases, as well as to reorient AI based decisions making to the level of human rights and humanitarian laws.

Transparency and Accountability

This brings the question of transparency into the discussion of how AI achieves its decisions in RSD because it renders choices which have a direct impact on individuals' lives (Gibbons, 2021). AI systems adopted in RSD are often operate as black boxes – this is why the reasoning behind their decisions is unclear not only to the parties employing the systems but also to the refugees themselves (Sandvik, 2021). Lack of transparency makes it hard for refugees and human rights campaigners to know if a decision made is fair whereby making it harder to find an error, appeal or challenge the system.

Transparency is therefore an important aspect when designing AI systems in RSD for enhance ability of the stakeholders to understand the reasoning behind the systems decision making processes (Cobbe et al., 2021). In this particular context, Explainable AI means creating models that offer understanding of how specific decisions at a specific outcome was arrived at. For example, if an AI tool evaluates particular factors as more relevant to define refugee status this reasoning must be available to case officers and, preferably, to the applicants to understand the process that defined their status. Such levels of transparency do not only help people trust Artificial Intelligence but also help refugees to be treated with the respect they deserve as they make life-changing decisions clear to them.

Accountability mechanisms must also be established to uphold ethical standards and ensure that AI developers, RSD authorities, and other stakeholders are responsible for maintaining the ethical integrity of the system (Saenz et al., 2023). Such mechanisms might comprise the creation of specific oversight committees to ensure the AI functioning in RSD with rights to review, question or alter the AI-generated decisions that seem ethically questionable to the board's members. Another measure is determining the procedures refugees can use to report a problem or appeal against a decision that has been made for them. This approach makes developers and operators of AI paid attention because their systems and decisions will be audited periodically. Further, regulation must define high standards of AI explaining and being accountable in RSD and reassert the principle of respect for human rights across the process (Lam et al., 2024).

Privacy Concerns

In RSD, AI depends on vast data accumulation and analysis to determination the refugee situation, ranging from demographic to biometric data and personal and/or sensitive data if any. This is quite alarming to refugee's privacy as they can hardly record, manage and have control over their data. Since a good percentage of refugees originate from parts of the world with little data protection, it becomes possible for the data to be misused hence exposing the individuals to enhanced unintended consequences in the line of their lives and privacy (Kaurin, 2019).

The question of ethics of data gathering in RSD is one of the most fundamental. Refugees have faith that the data that will be given by them will be used for deciding the legitimacy of their asylum request and will not endanger them (Schoemaker et al., 2020). However, if there are no strict measures against data protection, such data will be readily accessible, shared or utilised by wrongdoers putting the lives of those people, who have suffered so much, at the perpetrating agents' mercy. This is made worse by the possibility of data to be stored for an indefinite period or transfer to other agencies, directing the little control refugees have over their information, even further.

To address these privacy concerns, ethical frameworks must prioritize data minimization and security. Particularly, only the data required for the RSD process should be gathered, and once it is not used any more, it needs to be erased. Of them, biometric data deserve special protection because of the inherent risk and the impossibility to anonymize them(Archer et al., 2023). It is recommended that personal data of refugees should be encrypted, anonymized and stored securely by the developers of AI systems and the RSD authorities. Furthermore, refugees should know what data is collected, for what purpose, and their rights regarding data use, getting and changing the data, which increases trust in the given system.

The conflict of interest between national security and refugee's civil liberties can only be addressed by developing policies that will work for both host countries and refugee populations while respecting their rights(Wolff, 2021). For example, the integration of biometric identification into RSD and security can be helpful; nevertheless, this sort of application must be set into narrow constraints, prohibiting data exchange. Procedures for supervision and monitor for data collection should also be established in order to facilitate checking and assessing of data management from time to time with an aim of validating if the procedures used follow the provision of privacy.

COMPARATIVE INSIGHTS AND CASE STUDIES ON AI IN RSD

Incorporating Artificial Intelligence (AI) in Refugee Status Determination (RSD) raises ethical complexities, as highlighted by insights from Miss Sneha, Assistant Professor at Xavier Law School, XIM University. Collecting the case of RSD from a range of countries including Canada, the EU, UK and Australia, AI has revealed phenomenal possibilities of complexity reduction yet a range of ethical and practical issues. It is possible by analysing these cases to look at challenges that regions implement in these areas and learn how to address fairness, transparency, and privacy concerns.

Successes and Challenges in AI-Driven RSD Canada: AI in Immigration Triage with a Focus on Transparency Based in Canada's Immigration, Refugees and Citizenship Canada (IRCC), some facets have applied AI in RSD; although, in its application, AI is used in support roles to help humans make the decisions rather than making the decisions themselves(Nejadgholi et al., 2024). The current Canadian system employs the help of artificial intelligence to sort out cases on the basis of their difficulty with a view of facilitating distribution of resources as well as Case processing.

Canada's AI application has been used to eliminate backlog in straightforward cases whereby simple cases go through the process without intervention of human beings but, complicated cases require a the interference of human beings. According to the words of Miss Sneha, there should be transparency in such cases, especially when using artificial intelligence, with reference to vulnerable persons. Canada's way is completely clear about what it is doing with AI to applicants, and that is meeting the ethical principles of trust and autonomy since the decision-making process is clear and understood by anyone.

Nevertheless, efficiency comes with ethical concern on fairness in the implementation process. According to Miss Sneha, AI in RSD should not regress a bias that has already been designed. Although the use of the AI system means saving time, such patterns can sometimes missed certain cultural or personal circumstances of an applicant. This has made Canada develop a strong accountability system that makes sure that the use of AI in determining the triage does not bring prejudice in the rights and the dignity of the applicants, as Miss Sneha advised on continuity of checking fairness in the artificial intelligence model and advancing the socio-legal approach to the AI models.

European Union: AI in Border Management with iBorderCtrl The EU has developed iBorderCtrl, an AI-driven system designed to streamline immigration checks at borders(Sánchez-Monedero & Dencik, 2019). Utilizing facial recognition and emotion detection, iBorderCtrl is aimed at improving efficiency in screening individuals by quickly assessing the consistency of their responses.

By consequent, the utilisation of iBorderCtrl has proven to be useful in enhancing the efficiency of the borders specifically at the busy areas. Its usefulness is that it saves time to reveal contradictions as for applicants, and border agents, it frees time for cases with more suspicious activity(Chekovik & Achkoski, 2019). Due to overworking, Miss Sneha specifically focuses on the procedural organisation issues with an aim of enhancing responsiveness something that is in line with the European Union's aim of making the RSD process fast and more responsive.

The iBorderCtrl system has raised significant concerns over privacy and bias, as it involves complex personal data such as biometric and emotional assessments(Sánchez-Monedero & Dencik, 2020). According to Miss Sneha, transparency is essential in AI applications to maintain legitimacy and public trust. The system's lack of transparency has led to criticisms from human rights groups, with claims that applicants have little understanding or recourse if flagged by the system. iBorderCtrl's case reinforces Miss Sneha's view on the importance of open, accountable AI systems that respect the sensitive nature of data involved in RSD processes.

United Kingdom: Predictive Analytics and Case Prioritization in RSD The UK Home Office has explored the use of predictive analytics and AI to assist in prioritizing RSD cases. This approach aims to allocate resources efficiently by identifying high-priority cases and streamlining asylum application reviews.

The UK's AI model faced backlash due to concerns about racial biases, as it disproportionately flagged applicants from specific countries as high-risk(The UK is dropping an immigration algorithm that critics say is racist, 2020). Miss Sneha highlights that bias in AI systems is often a reflection of biased training data. The

UK's experience exemplifies this, as the AI model unintentionally reinforced discriminatory practices. Due to public criticism and calls for ethical accountability, the UK partially suspended its use of AI in prioritizing cases, prompting a re-evaluation of its system. The situation underscores Miss Sneha's recommendation for ongoing evaluation, fairness testing, and the need for socio-legal perspectives to ensure AI's equitable application.

Australia: Automation in Visa Processing for RSD Australia has deployed AI in visa processing, including RSD, as part of a pilot program aimed at improving processing times. The system automates certain administrative tasks, such as document checks and initial screening stages.

The Australian system encountered significant public backlash regarding the "black box" nature of its AI, where applicants had limited insight into the reasoning behind automated decisions(Weber & Gerard, 2024). Miss Sneha points out that transparency is vital for building trust and allowing applicants to contest decisions if needed. Public concerns about transparency led to a demand for greater human oversight and an appeals process, reinforcing Miss Sneha's stance on the necessity of open AI systems. The criticism prompted the Australian government to institute measures for human oversight, ensuring that AI decisions can be contested or reviewed if deemed unfair.

Analysis: The case studies from Canada, the EU, the UK, and Australia highlight patterns and distinct ethical issues that accompany the integration of AI in RSD. As Miss Sneha's interview reveals, ethical considerations—such as transparency, bias mitigation, and respect for applicant autonomy—are fundamental to building a reliable AI-based RSD system.

Canada's example illustrates the importance of transparency and maintaining human oversight over AI processes. Miss Sneha's insights on transparency and participant engagement support Canada's approach, where AI serves as an aid rather than a substitute for human judgment. Ensuring that applicants and their advocates understand the role of AI in RSD decisions, as Miss Sneha recommends, fosters accountability and trust.

Common Pitfalls: Bias in AI models, as seen in the UK's case, remains a central challenge across countries. Miss Sneha underscores that bias often originates from the data and assumptions used in AI training, which can inadvertently reinforce discriminatory outcomes. The UK's experience reveals the risks of using AI without comprehensive fairness assessments, highlighting the importance of fairness testing, as Miss Sneha suggests, to identify and mitigate any biased patterns in AI decision-making.

Lessons Learned: The various responses of these countries illustrate Miss Sneha's advocacy for rigorous accountability mechanisms and shared responsibility among policymakers, developers, and RSD authorities. The need for a participatory approach, as she suggests, ensures that the AI system remains aligned with ethical standards. Australia's move to introduce human oversight after public backlash demonstrates that trust and legitimacy in AI-driven RSD depend on transparency and ethical safeguards, affirming Miss Sneha's position on the importance of participatory, transparent AI models.

The application of AI in RSD across Canada, the EU, the UK, and Australia offers valuable lessons in aligning technological efficiency with ethical integrity(Kinchin & Mougouei, 2022). Miss Sneha's perspectives underscore the importance of transparency, fairness testing, and respect for privacy in designing AI for refugee determination processes. Balancing efficiency with human rights standards, as advocated by Miss Sneha, can lead to a more humane and effective RSD system that enhances public trust and safeguards the rights and dignity of refugees. Each case study demonstrates the potential of AI to support refugee management, while simultaneously emphasizing the need for careful oversight, accountability, and inclusivity to ensure that technological tools serve justice rather than undermine it.

In exploring these complexities, Miss Sneha underscore that clear regulatory guidance is essential to navigate the legal uncertainties and ethical challenges posed by AI systems, a perspective that holds significant relevance in the context of refugee management systems as well (Bag et al., 2024).

HUMAN RIGHTS AND AI IN RSD

The use of artificial intelligence in particular Refugee Status Determination and compliance with international human rights law comes with unique opportunities and challenges. RSD is not a legal process: indeed, it is the significant change of a personal status that is associated with rights and obligations, with the level of personal freedom, with the ability to control one's life(Refugee Status Determination, 2023). Bearing this in mind, integrating AI in such a delicate process means that human rights aspects must be at the heart of this process(Inverardi, 2021). This section considers how AI in the context of RSD intersects with the protection of international human rights law, analyses how RSD is affected in terms of the human rights of refugees and asylum seekers or the rights of the individual to be treated with respect, dignity, and fairness based on the right to non-discrimination, and finally, look at the ways in which AI's use in RSD can be best governed to ensure that the human rights standards are aligned.

AI and International Human Rights Law in RSD

The Universal Declaration of Human Rights (UDHR), alongside other core human rights treaties such as the International Covenant on Civil and Political Rights (ICCPR) and the 1951 Refugee Convention, establish the fundamental rights that must be upheld in any process impacting refugees(Gorlick, 2001). These frameworks enshrine rights such as the right to seek asylum, freedom from discrimination, the right to a fair hearing, and protection of personal dignity. In the broader applicability of the AI in RSD, must not function as an enhanced decision-making tool but as one that complied with the underpinning principles of mitigating unfairness in the determination process(Chohlas-Wood et al., 2023).

AI used in handling RSD needs to respect human rights laws that demand that a process should be bias free, equal and open. Since the AI models which have the potential of improving the effectiveness and reliability of the RSD processes have to conform if implemented they must be designed in a way that respects the principles of the protection of refugee rights. The absence of non-discrimination is particularly critical especially because the use of artificial intelligence systems means that the system will have to rely on data and reflect the discriminations captured in the same data. The ICCPR does not allow discrimination: this means in order to avoid discriminating AI outputs, systems in RSD have to have elements that correct for such trends(Nuenen et al., 2020). Moreover, the primary focus of the 1951 Refugee Convention is the right to seek asylum; therefore, the organisation's involvement in RSD should make access to a just procedure for seeking asylum easier.

Impact of AI on Refugee Rights: Dignity, Autonomy, and Fair Treatment

AI's influence on refugee rights is multi-dimensional, impacting critical areas such as personal dignity, autonomy, and the right to fair treatment(Kinchin, 2021). Each of these aspects reflects key values upheld by international human rights law, and the integration of AI in RSD poses unique challenges and opportunities in preserving them.

Dignity is a cornerstone of international human rights and directly relates to how individuals are treated within any legal framework, including RSD(McGregor et al., 2019). When AI systems are implemented without adequate safeguards, they risk treating individuals impersonally, potentially reducing refugees to data points rather than recognizing their complex individual needs and circumstances. Maintaining dignity requires that AI systems are developed with an understanding of the gravity of RSD decisions and an appreciation for the diverse backgrounds and vulnerabilities of refugee populations. By designing AI with a human-centric approach, developers can foster systems that honour each individual's unique context, ensuring that refugees are not stripped of their personal worth by an overly mechanized process.

Another typical area impacted by AI in RSD that belongs to human rights is autonomy(Aizenberg & Hoven, 2020). Refugees should not lose their subjectivity in the determination process even where they use these emerging technologies. The principles of autonomy in RSD are the rights of an individual to provide evidence, rights of an individual to have understanding and some influence on the factors forming the decision about his or her case and the right to appeal against the decision which seems unjust to the individual. AI's role in RSD must therefore safeguard decision making processes and ensure that decision making remains fully transparent and comprehensible to refugees as to ensure that refugees remain making decisions and not having decisions made for them. One of the main use cases of the explainable AI technologies, which can help to ensure that refugees keep control over decisions made for and on their behalf, or provide the means for AI truth, are the AI models, which independent of their accuracy and performance, allow for output results review, on what basis the decision were made, and direct interaction in case of applying for decision or request fulfilment.

The right to fair treatment is perhaps the most critical in the RSD process, where AI's efficiency benefits could be invaluable(Nazer et al., 2023). However, fair treatment cannot be compromised for efficiency; on the contrary, fairness must be a primary objective in the design and application of AI systems in RSD. Justification for the fair use of AI tools demands not only procedural openness but also reliable procedures for supervising and improving AI tools to guarantee that no forms of unfair bias occur. Further, refugees should be able to seek to challenge or appeal decisions and AI must work in this process as a supplement rather than the replacement for the fair and encompassing procedure.

Designing AI Systems to Uphold Human Rights Standards in RSD

To ensure that AI systems uphold human rights standards in RSD, certain core principles and design strategies must be implemented from the outset. These approaches focus on embedding respect for human rights into AI frameworks, ensuring that refugee rights are safeguarded and promoted in every aspect of the process.

Principle of Inclusivity and Non-Discrimination

AI systems should be designed with a proactive approach to inclusion in order that the data sources used in the development and training of the system are varied and involve minorities. This diversity is important in minimising biases that could occur due to data sameness or restricted cultural diversity since such analysis would likely work disadvantageously to refugees from certain origins. An inclusive AI design will not only give better results as compared to other forms of determinations, but it will be in compliance with the provision of equality or non-discrimination(Shams et al., 2023).

Transparency and Explainability

Transparency is fundamental to respecting human rights in AI-driven RSD. Explainable AI techniques should be used to make the decision-making process as transparent as possible to refugees, legal representatives, and humanitarian organizations(Prentzas et al., 2023). Explainability is essential to demystify AI operations and enable individuals to understand how determinations are made, promoting fairness and enhancing refugees' ability to participate actively in their cases. Regular audits and documentation of AI systems are further crucial to maintain transparency and accountability, ensuring that AI operations remain aligned with ethical standards(Primer: Defending the rights of refugees and migrants in the digital age, 2024).

Privacy and Data Protection

Given the sensitivity of refugee data, strict data protection measures are critical. AI systems in RSD should comply with data privacy regulations, minimizing the amount of data collected and limiting its use strictly to RSD purposes. Privacy concerns are heightened in RSD due to refugees' histories of persecution and displacement, and any mishandling of data could endanger their safety. Employing encryption, anonymization, and limiting data retention are all necessary steps to protect refugee privacy and align with human rights standards concerning personal data.

Mechanisms for Redress and Accountability

AI systems in RSD must be designed to allow for redress mechanisms, ensuring that refugees have pathways to challenge or appeal decisions that they believe to be incorrect or unfair. Establishing such mechanisms acknowledges refugees' right to fair treatment and recourse, upholding human rights principles in cases where AI-driven decisions fall short. Additionally, mechanisms for accountability—such as independent oversight bodies that monitor AI performance in RSD—provide an additional layer of protection, ensuring that AI systems adhere to human rights obligations and operate with integrity(Petković, 2023).

Human-Centric Design and Human Oversight

Finally, another distinguishing factor of AI in RSD is that human oversight is an essential part of AI and is not there to replace human judgement(Xu & Gao, 2023). In this way, the general approach of RSD that keeps the human at the centre of all processes allows using the potential of AI technologies to improve the efficiency and procedural justice of decisions and do not exclude the human factors of empathy and judgement. AI should be seen as an aid to the decision maker for areas that can be beneficial for them but also as a tool that can provide large quantities of information that might need to be looked at individually. The advantages of this approach are that it fosters maximum compliance with human rights principles of dignified treatment and non-discrimination.

POLICY RECOMMENDATIONS AND FUTURE DIRECTIONS FOR ETHICAL AI IN REFUGEE STATUS DETERMINATION (RSD)

In the pursuit of an effective and ethically responsible framework for AI in Refugee Status Determination (RSD), it is crucial to address multiple facets of AI design, collaboration, and governance. The issues with regard to AI in RSD, are deeply encompassing ranging from transparencies, fairness, issue of privacy, to the question of open, fair and suitable governance. When such issues are addressed under clear and emerging rules these countries like India can use of AI to enhance the RSD process while upholding the core tenets of human rights and refugee dignity without being a signatory to the 1951 Convention.

Emphasis on Transparency, Fairness, and Human Oversight

Implementing transparency, fairness, and human oversight are foundational to the ethical deployment of AI in RSD. Transparent AI systems allow for more accessible and accountable decision-making processes, which in turn builds trust among all stakeholders, particularly refugees and their advocates. Fairness involves ensuring that AI systems do not inherently favour any particular demographic or ethnic group, thereby reducing biases that could arise from skewed data sources or unrepresentative datasets. One of the initial steps involves pre-implementation testing on anonymized historical data to ensure algorithmic fairness across diverse variables, such as ethnicity, nationality, and religious background. This step mitigates risks of favouring certain groups over others, ensuring equitable treatment.

Equally important is the concept of fairness auditing of crowdsourcing AI systems on ongoing basis. To counteract the issue, a procedure called fairness audits becomes utilised in which similar claims are continuously reviewed for articles that are either discriminatorily biassed or inconsistently evaluated. This can be done by exploring the participation of human rights advocates and legal practitioners in auditing. Also, including human monitoring as one of the permanent components of AI projects in the RSD process allows avoiding artificially controlled decision-making by AI. For example, decisions made by AI might be subjected to human input especially in special situations where it is necessary to consider the human angle in the arguments of personal storeys besides historical data.

Development of Regulatory Frameworks Tailored to National Contexts

For countries like India, where diverse demographics influence various aspects of governance, regulatory frameworks for AI in RSD should be tailored to specific national needs. Currently, there are guidelines such as the G20 AI Principles, other guidelines such as the OECD AI Principles and UNESCO Recommendation on the Ethics of Artificial Intelligence all of which propose on the aspect of fairness, accountability and transparency(Corrêa et al., 2023). These frameworks can be suited to present analysis where refugee migration and status determination presents distinct challenges in the region as it fills India's AI vision of a human centric paradigm.

There should also be general principles regulating the amount of disclosure that national regulators require about AI's decision-making parameters. The key findings outlined here indicate that AI-based decision systems can be more transparent if the refugee applicants and their advocates, as well as the public, can see how the system came to certain decisions and why specific refugees gained entry to the United States. Moreover, the addition of an accountability system or a review board, for example, can help the experts have regular cheque-ups in the form of auditing moral correctness of the particular AI model.

Engagement of Technologists, Policymakers, and Human Rights Advocates

To foster a robust ethical governance model, interdisciplinary collaboration is essential. This includes participation of technologist, policymakers, legal expert and human rights advocates in the entire process starting from the development of the AI system(Fukuda-Parr & Gibbons, 2021). Such collaborations guarantee that the developed AI model is fluent in technical language, not only in brute mathematics but also in ethical paradigms such as narratology when interpreting intricate refugee storeys. This way, stakeholders introduce multiple viewpoints that will allow for the compliance of the system with both human rights and technological imperatives.

A primary strategy is to introduce a design framework that would permit open discussion of all parties involved and promote integrated goals and expectations. For example, human rights advocates can pronounce on aspects of ethical improbity that might not have come to the notice of the development team, and data scientists can speak to the development's technological viability and opportunities. This multidisciplinary approach not only strengthens the system's integrity but also promotes a well-rounded, ethical perspective that safeguards the dignity of refugees.

Fostering Dialogue and Interdisciplinary Approaches for AI Ethics

AI ethics or the implications of its application in RSD must include solution to ethical issues applying interdisciplinary approaches which involve inputs from law for the technological laws, sociology and psychology for the effects of the technology on human behaviour, and data science for the applicability of the AI technology. Such practise also may lead to the development of more realistic scenarios for the actual use of AI-decision of the refugees' fate, as well as include the refugees and advocates in the process to make sure all the outcomes are meeting the demand of the refugees' rights.

Stakeholders should encourage open dialogues through workshops, conferences, and advisory committees focused on AI ethics in RSD. These platforms enable policymakers, data scientists, and refugee advocates to exchange perspectives, share insights, and jointly address emerging challenges. For example, regular forums where stakeholders discuss case studies and identify potential biases can help refine the AI model's ethical parameters.

Strategies for Continuous Stakeholder Engagement and Feedback

Participatory design need to incorporate refugees, advocates and lawyers within the design process at every stage of the AI cycle. This method ensures that AI systems are aware of the cultural, social and legal aspects of the refugee matters. It was found that incorporating the views of the refugees during the design process will enable the system to have strong ethical and human centred principles as the construction of the system is sensitive to the practical concerns of applications(Oduntan & Ruthven, 2020).

In practical terms, participatory design revolves around a series of feedback cycles with representative participants from among refugees and their advocates. All of these stakeholders analyse and evaluate the different aspects that concern the AI model such as data acquisition as well as criteria for decision making. Such consultations promote openness and contributing to the development of the feeling that people is an essential part of the system because it involves them as refugees during these meetings.

Continuous feedback of refugees and advocates should also be established to ensure they can hold AI to account and make it continue to be relevant and receptive in AI-driven RSD. A web-based tool could be created to help applicants to share their impressions on the subject perceiving the AI's involvement as more or less fair and transparent. This platform could also act as the channel through which refugees can appeal potential bias or unfair decision made by the agents. Positive feedbacks maintain and adjust the ethical perspective in the system, in order to guarantee the respect to refugees' dignity.

To facilitate this, there should be a periodic feedback team that should review the submissions from the users, make an analysis of the common features found in the submissions to warrant a change in the AI model and then consult the developers of the AI to make the necessary changes. For example, if refugees from a given age group have substantiated complaints of unfair treatment in their feedback, then the feedback team can embark on a process of analysing the available data with data scientists to reduce biases.

Developing Data Quality and Privacy Standards for Ethical AI in RSD

Data quality is a critical component of ethical AI in RSD, as it directly affects the fairness and accuracy of the system's decisions. Diverse data sources are essential to avoid biases that may arise from limited or homogenous datasets. Data gathering should consider the background and the situation of refugees and should be aligned to provide fair decision making to the sufferers by the machine learning algorithms. This process therefore involves frequent changes and update of the data gathered in a bid to capture current refugee data in a biassed free manner.

Emulating norms of even international standard available such as those put forward by the United Nations High Commissioner for Refugees (UNHCR), India can develop its reliability in managing data. For instance, the pattern of data quality audits can remove historical biases and replenishment to the data can guarantee that the dataset reflects the current issues affecting refugees.

The sensitive nature of refugee data requires stringent privacy and data protection protocols to prevent misuse. India's RSD framework should ensure that the data in the framework is anonymized to ensure that the data is not accessed by unauthorised individuals. Data privacy procedures and protection confirm the information of refugees is secure and safe. If refugees are educated on how their information is utilised and measures put in place to safeguard the information, they feel comfortable with the RSD process.

Implementing data anonymization protocols aligned with international standards, such as the General Data Protection Regulation (GDPR), further supports privacy. Moreover, anonymizing data before analysis helps maintain confidentiality, safeguarding refugees' identities and ensuring their rights are protected throughout the RSD process(Ferrari, 2018).

Ongoing Legal and Ethical Training for AI Developers and RSD Officers

The gap between technology and ethical responsibility can only be closed if both AI developers and RSD officers engage in training that cuts across the disciplinary divide. Some suggested roles included ensuring that developers of technologies experienced fairness, non-discrimination and transparent methods, as well as making provision for RSD officers which would involve understanding and critical evaluation of the outputs of technologies(Widder et al., 2023). Thus, this combined knowledge aids in a coordinated process as to where biases exist so that they can be removed making decisions more ethical.

Training programs should focus on practical skills, such as bias detection and management, allowing developers and RSD officers to recognize and mitigate unintended ethical issues. Continuous learning sessions, involving case studies and hands-on exercises, enable teams to stay current with ethical standards and emerging best practices in AI and refugee law. Additionally, by fostering a shared understanding of ethical guidelines, these programs promote a unified commitment to upholding refugees' rights and human dignity within the RSD process.

CONCLUSION

In this paper, we have looked at the pertinent issue of artificial intelligence (AI) and Refugee Status Determination (RSD) specific to the ethical dimension that will enable proper utilisation of AI to amplify the positive impact of refugee migration management. With the infusion of AI into RSD processes, there is need to reassert the core values that govern the respect of human persons and their rights. In this paper, we synthesise the findings of this research and argue that there is a significant challenge of effectively and ethically deploying AI in RSD.

The research evidence reveals that, while the use of AI can enhance the effectiveness of RSD procedures, there remain numerous moral implications when using the technology: To prevent a decline in human value, application of AI should be limited by ethics addressing key issues of systematic unfairness. In this paper, we noted that transparency, fairness, and accountability are essential ethical principles due to possible risks in AI-driven RSD. In addition, this study also supports the view that the strategies for AI application in RSD has to be contingent and has to be adjusted over time to align with the prevailing cultural and legal trends. Integrated approaches to AI in migration scenarios can be based on the G20 Principles for AI and the SDG 16. These new international standards and guide rails for AI are focusing on issues such as transparency, accountability, and human rights in application of these intelligent technologies. Having embraced the following guidelines will be vital for shaping a responsible AI framework for RSD so as not to hazard the refugees' rights and dignity by technology advancements.

Moreover, the proposed approach of ethical decision-making framework in AI in RSD requires an integration of concepts of the conventional ethical theories with the requirements of effective AI-based refugee management. With adopting some practises of utilitarianism and deontological ethic this adapted framework will assist in the proper utilisation of AI that will hopefully make it efficient as well as encourage respect for human rights and fairness. To be specific, ethical usage of AI in RSD ensures that the clients' rights among displaced people are protected while doing so within a system that aligns humanism, humanitarianism with AI's operational utility.

Through conducting fairness cheques regularly and being transparent, RSD AI systems can develop trust within the public to guarantee refugees feel their cases are fairly handled. The ongoing issues of liable and ethics for fairer and more ethical governance indicate the imperative for a broad ethical framework appropriate for AI's application in RSD. It is necessary to achieve closer cooperation between technology developers, legislators, lawyers, and activists to close the information gap between the use of technology and its morality.

As a way forward toward implementing AI to its full extent in RSD but within an ethical context, there is the need for an appeal to enable all the stakeholders who are in refugee management, legal governance, and the providers of technologies. This entails a forward-looking approach to the application of ethical Artificial Intelligence, and the later, emphasising the protection of human worth and ethical re-consideration. The following key actions are recommended to promote an ethical framework for AI in RSD:

The integration of AI into RSD must have one primary goal, that is to promote human dignity and respect refugees' rights. This condition continues to demand that AI systems retain the status of tools, devoid of power to decide, especially where such decision would impact on people's lives. RSD processes are generally acute, as they entail people running from persecution, war, and violence; therefore, outcomes of these processes directly affect refugees' life paths. The other reason for rising to human rights principles in the operation of AI is to keep the automation process human enough.

There is the need to ensure that in order to maintain the dignity and rights of refugees AI must be used in a manner that will amplify their voices and storeys. This can be done where certain aspects of RSD are automated, thereby leaving other aspects of decision making preserved for officers equipped to understand personal and cultural differences. The Identification of ways that AI could be used ethically involves the development of rules that need to be followed in the deployment of AI in the business so as to safeguard the morality of the applications, thereby coming up with recommendations that insist on human involvement in any instance that the AI conducts preliminary assessments. Furthermore, by making it easier for applicants learn about the use of AI in the relevant processes, we are able to reduce the scepticism and give the applicants a certain sense of security knowing that their cases are being handled fairly.

The ethical map of AI in RSD is not fixed, established for a long time; it is always under process and conscious with the dynamism of AI technology. On the positive note future enhancements in AI like the machine learning and deep learning have possibilities of growth as well as have dangers. Ethical analysis needs to be performed pre-actively, so there must be means established to oversee, update, and report publicly on the AI systems regularly. This can involve having a fairness cheque occasionally, a transparency cheque, or engaging stakeholders, all of which cheque that AI models are aware of new ethic and social issues.

One recommended approach is to develop an independent oversight body tasked with monitoring AI applications in RSD. It would be up to this body to periodically scrutinise AI systems, scrutinise algorithms, and assess results for unfairness as well as bias. It would be a viable body that could work with human rights activist groups, lawyers and refugee activists in an effort to rein in the system while ensuring it is in tune with the ethical trends of the society. Creating these supervisory bodies shows concern for ethical action, as well as of the interest of the most vulnerable groups when it comes to RSD procedures.

One more critical dimension of the continued ethical assessment is the communication with the representatives of other disciplines. This is because of the insight gained from a mix of law from the technology point of view, human rights, and ethical consideration. By the activities of seminars, conferences, and research studies such as the current study, stakeholders can feedback experiences and new standards to design the ethical standards as the technology and norms develop. In addition, refugees themselves have been involved in the discussion on how AI decisions might affect their lives, thus making those most affected to also be the ones helping in setting the policies that guide the technology.

The current paper supports the further development of research and educational programmes to build an ethical culture for innovation in AI technology which is engaged with refugee management. Academic Institutions, legal bodies and technology firms could partner to fund specific projects and initiatives such as the calls for papers that target the establishment of new, School of Thought for permission to extend research into specific issues related to AI for refugees and other irregular migrants. When funding this kind of research, one can cultivate the empirical focus on the consequences of AI, as well as spot the opportunities for AI regulation.

Awareness and training sessions, specific to AI ethics should be created for migration management involving; policymakers, developers, and RSD officers. These programmes could offer important knowledge of how to apply AI ethically, focusing on the special factors of refugee context. Also, cross functional training assures that developers and RSD officers comprehend the technical and the ethical aspect concerning the provision of AI solutions. When decision makers are fully informed, they can combine their knowledge about RSD's and the application of AI in this field, with the objectives to protect human rights and stay updated with the latest technology available.

As the integration of Artificial Intelligence (AI) continues to advance across the globe, ethical principles for the governance of AI in RSD must therefore be established both at international and national level. There are well-established International Principles for the sake of proper organisation of business activities and relations of businesses with people, amongst which the main one is the United Nations Guiding Principles on Business and Human Rights that lay down the base for ethical usage of Artificial Intelligence. Nevertheless, these standards need to be make more context sensitive to the respective countries, because every country has its own social, cultural and legal environment.

For instance, India that has refugees from different countries establish a national protocol that follows the guidelines from the international standards in a bid to tackle certain peculiarities. This particular approach may establish benchmarks of best practise in terms of RSD transparency, accountability, and data privacy in order that the AI systems are capable of adapting to supporting the needs and rights of the refugees in the region. Therefore, the combination of collaboration on the international level allows nations to share the existing practises, provide support to one another in ethical situations, and ensure the development of a single approach aimed at ethical application of AI in RSD.

At a national level, implementing data protection laws and ethical AI policies can further reinforce AI's responsible use. These policies could mandate ethical reviews of AI applications, public reporting on AI outcomes, and avenues for grievances to ensure public trust and compliance. Establishing a comprehensive policy infrastructure promotes the long-term sustainability of ethical AI practices in RSD and strengthens the nation's commitment to human rights.

Based on the various recommendations of this paper, it is clear that a middle of the road approach to Artificial Intelligence in RSD is necessary to safeguard the dignity and rights of refugees. Advantages related to the usage of AI in improvement of RSD processes outcomes are evident; however, these opportunities can be achieved only within the compliance with appropriate ethical norms which are aimed at protection of people's rights and needs. The recommendations provided in this paper are meant as a way to help the stakeholders develop proper and ethical AI systems compatible with international human rights legislation.

A commitment to ethical AI in RSD must extend beyond mere compliance; it requires an ongoing dedication to respecting human rights, promoting fairness, and engaging with affected communities. By fostering an inclusive approach and upholding ethical principles, stakeholders can ensure that AI serves as a tool for positive transformation in refugee migration management. Through vigilance, accountability, and collaboration, the integration of AI in RSD can advance not only efficiency but also equity and justice for those most vulnerable.

REFERENCES

Aizenberg, E., & Hoven, J. V. D. (2020, July 1). Designing for human rights in AI. *Big Data & Society*, *7*(2), 205395172094956–205395172094956. DOI: 10.1177/2053951720949566

Alexander, M M. (1999, April 1). Refugee status determination conducted by UNHCR. Oxford University Press, 11(2), 251-289. DOI: 10.1093/ijrl/11.2.251

Algorithmic Fairness for Asylum Seekers and Refugees (AFAR). (2023, January 1). https://www.rsc.ox.ac.uk/research/algorithmic-fairness-for-asylum-seekers-and-refugees-afar

Archer, D. W., Pigem, B. D. B., Bogdanov, D., Craddock, M., Gascón, A., Jansen, R., Jug, M., Laine, K., McLellan, R. K., Ohrimenko, O., Raykova, M., Trask, A., & Wardley, S. (2023, January 1). UN Handbook on Privacy-Preserving Computation Techniques. Cornell University. https://doi.org/DOI: 10.48550/arXiv.2301

Artificial Intelligence & Human Rights. Opportunities & Risks. (n.d). https://papers.ssrn.com/sol3/papers.cfm?abstract_id=3259344

Automating Immigration and Asylum. (2023, April 13). https://www.rsc.ox.ac.uk/files/files-1/automating-immigration-and-asylum_afar_9-1-23.pdf

Bag, A., Chaturvedi, A., Sneha., & Tiwari, R. (2024, April 4). Navigating the Legal Landscape of AI-Induced Property Damage. IGI Global, undefined(undefined), 185-210. DOI: 10.4018/979-8-3693-1565-1.ch011

Bansak, K., Ferwerda, J., Hainmueller, J., Dillon, A., Hangartner, D., Lawrence, D., & Weinstein, J. M. (2018, January 18). Improving refugee integration through data-driven algorithmic assignment. *Science*, *359*(6373), 325–329. DOI: 10.1126/science.aao4408 PMID: 29348237

Barale, C. (2023, January 1). Refugee status determination: how cooperation with machine learning tools can lead to more justice. Cornell University. https://doi.org//arxiv.2308.11541DOI: 10.48550

Bhattacharjee, S. (2008, March 1). India Needs a Refugee Law. RELX Group (Netherlands). https://papers.ssrn.com/sol3/Delivery.cfm/SSRN_ID2066397_code396686.pdf?abstractid=1760460&mirid=1

Boswell, C. (2017, May 15). The Ethics of Refugee Policy. DOI: 10.4324/9781315239910

Cameron, H E., Goldfarb, A., & Morris, L. (2021, March 30). Artificial Intelligence for a Reduction of False Denials in Refugee Claims. Oxford University Press, 35(1), 493-510. DOI: 10.1093/jrs/feab054

Chatterjee, S. (2020, June 3). AI strategy of India: policy framework, adoption challenges and actions for government. Emerald Publishing Limited, 14(5), 757-775. DOI: 10.1108/TG-05-2019-0031

Chekovik, T., & Achkoski, J. (2019, January 1). Border Control and Using Analysis Tools due to the Humanitarian Aspect of the Immigrant Crisis. *SciPress Ltd.*, *85*, 1–13. . DOI: 10.18052/www.scipress.com/ILSHS.85.1

Chohlas-Wood, A., Coots, M., Goel, S., & Nyarko, J. (2023, July 24). Designing equitable algorithms. *Nature Computational Science*, *3*(7), 601–610. DOI: 10.1038/s43588-023-00485-4 PMID: 38177749

Cobbe, J., Lee, M S A., & Singh, J. (2021, March 1). Reviewable Automated Decision-Making. DOI: 10.1145/3442188.3445921

Corrêa, N K., Galvão, C., Santos, J W., Pino, C D., Pinto, E P., Barbosa, C., Massmann, D., Mambrini, R., Galvão, L., Terem, E., & Oliveira, N D. (2023, October 1). Worldwide AI ethics: A review of 200 guidelines and recommendations for AI governance. Elsevier BV, 4(10), 100857-100857. DOI: 10.1016/j.patter.2023.100857

Dehghani, F., Dibaji, M., Anzum, F., Dey, L., Basdemir, A., Bayat, S., Boucher, J., Drew, S., Eaton, S. E., Frayne, R., Ginde, G., Harris, A. E., Ioannou, Y., Lebel, C., Lysack, J., Arzuaga, L. S., Stanley, E A M., Souza, R., Santos, R. D. S., . . . Bento, M. (2024, August 28). Trustworthy and Responsible AI for Human-Centric Autonomous Decision-Making Systems. Cornell University. https://doi.org//arxiv .2408.15550DOI: 10.48550

Ferrara, E. (2023, December 26). Fairness and Bias in Artificial Intelligence: A Brief Survey of Sources, Impacts, and Mitigation Strategies. *Multidisciplinary Digital Publishing Institute*, *6*(1), 3–3. DOI: 10.3390/sci6010003

Ferrara, E. (2023, January 1). Fairness And Bias in Artificial Intelligence: A Brief Survey of Sources, Impacts, And Mitigation Strategies. Cornell University. https://doi.org/DOI: 10.48550/arXiv.2304

Ferrari, V. (2018, January 1). EU Blockchain Observatory and Forum Workshop on GDPR, Data Policy and Compliance. RELX Group (Netherlands). DOI: 10.2139/ssrn.3247494

Feuerriegel, S., Dolata, M., & Schwabe, G. (2020, May 11). Fair AI. *Business & Information Systems Engineering*, *62*(4), 379–384. DOI: 10.1007/s12599-020-00650-3

Freeman, S. (1994, October 1). Utilitarianism, Deontology, and the Priority of Right. Wiley, 23(4), 313-349. DOI: 10.1111/j.1088-4963.1994.tb00017.x

Freund, D., Lykouris, T., Paulson, E., Sturt, B., & Weng, W. (2023, July 7). Group fairness in dynamic refugee assignment. DOI: 10.1145/3580507.3597758

Fukuda-Parr, S., & Gibbons, E. (2021, June 19). Emerging Consensus on 'Ethical AI'. *Global Policy*, *12*(S6), 32–44. DOI: 10.1111/1758-5899.12965

Gerver, M. (2020, June 23). Sufficiency, Priority, and Selecting Refugees. Wiley, 37(5), 713-730. DOI: 10.1111/japp.12445

Gibbons, E. (2021, March 1). Toward a More Equal World: The Human Rights Approach to Extending the Benefits of Artificial Intelligence. *IEEE Technology and Society Magazine*, *40*(1), 25–30. DOI: 10.1109/MTS.2021.3056295

Gîrju, R. (2021, May 14). Adaptive Multimodal and Multisensory Empathic Technologies for Enhanced Human Communication. Cornell University. https://doi.org/ DOI: 10.48550/arXiv.2110

Gorlick, B. (2001, October 15). Human Rights and Refugees: Enhancing Protection Through International Human Rights Law. RELX Group (Netherlands). https://papers.ssrn.com/sol3/papers.cfm?abstract_id=2331753

Ibiricu, B., & Made, M L V D. (2020, June 10). Ethics by design: a code of ethics for the digital age. Emerald Publishing Limited, 30(3), 395-414. DOI: 10.1108/RMJ-08-2019-0044

India, A. I. 2023: Expert Group Report. (2023, October 14). https://www.meity.gov.in/content/indiaai-2023-expert-group-report-%E2%80%93-first-editionthe-ministry-electronics-and-information

Inverardi, P. (2021, November 23). The Challenge of Human Dignity in the Era of Autonomous Systems. *Springer Nature*, *25-29*, 25–29. Advance online publication. DOI: 10.1007/978-3-030-86144-5_4

Janssen, M., Hartog, M. H. D., Matheus, R., Ding, A. Y., & Kuk, G. (2020, December 28). Will Algorithms Blind People? The Effect of Explainable AI and Decision-Makers' Experience on AI-supported Decision-Making in Government. *Social Science Computer Review*, *40*(2), 478–493. DOI: 10.1177/0894439320980118

Kaur, N. (2013, January 1). Protection of Refugees in India: A Critical Analysis. RELX Group (Netherlands). DOI: 10.2139/ssrn.2214274

Kaurin, D. (2019, May 15). Data Protection and Digital Agency for Refugees. https://www.cigionline.org/sites/default/files/documents/WRC%20Research%20Paper%20no.12.pdf

Kinchin, N. (2021, September 3). Technology, Displaced? The Risks and Potential of Artificial Intelligence for Fair, Effective, and Efficient Refugee Status Determination. Cambridge University Press, 37(3). DOI: 10.26826/law-in-context.v37i3.157

Kinchin, N., & Mougouei, D. (2022, October 1). What Can Artificial Intelligence Do for Refugee Status Determination? A Proposal for Removing Subjective Fear. Oxford University Press, 34(3), 373-397. DOI: 10.1093/ijrl/eeac040

Lam, K T., Lange, B., Blili-Hamelin, B., Davidovic, J., Brown, S., & Hasan, A. (2024, June 3). A Framework for Assurance Audits of Algorithmic Systems., 38, 1078-1092. DOI: 10.1145/3630106.3658957

Ling, X., Gao, M., & Wang, D. (2020, November 6). Intelligent document processing based on RPA and machine learning., 33, 1349-1353. DOI: 10.1109/CAC51589.2020.9326579

Lippert-Rasmussen, K., & Vitikainen, A. (2020, January 1). The ethics of refugee prioritization: reframing the debate. Taylor & Francis, 13(1), 1-5. DOI: 10.1080/16544951.2020.1735019

McGregor, L., Murray, D., & Ng, V. (2019, April 1). International Human Rights Law as a Framework for Algorithmic Accountability. Cambridge University Press, 68(2), 309-343. DOI: 10.1017/S0020589319000046

Memon, A., Given-Wilson, Z., Ozkul, D., Richmond, K. M., Muraszkiewicz, J., Weldon, E., & Katona, C. (2024, January 30). Artificial Intelligence (AI) in the asylum system. *Medicine, Science, and the Law*, *64*(2), 87–90. DOI: 10.1177/00258024241227721 PMID: 38291676

Molnar, P. (2019, December 1). Technology on the margins: AI and global migration management from a human rights perspective. Edward Elgar Publishing, 8(2), 305-330. DOI: 10.4337/cilj.2019.02.07

Mühlhoff, R. (2023, January 1). Predictive privacy: Collective data protection in the context of artificial intelligence and big data. *Big Data & Society*, *10*(1), 2053 95172311668–205395172311668. DOI: 10.1177/20539517231166886

Nalbandian, L. (2022, August 3). An eye for an 'I:' a critical assessment of artificial intelligence tools in migration and asylum management. *Comparative Migration Studies*, *10*(1), 32. Advance online publication. DOI: 10.1186/s40878-022-00305-0 PMID: 35967601

Nazer, L., Zatarah, R., Waldrip, S., Ke, J. X. C., Moukheiber, M., Khanna, A. K., Hicklen, R. S., Moukheiber, L., Moukheiber, D., Ma, H., & Mathur, P. (2023, June 22). Bias in artificial intelligence algorithms and recommendations for mitigation. *PLOS Digital Health*, *2*(6), e0000278–e0000278. DOI: 10.1371/journal.pdig.0000278 PMID: 37347721

Nejadgholi, I., Molamohammadi, M., Missaghi, K., & Bakhtawar, S. (2024, September 2). Human-Centered AI Applications for Canada's Immigration Settlement Sector. Cornell University. https://doi.org//arxiv.2409.01461DOI: 10.48550

Nuenen, T. V., Ferrer, X., Such, J. M., & Coté, M. (2020, October 22). Transparency for Whom? Assessing Discriminatory Artificial Intelligence. *Computer*, *53*(11), 36–44. DOI: 10.1109/MC.2020.3002181

Oduntan, O., & Ruthven, I. (2020, May 10). People and places: Bridging the information gaps in refugee integration. Wiley, 72(1), 83-96. DOI: 10.1002/asi.24366

Petković, D. (2023, January 30). It is Not "Accuracy vs. Explainability"—We Need Both for Trustworthy AI Systems. *IEEE Transactions on Technology and Society*, *4*(1), 46–53. DOI: 10.1109/TTS.2023.3239921

Policy on the Protection of Personal Data of Persons of Concern to UNHCR. (2023, February 6). https://www.refworld.org/docid/55643c1d4.html

Prentzas, N., Kakas, A. C., & Pattichis, C. S. (2023, January 1). Explainable AI applications in the Medical Domain: a systematic review. Cornell University. https://doi.org/DOI: 10.48550/arXiv.2308

Primer: Defending the rights of refugees and migrants in the digital age. (2024, February 5). https://www.amnesty.org/en/documents/pol40/7654/2024/en/

Refugee protection in the artificial intelligence era. (2022, September 7). https://www.chathamhouse.org/2022/09/refugee-protection-artificial-intelligence-era

Refugee Status Determination. (2023, February 6). https://www.refworld.org/rsdproceduralstandards.html

Robertson, M., Morris, K., & Walter, G. (2007, September 7). Overview of Psychiatric Ethics V: Utilitarianism and the Ethics of Duty. *Australasian Psychiatry*, *15*(5), 402–410. DOI: 10.1080/10398560701439640 PMID: 17828638

Rupp, G. (2016, September 1). Rethinking the Assumptions of Refugee Policy: Beyond Individualism to the Challenge of Inclusive Communities. *Journal on Migration and Human Security*, *4*(3), 76–82. DOI: 10.1177/233150241600400303

Saenz, A., Harned, Z., Banerjee, O., Abràmoff, M. D., & Rajpurkar, P. (2023, October 6). Autonomous AI systems in the face of liability, regulations and costs. *NPJ Digital Medicine*, *6*(1), 185. Advance online publication. DOI: 10.1038/s41746-023-00929-1 PMID: 37803209

Sánchez-Monedero, J., & Dencik, L. (2019, January 1). The politics of deceptive borders: 'biomarkers of deceit' and the case of iBorderCtrl. Cornell University. https://doi.org//arxiv.1911.09156DOI: 10.48550

Sánchez-Monedero, J., & Dencik, L. (2020, August 3). The politics of deceptive borders: 'biomarkers of deceit' and the case of iBorderCtrl. Routledge, 25(3), 413-430. DOI: 10.1080/1369118X.2020.1792530

Sandvik, K B. (2021, September 30). Digital Refugee Lawyering: Risk, Legal Knowledge, and Accountability. Oxford University Press, 40(4), 414-432. DOI: 10.1093/rsq/hdab013

Schoemaker, E., Baslan, D., Pon, B., & Dell, N. (2020, July 1). Identity at the margins: data justice and refugee experiences with digital identity systems in Lebanon, Jordan, and Uganda. Taylor & Francis, 27(1), 13-36. DOI: 10.1080/02681102.2020.1785826

Shams, R. A., Zowghi, D., & Bano, M. (2023, January 1). Challenges and Solutions in AI for All. Cornell University. https://doi.org/DOI: 10.48550/arXiv.2307

Shneiderman, B. (2020, October 16). Bridging the Gap Between Ethics and Practice. *ACM Transactions on Interactive Intelligent Systems*, *10*(4), 1–31. DOI: 10.1145/3419764

Smrkolj, M. (2009, October 14). International Institutions and Individualized Decision-Making: An Example of UNHCR's Refugee Status Determination. Springer Nature, 165-193. DOI: 10.1007/978-3-642-04531-8_7

Talhouk, R., Montague, K., Garbett, A., Araújo-Soares, V., Akik, C., Ghattas, H., Ahmad, B., & Balaam, M. (2019, June 3). A Call For Embedding Dignity In Humanitarian Technologies., undefined(undefined). DOI: 10.1145/3328320.3328373

The UK is dropping an immigration algorithm that critics say is racist. (2020, August 5)., undefined(undefined). Retrieved October 29, 2024 from https://www.technologyreview.com/2020/08/05/1006034/the-uk-is-dropping-an-immigration-algorithm-that-critics-say-is-racist/

Vorster, J. (2009, July 26). The contribution of deontological Christian ethics to the contemporary human rights discourse. *In die Skriflig*, *43*(3). Advance online publication. DOI: 10.4102/ids.v43i3.234

Weber, L., & Gerard, A. (2024, April 10). Robodeport or surveillance fantasy?: how automated is automatic visa cancellation in Australia?. Frontiers Media, 9(undefined). DOI: 10.3389/fsoc.2024.1336160

Weber, R H. (2019, November 15). Socio-ethical values and legal rules on automated platforms: The quest for a symbiotic relationship. Elsevier BV, 36, 105380-105380. DOI: 10.1016/j.clsr.2019.105380

Widder, D G., Zhen, D., Dabbish, L., & Herbsleb, J D. (2023, June 12). It's about power: What ethical concerns do software engineers have, and what do they (feel they can) do about them?. DOI: 10.1145/3593013.3594012

Wolff, J. (2006, May 1). Making the World Safe for Utilitarianism. Cambridge University Press, 58(undefined), 1-22. DOI: 10.1017/S1358246106058012

Wolff, S. (2021, May 13). The Security Sector Governance–Migration Nexus. DOI: 10.2307/j.ctv1v3gqt4

Xu, W., & Gao, Z. (2023, January 1). Enabling Human-Centered AI: A Methodological Perspective. Cornell University. https://doi.org//arxiv.2311.06703DOI: 10.48550

Young, M., Himmelreich, J., Honcharov, D., & Soundarajan, S. (2021, June 9). The Right Tool for The Job? Assessing the Use of Artificial Intelligence for Identifying Administrative Errors. DOI: 10.1145/3463677.3463714

ADDITIONAL READING

Akhmetova, R., & Harris, E. (2021). Politics of technology: The use of artificial intelligence by US and Canadian immigration agencies and their impacts on human rights. In *Digital identity, virtual borders and social media*. Edward Elgar Publishing., DOI: 10.4337/9781789909159.00008

Berlyand, Y., Raja, A. S., Dorner, S. C., Prabhakar, A. M., Sonis, J. D., Gottumukkala, R. V., Succi, M. D., & Yun, B. J. (2018). How artificial intelligence could transform emergency department operations. *The American Journal of Emergency Medicine*, *36*(8), 1515–1517. DOI: 10.1016/j.ajem.2018.01.017 PMID: 29321109

Chen, D. L., & Eagel, J. (2017). Can machine learning help predict the outcome of asylum adjudications? In *Proceedings of the 16th edition of the International Conference on Artificial Intelligence and Law (ICAIL '17)* (pp. 237–240). Association for Computing Machinery. DOI: 10.1145/3086512.3086538

Hamlin, R. (2014). *Let me be a refugee: Administrative justice and the politics of asylum in the United States, Canada, and Australia*. Academic., DOI: 10.1093/acprof:oso/9780199373307.001.0001

Iyer, S., Meng, X.-L., & Vittert, L. (2023). Close to refuge: Integrating AI and human insights for intervention and prevention: A conversation with Seema Iyer. *Harvard Data Science Review*, *5*(4). Advance online publication. DOI: 10.1162/99608f92.1a58d824

Madianou, M. (2019). Technocolonialism: Digital innovation and data practices in the humanitarian response to refugee crises. *Social Media + Society*, *5*(3), 2056305119863146. Advance online publication. DOI: 10.1177/2056305119863146

KEY TERMS AND DEFINITIONS

Refugee Status Determination: The legal process by which authorities decide if an asylum seeker qualifies for refugee protection under international law.

Ethical AI Governance: Frameworks and principles guiding the responsible and fair deployment of AI systems to ensure transparency and accountability.

Human Dignity in AI: The principle that AI systems should respect and uphold the inherent worth and rights of individuals in decision-making processes.

Bias in AI Systems: The risk that AI algorithms may perpetuate or exacerbate societal biases, leading to unfair outcomes in automated decision-making.

Comparative Analysis of AI in Migration: A study comparing how different countries apply AI technologies in their migration and refugee management systems.

Privacy and Data Protection: Safeguarding personal and sensitive information in AI systems, particularly in refugee cases, to prevent misuse and breaches.

AI in Humanitarian Contexts: The use of AI technologies to assist in humanitarian efforts, such as migration management, while ensuring ethical considerations are met.

Refugee Rights and Technology: The intersection of refugee protection laws with the use of emerging technologies like AI, ensuring that human rights are upheld.

Chapter 13
AI and Environmental Ethics:
The Role of AI in Addressing Climate Change Ethical Considerations in AI-Driven Environmental Solutions

A. Soundariya
B.S. Abdur Rahman Crescent Institute of Science and Technology, India

Niyati Kumari Behera
https://orcid.org/0000-0002-9384-2949
B.S. Abdur Rahman Crescent Institute of Science and Technology, India

P. Selvakumar
https://orcid.org/0000-0002-3650-4548
Science and Humanities, Nehru Institute of Technology, India

Pankaj Singh Chandel
https://orcid.org/0000-0002-4135-8194
Dev Sanskriti Vishwavidyalaya, India

Sandeep Gupta
https://orcid.org/0000-0002-3734-3723
Graphic Era University (Deemed), Dehradun, India

T. C. Manjunath
https://orcid.org/0000-0003-2545-9160
Rajarajeswari College of Engineering, India

ABSTRACT

The intersection of artificial intelligence (AI) and environmental ethics represents a critical and increasingly relevant area of inquiry as technology advances and environmental concerns escalate. AI, with its transformative potential, offers substantial benefits for environmental management, including enhanced data analysis, predictive modeling, and optimization of resource use. For example, AI-driven tools can enhance climate modeling, predict natural disasters, and optimize energy consumption. In agriculture, AI can be used to monitor crop health, manage water

DOI: 10.4018/979-8-3693-9035-1.ch013

usage, and reduce waste through precision farming techniques. Additionally, AI applications in environmental monitoring can detect pollution levels, track wildlife populations, and support conservation efforts. These capabilities offer the promise of more effective and data-driven approaches to addressing environmental challenges and achieving sustainability goals. The benefits of AI in environmental management may not be evenly distributed across different regions and communities.

INTRODUCTION

The intersection of artificial intelligence (AI) and environmental ethics represents a critical and increasingly relevant area of inquiry as technology advances and environmental concerns escalate. AI, with its transformative potential, offers substantial benefits for environmental management, including enhanced data analysis, predictive modeling, and optimization of resource use. For example, AI-driven tools can enhance climate modeling, predict natural disasters, and optimize energy consumption. In agriculture, AI can be used to monitor crop health, manage water usage, and reduce waste through precision farming techniques. Additionally, AI applications in environmental monitoring can detect pollution levels, track wildlife populations, and support conservation efforts. These capabilities offer the promise of more effective and data-driven approaches to addressing environmental challenges and achieving sustainability goals. The benefits of AI in environmental management may not be evenly distributed across different regions and communities.

The intersection of AI and environmental ethics also requires a holistic approach to integrating technology with broader environmental values and goals. Ethical AI practices should be aligned with principles of environmental sustainability, conservation, and respect for natural ecosystems (Abdalla AN, et al., 2021). This involves considering the long-term impacts of AI technologies on environmental health and ensuring that their deployment supports rather than undermines sustainability objectives. Engaging with diverse stakeholders, including environmental organizations, policymakers, and the public, can help ensure that AI applications in environmental management reflect a broad range of perspectives and values. In addition, fostering interdisciplinary collaboration is key to addressing the complex ethical issues at the intersection of AI and environmental ethics (Ramakrishnan T., et al., 2022). Collaborations between technologists, environmental scientists, ethicists, and policymakers can facilitate the development of AI solutions that are both effective and ethically sound. By bringing together expertise from different fields, stakeholders can identify potential risks, address ethical dilemmas, and design AI systems that contribute positively to environmental sustainability. In conclusion, the intersection of AI and environmental ethics is a critical area of exploration as technology and environmental concerns increasingly intersect. While AI offers significant potential for advancing environmental management and sustainability, it also presents ethical challenges related to its environmental impact, equity, transparency, and alignment with broader environmental values. By fostering interdisciplinary collaboration, ensuring transparency and accountability, and prioritizing sustainability, we can navigate the complexities of AI and environmental ethics and work towards a future where technological advancements contribute to the health and well-being of both people and the planet.

AI FOR CLIMATE CHANGE MITIGATION: OPPORTUNITIES AND CHALLENGES

As climate change becomes an increasingly urgent global challenge, AI technologies present opportunities to accelerate efforts towards sustainability and resilience (Abduljabbar R, et al.,2019). However, alongside these opportunities are significant challenges that must be addressed to ensure that AI contributes effectively and ethically to climate action. Machine learning algorithms can process complex climate models, predict weather patterns, and assess environmental impacts with greater accuracy and efficiency than traditional methods. For example, AI can enhance climate forecasting by integrating data from satellites, sensors, and historical records, enabling more precise predictions of extreme weather events and long-term climate trends (Sugumaran K. M, D., et al., 2024). This improved forecasting capability can help governments, businesses, and communities prepare for and respond to climate-related risks, ultimately supporting more effective mitigation strategies. In addition to forecasting, AI can optimize energy use and reduce emissions through advancements in energy management. In industrial settings, AI can optimize processes to minimize energy consumption and emissions, contributing to overall reductions in carbon footprints. Moreover, the success of AI in climate change mitigation depends on interdisciplinary collaboration and stakeholder engagement. Addressing climate change effectively requires input from various fields, including environmental science, technology, policy, and ethics. Collaborative efforts between technologists, climate scientists, policymakers, and communities can facilitate the development of AI solutions that are both technically robust and aligned with broader climate goals. Engaging diverse stakeholders ensures that AI-driven climate actions are comprehensive, inclusive, and responsive to the needs of different communities. In conclusion, AI presents significant opportunities for advancing climate change mitigation through improved forecasting, optimized energy management, and enhanced environmental monitoring. By harnessing AI's capabilities, we can make substantial progress towards reducing emissions and building resilience against climate impacts.

Ethical AI for Sustainable Development: Aligning AI with Environmental Values

As AI technologies proliferate and integrate into various sectors, the ethical deployment of these systems becomes crucial for advancing environmental sustainability and addressing global ecological challenges (Bacco M, et al., 2018). This alignment involves a multifaceted approach that considers the environmental impact of AI itself, integrates sustainability into AI development and application, and ensures that AI supports and amplifies environmental values rather than undermining them. Data centers, which power AI applications, are known for their high energy consumption and carbon footprints. To address these issues, it is essential to adopt practices that reduce the environmental impact of AI infrastructure. Moreover, integrating environmental values into the design and application of AI systems is crucial for promoting sustainability. AI technologies should be developed with a focus on supporting environmental conservation and sustainable practices. This involves incorporating environmental considerations into AI algorithms and models to ensure that they promote efficient resource use, reduce waste, and minimize ecological harm. Additionally, establishing robust governance frameworks and oversight mechanisms can help address potential biases or unintended consequences associated with AI technologies. By promoting transparency and accountability, we can ensure that AI systems are used responsibly and ethically in ways that support environmental sustainability. The ethical deployment of AI also involves addressing the broader social and environmental impacts of AI technologies (Vijayakumar G, et al., 2024). AI systems should be designed and applied in ways that support equitable and inclusive outcomes, avoiding exacerbation of existing inequalities.

Furthermore, interdisciplinary collaboration and stakeholder engagement are essential for aligning AI with environmental values (Baduge SK, et al.,2022). Effective integration of AI into sustainable development requires input from diverse fields, including environmental science, technology, ethics, and policy. Collaborations between technologists, environmental experts, policymakers, and communities can facilitate the development of AI solutions that are both technically advanced and ethically sound. By incorporating environmental ethics into AI education and professional development, we can build a workforce that is aware of the ecological implications of their work and committed to aligning AI technologies with sustainability goals. This education should emphasize the role of AI in promoting environmental stewardship, addressing climate change, and supporting sustainable development. In conclusion, aligning AI with environmental values is crucial for ensuring that technological advancements contribute to sustainable development and address global ecological challenges. By minimizing the environmental footprint of AI technologies, integrating sustainability into AI design and application, and promoting transparency and accountability, we can ensure that AI supports and amplifies environmental values.

AI-Driven Environmental Monitoring: Surveillance, Detection, and Prediction

AI-driven environmental monitoring represents a transformative advancement in how we observe, analyze, and respond to environmental changes and challenges. By harnessing the power of artificial intelligence, we can significantly enhance our capabilities in surveillance, detection, and prediction of environmental phenomena. These capabilities are crucial for managing natural resources, protecting ecosystems, and addressing pressing issues such as climate change and pollution. However, while AI offers substantial benefits for environmental monitoring, it also raises important considerations related to privacy, data management, and ethical use. Surveillance and Detection AI-powered surveillance systems have revolutionized environmental monitoring by providing more comprehensive and real-time data about environmental conditions. Traditional monitoring methods, which often rely on manual observation and limited data collection, are being supplemented by advanced AI technologies that utilize satellite imagery, remote sensing, and IoT sensors. For instance, AI algorithms can process satellite images to detect changes in land use, deforestation, and urban expansion with high precision (Carranza-García M, et al., 2019). This capability is invaluable for monitoring large and remote areas that are otherwise difficult to assess. For example, AI systems can analyze data from air quality sensors to detect pollutants and identify sources of contamination, providing timely information to regulatory agencies and the public (Paramasivam Selvakumar, et al., 2024). Similarly, underwater sensors equipped with AI can monitor marine environments, detect pollution events, and track the health of coral reefs and other marine ecosystems. This enhanced surveillance capability allows for more proactive and informed responses to environmental issues.

Prediction and Forecasting. For instance, AI algorithms can forecast extreme weather events such as hurricanes, floods, and heatwaves by analyzing meteorological data and climate models. These predictions enable governments, businesses, and communities to prepare for and mitigate the impacts of extreme weather, ultimately enhancing resilience and reducing vulnerability.

AI is also instrumental in predicting environmental changes related to climate change, such as shifts in temperature patterns, sea-level rise, and changes in biodiversity. By integrating data from various sources, including climate models, historical records, and current observations, AI systems can provide insights into potential future scenarios and help guide climate adaptation strategies (Chen L, et al.,2022). For example, AI can model the impacts of different greenhouse gas emission scenarios on global temperatures and sea levels, informing policy decisions and climate action plans. Future Directions Looking ahead, the future of AI-driven environmental monitoring will likely involve further advancements in technology and integration. Emerging technologies such as edge computing, which enables data processing at the source, and advanced sensor networks will enhance the capabilities of AI systems and improve real-time monitoring. Additionally, the integration of AI with other technologies, such as blockchain for data integrity and automated decision-making systems, has the potential to further enhance environmental monitoring and management. Collaboration between researchers, technologists, policymakers, and communities will be crucial for advancing AI-driven environmental monitoring (Das U, et al.,2018). In conclusion, AI-driven environmental monitoring represents a significant advancement in our ability to observe, detect, and predict environmental changes. Through enhanced surveillance, detection, and forecasting capabilities, AI technologies offer valuable tools for managing natural resources, protecting ecosystems, and addressing environmental challenges.

The Ethics of AI-Optimized Resource Management: Efficiency and Justice

The ethics of AI-optimized resource management involves a nuanced examination of how artificial intelligence (AI) can balance efficiency with justice in the allocation and utilization of resources. As AI technologies become increasingly integral to managing resources such as energy, water, and raw materials, their impact on both operational efficiency and social equity requires careful consideration (Ding Z, et al., 2022). The promise of AI in optimizing resource management is substantial, offering improvements in efficiency, cost savings, and sustainability. However, this potential must be weighed against ethical concerns related to fairness, transparency, and inclusivity. Similarly, AI can optimize water usage in agriculture by analyzing weather patterns, soil conditions, and crop needs, leading to significant water savings and improved crop yields. In manufacturing, AI can streamline production processes, minimize waste, and optimize supply chains, contributing to overall resource efficiency. Despite these advantages, the ethical implications of AI-optimized resource management are complex. One significant concern is the potential for exacerbating existing inequalities. Resource management systems that rely on AI must be designed to ensure that benefits are distributed fairly and that vulnerable communities are not disproportionately disadvantaged. For instance, while AI-driven energy management can lead to significant efficiency gains, there is a risk that the benefits may be concentrated in wealthier regions or sectors, leaving underserved communities with fewer improvements and potentially higher costs. Ensuring that AI applications in resource management are inclusive and equitable is crucial for addressing social justice issues and promoting fairness. Transparency and accountability are also key ethical considerations in AI-optimized resource management. AI systems that manage resources must operate in a transparent manner, with clear explanations of how decisions are made and how data is used. Furthermore, the ethical deployment of AI in resource management involves considering the environmental impact of the technologies themselves. This includes involving representatives from marginalized communities, local governments, and environmental organizations in the design and deployment process (Selvakumar, P.,2023). By incorporating a wide range of viewpoints, AI systems can be better tailored to address the needs of various stakeholders and contribute to more equitable resource management outcomes. In conclusion, the ethics of AI-optimized resource management involve balancing the benefits of efficiency with the imperative of justice. While AI technologies offer substantial improvements in resource management efficiency, it is crucial to address ethical considerations related to fairness, transparency, inclusivity, and environmental impact.

AI-Assisted Conservation: Protecting Biodiversity and Ecosystems

AI-assisted conservation represents a groundbreaking advancement in the efforts to protect biodiversity and ecosystems. By leveraging artificial intelligence, conservationists can enhance their ability to monitor, analyze, and manage natural environments more effectively than traditional methods allow (Ebrahimi M, et al., 2019). This integration of AI into conservation strategies holds immense promise for safeguarding endangered species, managing habitats, and addressing the complex challenges posed by environmental changes. However, while AI offers powerful tools for conservation, its application must be carefully managed to address potential ethical, practical, and ecological concerns. Enhanced Monitoring and Data Analysis AI-driven tools are also instrumental in monitoring wildlife populations and behaviors. Camera traps equipped with AI can automatically identify and classify animal species from images, reducing the need for manual data review and increasing the accuracy of species identification. This technology allows for more efficient tracking of endangered species, monitoring of migratory patterns, and detection of poaching activities. By improving the granularity and timeliness of wildlife data, AI assists conservationists in making informed decisions and implementing targeted conservation strategies. Predictive Modeling and Risk Assessment. For example, AI can predict the impacts of climate change on biodiversity by modeling shifts in temperature, precipitation, and habitat availability. These predictions enable conservationists to anticipate and prepare for future challenges, such as the effects of climate change on species distribution and ecosystem health.AI also supports risk assessment by identifying areas of high conservation priority and potential threats.

Machine learning models can analyze patterns in environmental data to highlight regions at risk of ecological degradation, such as areas prone to illegal logging or pollution. By focusing conservation efforts on high-risk areas, resources can be allocated more efficiently, and proactive measures can be taken to mitigate potential threats before they cause significant harm. Habitat Management and Restoration. AI-assisted conservation extends to habitat management and restoration efforts. AI technologies can optimize habitat management practices by analyzing data on soil health, vegetation cover, and water resources. For example, AI can help design and implement reforestation projects by identifying optimal planting locations and species based on environmental conditions and historical data. This targeted approach improves the chances of successful restoration and enhances ecosystem resilience.

Furthermore, AI can support habitat connectivity efforts by analyzing landscape fragmentation and identifying potential corridors for wildlife movement (Elahi E, et al.,2019). By understanding how different land use patterns affect habitat connectivity, conservationists can design strategies to mitigate fragmentation and support species migration. This helps maintain ecological processes and promotes the long-term viability of ecosystems. Ethical and Practical Considerations. While the potential benefits of AI-assisted conservation are substantial, there are ethical and practical considerations that must be addressed. One concern is the potential for over-reliance on technology, which may overshadow the importance of on-the-ground conservation efforts and local knowledge. AI should complement, rather than replace, traditional conservation practices and community engagement (Facchinetti D, et al.,2021). Collaboration between technologists, conservationists, and local communities is essential for ensuring that AI solutions are effectively integrated into conservation strategies and that they address the needs and priorities of those directly involved in conservation work. Ongoing research and development, as well as collaboration between technology providers and conservation organizations, are essential for overcoming these challenges and maximizing the impact of AI in conservation. Future Directions. Looking ahead, the future of AI-assisted conservation will likely involve further advancements in technology and increased integration with other emerging fields, such as genomics and remote sensing. Additionally, greater collaboration between researchers, practitioners, policymakers, and local communities will be key to addressing the complex challenges of conservation. In conclusion, AI-assisted conservation offers significant opportunities for protecting biodiversity and managing ecosystems more effectively. Through enhanced monitoring, predictive modeling, and habitat management, AI technologies can contribute to more informed and efficient conservation strategies.

THE CARBON FOOTPRINT OF AI: ASSESSING THE ENVIRONMENTAL IMPACT OF AI DEVELOPMENT

Energy Consumption and Carbon Emissions

Data centers, which host the servers and infrastructure necessary for AI operations, are major contributors to AI's carbon footprint. These facilities house large numbers of GPUs and CPUs, which perform the computations required for AI training and inference (Fang B, et al.,2023). The energy demands of these data centers are considerable, and their environmental impact depends heavily on the energy mix used to power them. Techniques such as model pruning, quantization, and transfer learning can significantly decrease the amount of computation needed without sacrificing performance. Additionally, optimizing the hardware used for AI, such as developing energy-efficient processors and employing techniques like specialized AI chips, can reduce the overall energy consumption.

Lifecycle Analysis and Transparency. This analysis includes not only the direct energy consumption and emissions but also the embedded carbon associated with manufacturing hardware, data center construction, and the end-of-life disposal of electronic components. Lifecycle analysis provides a holistic view of the environmental impact and identifies opportunities for improvement at various stages of the AI lifecycle.

Transparency in reporting and communicating the carbon footprint of AI is vital for fostering accountability and driving improvements. Organizations developing and deploying AI systems should disclose their energy consumption, emissions data, and the steps they are taking to mitigate their environmental impact (Gautam K, et al.,2023). Transparent reporting helps stakeholders make informed decisions about the environmental implications of AI technologies and encourages industry-wide efforts to adopt sustainable practices. Ethical and Regulatory Considerations. The ethical and regulatory dimensions of AI's carbon footprint are also important considerations. As the environmental impact of AI becomes more apparent, there is a growing call for regulations and standards that address the carbon emissions associated with AI development and deployment. Policymakers and industry leaders must work together to establish guidelines that promote energy efficiency, encourage the use of renewable energy, and incentivize sustainable practices in AI development. In addition to regulatory measures, ethical considerations involve addressing the broader implications of AI's carbon footprint for global climate goals. Ensuring that AI technologies support rather than hinder efforts to combat climate change is crucial. For example, while AI can contribute to energy efficiency and environmental monitoring, its development and deployment must be managed in a way that aligns with global efforts to reduce carbon emissions and mitigate climate change. Future Directions Looking ahead, the future of AI and its carbon footprint will likely involve continued innovation in both technology and practices. Advances in energy-efficient AI algorithms, sustainable data center design, and renewable energy adoption are expected to play key roles in reducing the carbon footprint of AI. Additionally, integrating AI with climate modeling and environmental monitoring can help optimize resource use and support efforts to address climate change. Collaboration between AI researchers, technology developers, policymakers, and environmental advocates will be essential for addressing the carbon footprint of AI and ensuring that technological advancements contribute positively to sustainability. In conclusion, assessing the carbon footprint of AI involves a comprehensive examination of energy consumption, emissions, and efficiency throughout the lifecycle of AI systems. By improving algorithmic and hardware efficiency, adopting sustainable data center practices, and promoting transparency and regulatory measures, the environmental impact of AI can be effectively managed. As AI continues to advance, it is crucial to integrate sustainability considerations into its development and deployment, ensuring that the technology supports global climate goals and contributes to a more sustainable future.

AI and Environmental Justice: Addressing Disproportionate Impacts AI's intersection with environmental justice is a critical area of concern as it holds the potential to either exacerbate or mitigate disproportionate environmental impacts across different communities (Ghadami N, et al.,2021). As AI technologies become increasingly influential in managing environmental resources, monitoring pollution, and predicting climate impacts, their deployment must be carefully managed to avoid reinforcing existing inequalities and to support fair and equitable outcomes.

Disproportionate Impacts of AI Deployment.AI systems can influence environmental outcomes in ways that disproportionately affect marginalized communities. For example, AI-driven environmental monitoring and management tools can improve pollution detection and resource allocation, yet if these technologies are implemented without considering local contexts, they may not address the specific needs of the most affected communities. In some cases, high-income or urban areas might benefit from advanced monitoring and intervention technologies, while low-income or rural communities may receive fewer resources or less attention, exacerbating existing disparities. For instance, if historical data used to train AI models reflect existing inequalities or gaps in environmental protection, the resulting AI systems might reinforce these biases. This could lead to inadequate monitoring of pollution or environmental hazards in underserved areas, further disadvantaging communities that are already exposed to higher environmental risks.AI for Equitable Resource Distribution. To address these challenges, AI must be employed in ways that promote equitable resource distribution and ensure that environmental benefits are accessible to all communities. One approach is to ensure that AI-driven environmental monitoring and management systems are designed with an emphasis on inclusivity. This involves incorporating diverse data sources and community input to ensure that the needs and conditions of all populations are considered. Moreover, AI can be leveraged to address environmental justice issues by targeting interventions and resources where they are most needed (Hanifa M, et al.,2023). By using AI to focus resources and actions on areas with the greatest need, policymakers and organizations can work towards more equitable environmental outcomes. Ethical Considerations and Community Engagement Ethical considerations are central to ensuring that AI supports environmental justice. This involves actively working to prevent and address any potential biases in AI systems and ensuring that the technology contributes positively to the well-being of all communities. This can be achieved through public consultations, partnerships with local organizations, and ongoing dialogue with affected populations. By prioritizing community engagement, stakeholders can build trust and ensure that AI solutions are implemented in ways that promote environmental justice. Future Directions and Policy Implications. Looking forward, integrating AI with environmental justice considerations will require ongoing research, collaboration, and policy development. This includes creating AI systems that are designed to identify and address disparities in environmental impacts and to support equitable access to environmental benefits. Policy implications include the need for regulations and guidelines that ensure AI technologies are used in ways that promote fairness and equity. Governments and organizations should work together to establish standards for AI deployment that consider environmental justice and to provide support for communities affected by environmental disparities. Additionally, fostering interdisciplinary collaboration between technologists, environmental scientists, social scientists, and justice advocates will be crucial for developing and implementing effective AI solutions that address both environmental and social challenges. By prioritizing transparency, accountability, community engagement, and ethical considerations, we can harness the power of AI to address disproportionate environmental impacts and support a more just and equitable approach to environmental protection.

Transparency and Accountability in AI-Driven Environmental Decision-Making

As AI technologies become increasingly integrated into environmental management, resource allocation, and policy-making, their impact on ecosystems and communities must be carefully monitored (Hannan MA, et al.,2021). Ensuring that these systems operate transparently and are held accountable helps to build trust, foster ethical practices, and promote better outcomes for both people and the environment. Transparency in AI Systems Transparency in AI systems involves making the operations, decisions, and data used by these systems open and understandable to stakeholders. For AI-driven environmental decision-making, this means providing clear explanations of how AI models process data, make predictions, and derive conclusions. Transparency is essential for several reasons, understanding and Trust By making AI processes transparent, stakeholders—including policymakers, environmental organizations, and the public—can understand how decisions are made and assess the validity of AI outputs. This understanding fosters trust in AI systems, which is crucial for their acceptance and effective use in environmental management. Explainability AI models, especially complex ones like deep learning networks, can often function as black boxes" where the decision-making process is not easily understood. Explainable AI (XAI) aims to make these models more interpretable by providing insights into how inputs are transformed into outputs. This is crucial for understanding the rationale behind environmental decisions and ensuring that they align with ecological goals and policies.

Accountability in AI Systems. Accountability in AI-driven environmental decision-making involves ensuring that AI systems are responsible for their outcomes and that there are mechanisms in place to address any adverse impacts or errors. Key aspects of accountability include Responsibility for Decisions Accountability requires clear attribution of responsibility for decisions made by AI systems. It is essential to identify who is accountable for the design, implementation, and deployment of AI technologies, as well as for the decisions and actions taken based on AI outputs. This helps ensure that there are mechanisms for addressing mistakes, inaccuracies, or unintended consequences.

Ethical Standards and Compliance Implementing: Ethical standards and regulatory compliance is vital for maintaining accountability (Ise T, et al.,2019). Organizations using AI for environmental decision-making should adhere to established guidelines and best practices that govern the ethical use of AI. This includes following environmental regulations, data protection laws, and industry standards to ensure that AI systems are used responsibly and in compliance with legal requirements.

Monitoring and Evaluation: Continuous monitoring and evaluation of AI systems are crucial for maintaining accountability (Jahanger A, et al.,2023). Regular assessments help identify any issues or discrepancies in AI performance and ensure that systems remain effective and aligned with environmental goals. Monitoring also involves tracking the impact of AI decisions on ecosystems and communities to ensure that any negative effects are promptly addressed.

Addressing Bias and Fairness. A significant aspect of transparency and accountability is addressing biases and ensuring fairness in AI systems (Kaack LH, et al.,2022). AI models can inadvertently perpetuate or amplify existing biases if they are trained on skewed or unrepresentative data. This can lead to unfair or harmful outcomes, particularly in environmental decision-making where the stakes can be high for vulnerable communities and ecosystems. Stakeholder Involvement Engaging a broad range of stakeholders, including local communities, environmental experts, and advocacy groups, helps ensure that AI systems reflect diverse perspectives and needs. Public Engagement and Communication Effective public engagement and communication are key components of transparency and accountability. Engaging with the public and stakeholders helps ensure that AI-driven environmental decisions are made with input from those affected by or interested in environmental outcomes. Collaboration between AI developers, environmental scientists, policymakers, and community organizations will be crucial for addressing challenges and advancing best practices. In conclusion, transparency and accountability are fundamental to the ethical and effective use of AI in environmental decision-making. By ensuring that AI systems are open, understandable, and responsible, stakeholders can build trust, address biases, and promote fair and equitable environmental management. As AI technologies continue to evolve, ongoing efforts to enhance transparency and accountability will be essential for achieving positive environmental outcomes and supporting sustainable development.

AI, Energy, and Resource Consumption: Optimizing for Sustainability

As AI technologies become more integral to various sectors, their potential to influence energy usage and resource management must be carefully evaluated to maximize environmental benefits while minimizing adverse impacts (Kadow C, et al.,2020). Optimizing AI for sustainability involves leveraging AI's capabilities to improve efficiency, reduce waste, and support the transition to more sustainable practices across industries.

Energy Consumption in AI Systems AI systems, particularly those involving machine learning and deep learning, are known for their high energy consumption. Efforts to optimize AI for sustainability must address the high energy demands of AI training and deployment. The potential for AI to inadvertently contribute to environmental harm must be carefully managed. For example, the production and disposal of AI hardware can have environmental impacts, including resource extraction and electronic waste. Sustainable practices in hardware manufacturing, such as using recycled materials and designing for longevity, can help mitigate these effects. Ethical considerations also involve ensuring that AI-driven solutions for sustainability are equitable and do not disproportionately impact marginalized communities (Kadow C, et al.,2020). Additionally, advancements in energy storage and renewable energy technologies will complement AI's capabilities, further supporting sustainable practices. Collaboration between AI researchers, energy experts, policymakers, and industry stakeholders will be crucial for driving progress in this area. By working together, these groups can develop and implement strategies that maximize the benefits of AI while addressing its environmental impacts. Establishing standards and best practices for sustainable AI development and deployment will help ensure that technological advancements contribute positively to environmental goals. In conclusion, optimizing AI for sustainability involves addressing the energy consumption and environmental impacts associated with AI systems while leveraging AI's capabilities to improve resource management and support sustainable practices. By enhancing the efficiency of AI algorithms, improving data center operations, and applying AI to energy and resource management, stakeholders can work towards a more sustainable future. Ethical considerations and collaboration will be key to ensuring that AI technologies contribute positively to environmental goals and support equitable and responsible resource use.

The Role of AI in Sustainable Agriculture and Food Systems

Artificial intelligence (AI) is transforming agriculture and food systems by introducing innovative solutions that enhance sustainability, efficiency, and resilience. As the global population grows and the demand for food increases, AI technologies offer promising avenues to address the challenges facing agriculture, such as resource scarcity, environmental degradation, and food security. The integration of AI into agriculture involves optimizing resource use, improving crop yields, and developing sustainable practices that support the long-term health of ecosystems and communities. Optimizing Resource Use One of the primary ways AI contributes to sustainable agriculture is through the optimization of resource use. Traditional farming methods often involve blanket applications of water, fertilizers, and pesticides, which can lead to inefficiencies and environmental harm, such as water pollution and soil degradation. AI technologies, particularly in the form of precision agriculture, offer a more targeted approach (Lee WJ, et al.,2019). Similarly, AI can assist in precision fertilization, where nutrient application is tailored to specific areas of a field based on soil and crop data, reducing runoff and nutrient pollution. Enhancing Crop Yields and Quality Furthermore, AI can support crop breeding programs by analyzing genetic data and predicting the performance of different crop varieties under various conditions. This helps farmers implement practices that enhance soil quality, reduce erosion, and support biodiversity In addition, AI can be used to optimize the management of livestock, contributing to more sustainable animal husbandry. This helps reduce overproduction and spoilage, ensuring that food is distributed more efficiently and reaching consumers in a timely manner. Supporting Climate Resilience AI-driven climate models can analyze historical weather data and current climate trends to predict future conditions, helping farmers plan for and adapt to changes in their growing environment (Ma W, et al.,2018). For instance,

AI can assist in selecting crop varieties that are better suited to future climate conditions, or in developing strategies for managing water resources under changing precipitation patterns (Zhang R, et al., 2019). By providing actionable insights and supporting adaptive practices, AI helps farmers build resilience to climate variability and reduce the risks associated with extreme weather events. Ethical and Social Considerations While AI offers numerous benefits for sustainable agriculture, it is important to address ethical and social considerations to ensure that its adoption supports equitable outcomes. Issues such as data privacy, access to technology, and the potential displacement of traditional farming practices must be carefully managed (Nawaz R, et al.,2021).

AI-Driven Waste Reduction and Management: Circular Economy Solutions

AI-driven technologies play a pivotal role in revolutionizing waste management by optimizing various stages of the waste hierarchy—prevention, reduction, reuse, recycling, and recovery—thereby fostering a sustainable circular economy.AI algorithms and machine learning models significantly enhance waste segregation processes.. High-resolution cameras and sensors analyze the composition of waste, enabling automated sorting with greater precision. For instance, AI systems can distinguish between different types of plastics, metals, and organics, ensuring that recyclables are properly separated and reducing contamination in recycling streams. This predictive capability helps municipalities and waste management companies to plan more efficient waste collection schedules, reducing operational costs and minimizing the environmental impact of waste collection activities. AI also facilitates dynamic adjustments based on changing patterns, such as seasonal variations or special events, further improving efficiency. AI's role extends to enhancing recycling processes through advanced sorting technologies (Newman ME, et al.,2014). For instance, AI can be used to identify and recover valuable materials from electronic waste, a category that poses significant challenges due to its complex composition. By automating the extraction of precious metals and rare materials, AI-driven systems not only reduce the volume of e-waste but also recover resources that would otherwise be lost. This contributes to a more sustainable lifecycle for electronic products and supports the circular economy by promoting resource efficiency.AI-driven waste reduction and management systems also support the circular economy by enabling the efficient use of secondary materials. Through sophisticated algorithms, AI can track and manage the flow of recycled materials, ensuring they are reintegrated into the production cycle effectively (Olabi AG, et al.,2023). This not only reduces the reliance on virgin resources but also supports the creation of a closed-loop system where materials are perpetually cycled through the economy.

In conclusion, AI-driven waste reduction and management solutions represent a significant advancement in the pursuit of a circular economy. By enhancing waste segregation, optimizing collection and recycling processes, and supporting sustainable product design, AI contributes to a more efficient and environmentally responsible waste management system.

The Intersection of AI, Policy, and Environmental Regulation

The intersection of Artificial Intelligence (AI), policy, and environmental regulation represents a crucial nexus for advancing sustainable development and addressing pressing environmental challenges. As AI technologies evolve and proliferate, they offer unprecedented opportunities to enhance environmental management, improve regulatory compliance, and drive policy innovation. AI's integration into environmental regulation is transforming the way governments and organizations monitor, enforce, and develop policies aimed at protecting natural resources and ecosystems. This predictive capability is invaluable for policymakers, as it allows them to evaluate the potential effectiveness of regulations before implementation. For instance, AI-driven models can simulate the impact of proposed emission reduction targets on air quality and public health, helping policymakers design more effective and evidence-based environmental regulations (Partel V, et al.,2019).AI also plays a crucial role in regulatory compliance and enforcement. Traditional compliance monitoring often relies on manual inspections and reporting, which can be resource-intensive and prone to errors.

To address this, policies and regulations should mandate clear documentation of AI methodologies and decision-making processes, as well as mechanisms for auditing and reviewing AI systems to ensure they operate fairly and effectively. The rapid pace of AI innovation also poses a challenge for environmental regulation, as existing regulatory frameworks may struggle to keep up with technological advancements. Policymakers need to adopt flexible and adaptive regulatory approaches that can accommodate the evolving nature of AI technologies (Raj EFI,et al., 2021). This may involve creating regulatory sandboxes where new AI applications can be tested and evaluated in a controlled environment before broader deployment. Additionally, international collaboration and harmonization of AI-related regulations can help address cross-border environmental issues and ensure consistent standards for AI applications in environmental management. Furthermore, integrating AI into environmental regulation requires a multidisciplinary approach, involving collaboration between AI experts, environmental scientists, policymakers, and regulatory authorities. Such collaboration ensures that AI solutions are developed with a deep understanding of environmental contexts and regulatory requirements. It also facilitates the exchange of knowledge and best practices, helping to overcome technical and policy challenges. In conclusion, the intersection of AI, policy, and environmental regulation holds significant promise for advancing environmental protection and sustainable development (Yan B, et al.,2021)..

AI and Environmental Education: Empowering Informed Decision-Making

The integration of Artificial Intelligence (AI) into environmental education represents a transformative advancement, enabling more effective, personalized, and impactful learning experiences. As environmental challenges become increasingly complex, AI has the potential to significantly enhance how individuals and communities understand and engage with environmental issues. By leveraging AI technologies, educators can provide deeper insights into ecological systems, foster informed decision-making, and promote sustainable practices among students and the public alike.AI-driven tools offer a range of applications in environmental education, from personalized learning experiences to sophisticated simulations that enhance understanding of complex environmental processes (Reddy KSP, et al., 2020). Adaptive learning platforms powered by AI can tailor educational content to individual learners' needs, providing customized lessons based on their prior knowledge, learning pace, and areas of interest. For instance, an AI system might analyze a student's performance and adapt the curriculum to focus more on areas where they struggle, or offer advanced materials in topics they excel in. This personalized approach helps ensure that learners acquire a thorough understanding of environmental concepts at their own pace, making education more engaging and effective.

Environmental education. These tools can provide on-demand information, answer questions, and guide learners through educational materials. For instance, an AI chatbot could assist students with inquiries about specific environmental topics, offer explanations, and recommend additional resources based on their interests. This 24/7 availability of information helps reinforce learning and provides a valuable resource for learners seeking to deepen their understanding of environmental issues outside of traditional classroom settings.AI can also enhance community engagement and public awareness through targeted educational campaigns. By analyzing demographic data and online behaviors, AI can help design and deliver customized educational content to specific audiences. For example, an AI system might identify key environmental concerns within a particular community and tailor educational materials and outreach efforts to address those concerns. This targeted approach ensures that educational initiatives are relevant and impactful, increasing the likelihood of behavioral change and community involvement in environmental conservation efforts. In addition, AI can support the development of innovative teaching methods and tools (Saheb T, et al.,2022). For example, AI can assist in creating interactive educational games and applications that teach environmental concepts through experiential learning. These tools can simulate real-world environmental challenges and encourage learners to devise solutions, fostering problem-solving skills and a deeper understanding of ecological principles (Xiang X, et al., 2021). By making learning interactive and engaging, AI helps cultivate a generation of environmentally conscious individuals equipped to tackle future challenges. Additionally, educators must be trained to effectively use AI tools and interpret their outputs, ensuring that technology enhances rather than replaces human judgment and pedagogical expertise. Moreover, while AI can provide valuable insights and support, it is essential to complement technological advancements with traditional educational approaches that emphasize critical thinking, ethical considerations, and hands-on experiences.

AI, Environmental Ethics, and a Sustainable Future - Directions and Responsibilities

The convergence of Artificial Intelligence (AI) and environmental ethics signifies a pivotal development in the quest for a sustainable future. As AI technologies become increasingly integral to various aspects of environmental management, their role in shaping and upholding ethical standards is paramount. This intersection underscores the need for a nuanced approach that balances technological innovation with moral considerations, ensuring that AI contributes positively to environmental sustainability while respecting ethical principles and responsibilities (Tomazzoli C, et al.,2020).AI holds tremendous promise for advancing environmental sustainability through applications such as resource management, pollution control, and climate modeling. For instance, AI algorithms can optimize energy usage in smart grids, predict natural disasters, and monitor deforestation, all of which contribute to more effective environmental stewardship. However, the deployment of AI in these domains also raises critical ethical questions about how these technologies are designed, implemented, and governed. One fundamental ethical consideration is the principle of fairness. AI systems must be designed to avoid exacerbating existing inequalities or creating new ones.

Moreover, accountability mechanisms must be in place to address any adverse outcomes or unintended consequences resulting from AI deployment. Ensuring that AI technologies are subject to rigorous oversight and that there are clear channels for redress is crucial for maintaining ethical standards and fostering public trust. Protecting this data from misuse and ensuring that privacy rights are respected is essential. Ethical AI practices involve implementing robust data protection measures and being transparent about how data is collected, used, and shared. This not only safeguards individual privacy but also helps prevent potential abuses of AI technologies. The environmental impact of AI itself is another important ethical consideration (Wei MCF, et al., 2020). While AI can contribute to sustainability, the technologies underlying AI, such as data centers and computational resources, have their own environmental footprints. Moreover, there is a responsibility to consider the long-term implications of AI on environmental systems and ecosystems. AI technologies should be designed with foresight, considering potential future impacts on ecological balance and biodiversity. This involves adopting a precautionary approach and conducting thorough impact assessments to anticipate and mitigate any negative effects that AI might have on natural environments. Ethical AI deployment requires a commitment to safeguarding the integrity of ecosystems and ensuring that technological innovations do not compromise environmental health. Collaboration and interdisciplinary approaches are essential for addressing the ethical dimensions of AI in environmental contexts.

CONCLUSION

In the contemporary landscape of technological and environmental challenges, Artificial Intelligence (AI) emerges as both a formidable tool and a subject of ethical scrutiny. Its role in addressing climate change is undeniably transformative, offering solutions ranging from predictive modeling and optimization to real-time monitoring and resource management. However, the integration of AI into environmental strategies necessitates a deep engagement with ethical considerations to ensure that its deployment contributes positively without compromising fundamental values. AI's capacity to address climate change is vast. Through sophisticated algorithms, AI enhances the precision of climate models, improving our understanding of environmental dynamics and enabling more effective policy responses. AI can optimize energy consumption, advance renewable energy technologies, and support sustainable agricultural practices, contributing significantly to mitigation and adaptation efforts. For instance, AI-driven systems can forecast extreme weather events with greater accuracy, aiding in preparedness and response strategies. Additionally, AI's ability to analyze vast datasets can reveal insights into emissions patterns, enabling targeted interventions and policy formulation. Yet, with these capabilities come profound ethical responsibilities. Ensuring fairness in AI-driven environmental solutions is paramount. As AI technologies are integrated into climate strategies, it is crucial to consider their impact on different populations, particularly marginalized and vulnerable communities. Ethical AI deployment demands that benefits be distributed equitably and that any negative consequences be addressed to avoid exacerbating existing inequalities. For example, while AI can optimize resource allocation, it must not disadvantage those who are already at a disadvantage or disproportionately affected by environmental policies.

This transparency fosters accountability, enabling stakeholders to scrutinize AI-driven outcomes and ensure that they align with environmental and social goals. The environmental impact of AI itself is another critical ethical concern. While AI holds the potential to advance sustainability, its development and deployment come with an environmental footprint. Data centers, computational resources, and energy consumption associated with AI technologies can contribute to carbon emissions and resource depletion. Addressing these impacts requires a commitment to sustainable AI practices, such as optimizing algorithms for energy efficiency, using renewable energy sources, and minimizing waste. By incorporating sustainability into the development lifecycle of AI technologies, we can mitigate their environmental footprint and align their use with broader sustainability goals. Ethical AI deployment involves not only leveraging technology for immediate gains but also ensuring that it supports long-term ecological health and resilience. Collaboration and interdisciplinary approaches are key to navigating the ethical landscape of AI in environmental solutions. Engaging ethicists, environmental scientists, technologists, and policymakers in dialogue helps to align AI technologies with ethical principles and sustainability objectives. Such collaboration ensures that diverse perspectives are considered, leading to more comprehensive and responsible approaches to AI deployment. In conclusion, AI has the potential to play a pivotal role in addressing climate change and advancing environmental sustainability.

REFERENCES

Abdalla, A. N., Nazir, M. S., Tao, H., Cao, S., Ji, R., Jiang, M., & Yao, L. (2021). Integration of energy storage system and renewable energy sources based on artificial intelligence: An overview. *Journal of Energy Storage*, *40*, 102811. DOI: 10.1016/j.est.2021.102811

Abduljabbar, R., Dia, H., Liyanage, S., & Bagloee, S. A. (2019). Applications of artificial intelligence in transport: An overview. *Sustainability (Basel)*, *11*(1), 189. DOI: 10.3390/su11010189

Bacco M, Berton A, Ferro E, Gennaro C, Gotta A, Matteoli S, Paonessa F, Ruggeri M, Virone G, Zanella A (2018) Smart farming: opportunities, challenges and technology enablers. 2018 IoT Vert Top Summit Agric Tuscany (IOT Tuscany). 1–6. https://doi.org/DOI: 10.1109/IOT-TUSCANY.2018.8373043

Baduge, S. K., Thilakarathna, S., Perera, J. S., Arashpour, M., Sharafi, P., Teodosio, B., Shringi, A., & Mendis, P. (2022). Artificial intelligence and smart vision for building and construction 4.0: Machine and deep learning methods and applications. *Automation in Construction*, *141*, 104440. DOI: 10.1016/j.autcon.2022.104440

Carranza-García, M., García-Gutiérrez, J., & Riquelme, J. C. (2019). A framework for evaluating land use and land cover classification using convolutional neural networks. *Remote Sensing (Basel)*, *11*(3), 274. DOI: 10.3390/rs11030274

Chen, L., Msigwa, G., Yang, M., Osman, A. I., Fawzy, S., Rooney, D. W., & Yap, P.-S. (2022b). Strategies to achieve a carbon neutral society: A review. *Environmental Chemistry Letters*, *20*(4), 2277–2310. DOI: 10.1007/s10311-022-01435-8 PMID: 35431715

Das, U., Pathak, P., Meena, M., & Mallikarjun, N. (2018). Precision farming a promising technology in horticulture: A review. *Int J Pure Appl Biosci*, *6*(1), 1596–1606. DOI: 10.18782/2320-7051.3088

Ding, Z., Chen, Z., Liu, J., Evrendilek, F., He, Y., & Xie, W. (2022). Co-combustion, life-cycle circularity, and artificial intelligence-based multi-objective optimization of two plastics and textile dyeing sludge. *Journal of Hazardous Materials*, *426*, 128069. DOI: 10.1016/j.jhazmat.2021.128069 PMID: 34959215

Ebrahimi, M., Sarikhani, M. R., Safari Sinegani, A. A., Ahmadi, A., & Keesstra, S. (2019). Estimating the soil respiration under different land uses using artificial neural network and linear regression models. *Catena*, *174*, 371–382. DOI: 10.1016/j.catena.2018.11.035

Elahi, E., Weijun, C., Zhang, H., & Abid, M. (2019a). Use of artificial neural networks to rescue agrochemical-based health hazards: A resource optimisation method for cleaner crop production. *Journal of Cleaner Production*, *238*, 117900. DOI: 10.1016/j.jclepro.2019.117900

Facchinetti, D., Santoro, S., Galli, L. E., Fontana, G., Fedeli, L., Parisi, S., Bonacchi, L. B., Šušnjar, S., Salvai, F., Coppola, G., Matteucci, M., & Pessina, D. (2021). Reduction of pesticide use in fresh-cut salad production through artificial intelligence. *Applied Sciences (Basel, Switzerland)*, *11*(5), 1992. DOI: 10.3390/app11051992

Fang, B., Yu, J., Chen, Z., Osman, A. I., Farghali, M., Ihara, I., Hamza, E. H., Rooney, D. W., & Yap, P.-S. (2023). Artificial intelligence for waste management in smart cities: A review. *Environmental Chemistry Letters*, *21*(4), 1959–1989. Advance online publication. DOI: 10.1007/s10311-023-01604-3 PMID: 37362015

Gautam, K., Sharma, P., Dwivedi, S., Singh, A., Gaur, V. K., Varjani, S., Srivastava, J. K., Pandey, A., Chang, J.-S., & Ngo, H. H. (2023). A review on control and abatement of soil pollution by heavy metals: Emphasis on artificial intelligence in recovery of contaminated soil. *Environmental Research*, *225*, 115592. DOI: 10.1016/j.envres.2023.115592 PMID: 36863654

Ghadami, N., Gheibi, M., Kian, Z., Faramarz, M. G., Naghedi, R., Eftekhari, M., Fathollahi-Fard, A. M., Dulebenets, M. A., & Tian, G. (2021). Implementation of solar energy in smart cities using an integration of artificial neural network, photovoltaic system and classical Delphi methods. *Sustainable Cities and Society*, *74*, 103149. DOI: 10.1016/j.scs.2021.103149

Hanifa M, Agarwal R, Sharma U, Thapliyal PC, Singh LP (2023) A review on CO2 capture and sequestration in the construction industry: emerging approaches and commercialised technologies. J Co2 Util 67:102292. DOI: 10.1016/j.jcou.2022.102292

Hannan, M. A., Al-Shetwi, A. Q., Ker, P. J., Begum, R. A., Mansor, M., Rahman, S. A., Dong, Z. Y., Tiong, S. K., Mahlia, T. M. I., & Muttaqi, K. M. (2021). Impact of renewable energy utilization and artificial intelligence in achieving sustainable development goals. *Energy Reports*, *7*, 5359–5373. DOI: 10.1016/j.egyr.2021.08.172

Ise, T., & Oba, Y. (2019). Forecasting climatic trends using neural networks: An experimental study using global historical data. *Frontiers in Robotics and AI*, *6*, 32. DOI: 10.3389/frobt.2019.00032 PMID: 33501048

Jahanger, A., Ozturk, I., Chukwuma Onwe, J., Joseph, T. E., & Razib Hossain, M. (2023). Do technology and renewable energy contribute to energy efficiency and carbon neutrality? Evidence from top ten manufacturing countries. *Sustainable Energy Technologies and Assessments*, *56*, 103084. DOI: 10.1016/j.seta.2023.103084

Kaack, L. H., Donti, P. L., Strubell, E., Kamiya, G., Creutzig, F., & Rolnick, D. (2022). Aligning artificial intelligence with climate change mitigation. *Nature Climate Change*, *12*(6), 518–527. DOI: 10.1038/s41558-022-01377-7

Kadow, C., Hall, D. M., & Ulbrich, U. (2020). Artificial intelligence reconstructs missing climate information. *Nature Geoscience*, *13*(6), 408–413. DOI: 10.1038/s41561-020-0582-5

Lee, J., & Yoo, H. J. (2021). An overview of energy-efficient hardware accelerators for on-device deep-neural-network training. *IEEE Open Journal of the Solid-State Circuits Society*, *1*, 115–128. DOI: 10.1109/OJSSCS.2021.3119554

Lee, W. J., Wu, H., Yun, H., Kim, H., Jun, M. B. G., & Sutherland, J. W. (2019) Predictive maintenance of machine tool systems using artificial intelligence techniques applied to machine condition data. *Proc CIRP 80*:506–511. https://doi.org/ DOI: 10.1016/j.procir.2018.12.019

Ma, W., Qiu, Z., Song, J., Li, J., Cheng, Q., Zhai, J., & Ma, C. (2018). A deep convolutional neural network approach for predicting phenotypes from genotypes. *Planta*, *248*(5), 1307–1318. DOI: 10.1007/s00425-018-2976-9 PMID: 30101399

Nawaz, R., Akhtar, R., Shahid, M. A., Qureshi, I. M., & Mahmood, M. H. (2021). Machine learning based false data injection in smart grid. *International Journal of Electrical Power & Energy Systems*, *130*, 106819. DOI: 10.1016/j.ijepes.2021.106819

Newman, M. E., McLaren, K. P., & Wilson, B. S. (2014). Assessing deforestation and fragmentation in a tropical moist forest over 68 years; the impact of roads and legal protection in the Cockpit Country, Jamaica. *Forest Ecology and Management*, *315*, 138–152. DOI: 10.1016/j.foreco.2013.12.033

Olabi, A. G., Abdelghafar, A. A., Maghrabie, H. M., Sayed, E. T., Rezk, H., Radi, M. A., Obaideen, K., & Abdelkareem, M. A. (2023). Application of artificial intelligence for prediction, optimization, and control of thermal energy storage systems. *Thermal Science and Engineering Progress*, *39*, 101730. DOI: 10.1016/j.tsep.2023.101730

Partel, V., Charan Kakarla, S., & Ampatzidis, Y. (2019). Development and evaluation of a low-cost and smart technology for precision weed management utilizing artificial intelligence. *Computers and Electronics in Agriculture*, *157*, 339–350. DOI: 10.1016/j.compag.2018.12.048

Raj, E. F. I., Appadurai, M., & Athiappan, K. (2021). Precision farming in modern agriculture. *Smart Agriculture Automation Using Advanced Technologies*, *65*, 294–329. DOI: 10.1007/978-981-16-6124-2_4

Ramakrishnan, T., Mohan Gift, M. D., Chitradevi, S., Jegan, R., & Subha Hency Jose, P. Nagaraja, Rajneesh Sharma H.N., Selvakumar P., Sintayehu Mekuria Hailegiorgis, (2022). "Study of Numerous Resins Used in Polymer Matrix Composite Materials", Advances in Materials Science and Engineering, vol. 2022, Article ID 1088926, 8 pages, 2022. https://doi.org/DOI: 10.1155/2022/1088926

Reddy, K. S. P., & Roopa, Y. M. L.N KR, Nandan NS (2020) IoT based smart agriculture using machine learning. In: 2020 Second international conference on inventive research in computing applications (ICIRCA), pp 130–134. https://doi.org/DOI: 10.1109/ICIRCA48905.2020.918337

Saheb, T., Dehghani, M., & Saheb, T. (2022). Artificial intelligence for sustainable energy: A contextual topic modeling and content analysis. *Sustainable Computing : Informatics and Systems*, *35*, 100699. DOI: 10.1016/j.suscom.2022.100699

Selvakumar, P. (2023). Biomedical applications of polymers in biosensors, cancer vaccines and drug delivery systems. *Biopolimery i Kletka*, *39*(4), 245–256. DOI: 10.7124/bc.000AA0

Selvakumar, P., Mariappan, P., Shunmugam, S., Paul, R., & Jabez, S. (2024). Removal of chemicals from effluent using photobioreactor technology to improve environmental and health impacts. *Zeitschrift für Physikalische Chemie*, *2024*(7), 1295–1302. Advance online publication. DOI: 10.1515/zpch-2023-0450

Sugumaran, K. M. D., Pandiyarajan V., Y. B S, Selvakumar P. and M. S, (2024). "Analysis of Electromyography Signals for Control of Mechanical Prosthesis using Machine Learning Techniques," 2024 2nd International Conference on Intelligent Data Communication Technologies and Internet of Things (IDCIoT), Bengaluru, India, 2024, pp. 1059-1067, DOI: 10.1109/IDCIoT59759.2024.10467707

Tomazzoli, C., Scannapieco, S., & Cristani, M. (2020). Internet of Things and artificial intelligence enable energy efficiency. *Journal of Ambient Intelligence and Humanized Computing*, *14*(5), 4933–4954. DOI: 10.1007/s12652-020-02151-3

Vijayakumar, G., Rajkumar, M., Rajiv Chandar, N., Selvakumar, P., & Duraisamy, R. (2024). Muniyandi Rajkumar, Rajiv Chandar N, Selvakumar P and Ramesh Duraisamy, (2024). Environmentally friendly TDS removal from waste water by electrochemical ion exchange batch-type recirculation (EIR) technique. *Environmental Science. Water Research & Technology*, *10*(4), 826–835. DOI: 10.1039/D3EW00793F

Wang, Z., & Srinivasan, R. S. (2017). A review of artificial intelligence based building energy use prediction: Contrasting the capabilities of single and ensemble prediction models. *Renewable & Sustainable Energy Reviews*, *75*, 796–808. DOI: 10.1016/j.rser.2016.10.079

Wei, M. C. F., Maldaner, L. F., Ottoni, P. M. N., & Molin, J. P. (2020). Carrot yield mapping: A precision agriculture approach based on machine learning. *AI*, *1*(2), 229–241. DOI: 10.3390/ai1020015

Xiang, X., Li, Q., Khan, S., & Khalaf, O. I. (2021). Urban water resource management for sustainable environment planning using artificial intelligence techniques. *Environmental Impact Assessment Review*, *86*, 106515. DOI: 10.1016/j.eiar.2020.106515

Yan, B., Hao, F., & Meng, X. (2021). When artificial intelligence meets building energy efficiency, a review focusing on zero energy building. *Artificial Intelligence Review*, *54*(3), 2193–2220. DOI: 10.1007/s10462-020-09902-w

Zhang, R., Chen, Z.-Y., Xu, L.-J., & Ou, C.-Q. (2019a). Meteorological drought forecasting based on a statistical model with machine learning techniques in Shaanxi province, China. *The Science of the Total Environment*, *665*, 338–346. DOI: 10.1016/j.scitotenv.2019.01.431 PMID: 30772564

Compilation of References

Aayog, N. I. T. I. (2018). *National strategy for artificial intelligence*. Retrieved November 27, 2024 from https://www.niti.gov.in/sites/default/files/2023-03/National-Strategy-for-Artificial-Intelligence.pdf

Abdalla, A. N., Nazir, M. S., Tao, H., Cao, S., Ji, R., Jiang, M., & Yao, L. (2021). Integration of energy storage system and renewable energy sources based on artificial intelligence: An overview. *Journal of Energy Storage*, *40*, 102811. DOI: 10.1016/j.est.2021.102811

Abduljabbar, R., Dia, H., Liyanage, S., & Bagloee, S. A. (2019). Applications of artificial intelligence in transport: An overview. *Sustainability (Basel)*, *11*(1), 189. DOI: 10.3390/su11010189

Adams, C., Pente, P., Lemermeyer, G., & Rockwell, G. (2023). Ethical principles for artificial intelligence in K-12 education. *Computers and Education: Artificial Intelligence*, *4*, 100131. Advance online publication. DOI: 10.1016/j.caeai.2023.100131

African Union. (2024). *Continental artificial intelligence strategy*. Retrieved March 15, 2025 from https://au.int/sites/default/files/documents/44004-doc-EN-_Continental_AI_Strategy_July_2024.pdf

Aizenberg, E., & Hoven, J. V. D. (2020, July 1). Designing for human rights in AI. *Big Data & Society*, *7*(2), 205395172094956–205395172094956. DOI: 10.1177/2053951720949566

Ajunwa, I., Friedler, S. A., Scheidegger, C., & Venkatasubramanian, S. (2016, March 10). *Hiring by Algorithm: Predicting and Preventing Disparate Impact*. https://www.semanticscholar.org/paper/Hiring-by-Algorithm%3A-Predicting-and-Preventing-Ajunwa-Friedler/bd31ad5e998629998f35db9a10d858b36e603248

Albrecht, J. P. (2016). How the GDPR Will Change the World. *European Data Protection Law Review*, *2*(3), 287–289. DOI: 10.21552/EDPL/2016/3/4

Alexander, M M. (1999, April 1). Refugee status determination conducted by UNHCR. Oxford University Press, 11(2), 251-289. DOI: 10.1093/ijrl/11.2.251

Algorithmic Fairness for Asylum Seekers and Refugees (AFAR). (2023, January 1). https://www.rsc.ox.ac.uk/research/algorithmic-fairness-for-asylum-seekers-and-refugees-afar

Al-kfairy, M., Mustafa, D., Kshetri, N., Insiew, M., & Alfandi, O. (2024). Ethical Challenges and Solutions of Generative AI: An Interdisciplinary Perspective. *Informatics (MDPI)*, *11*(3), 3. Advance online publication. DOI: 10.3390/informatics11030058

Allen, L. E., & Saxon, C. S. (1995). Better language, better thought, better communication: The A-Hohfeld language for legal analysis. In *Proceedings of the Fifth International Conference on Artificial Intelligence and Law* (p. 219). ACM Press. DOI: 10.1145/222092.222245

Almeida, P., Santos, C., & Farias, J. (2020). Artificial intelligence regulation: a meta-framework for formulation and governance.. DOI: 10.24251/HICSS.2020.647

Alston, P. (2019). Extreme poverty and human rights: Seventy-fourth session Agenda item 70(b) Promotion and protection of human rights: Human rights questions, including alternative approaches for improving the effective enjoyment of human rights and fundamental freedoms (A/74/493). United Nations General Assembly. https://undocs.org/en/A/74/493

Altmiller, G., & Pepe, L. H. (2022). Influence of Technology in Supporting Quality and Safety in Nursing Education. *The Nursing Clinics of North America*, *57*(4), 551–562. DOI: 10.1016/j.cnur.2022.06.005 PMID: 36280294

Anderson, E. (2023). Dewey's moral philosophy. In Zalta, E. N., & Nodelman, U. (Eds.), *The Stanford encyclopedia of philosophy*. Stanford University., https://plato.stanford.edu/archives/spr2023/entries/dewey-moral/

Angwin, J., Larson, J., Mattu, S., & Kirchner, L. (2016). *Machine Bias: There's Software Used Across the Country to Predict Future Criminals*. And It's Biased Against Blacks. ProPublica.

Angwin, J., Larson, J., Mattu, S., & Kirchner, L. (2022). Machine bias. In *Ethics of data and analytics* (pp. 254–264). Auerbach Publications. DOI: 10.1201/9781003278290-37

Archer, D. W., Pigem, B. D. B., Bogdanov, D., Craddock, M., Gascón, A., Jansen, R., Jug, M., Laine, K., McLellan, R. K., Ohrimenko, O., Raykova, M., Trask, A., & Wardley, S. (2023, January 1). UN Handbook on Privacy-Preserving Computation Techniques. Cornell University. https://doi.org/DOI: 10.48550/arXiv.2301

Arciniegas Gome, M. (2023). *Reframing privacy in the digital age: The shift to data use control* (Doctoral dissertation, Carnegie Mellon University).

Arnheim, R. (1992). Deus ex machina. *British Journal of Aesthetics*, *32*(3), 221–226. DOI: 10.1093/bjaesthetics/32.3.221

Arnold, T., Lee, M., & Yampolskiy, R. (2019). The Ethics of AI: Challenges and Opportunities. *Ethics and Information Technology*.

Artificial Intelligence & Human Rights. Opportunities & Risks. (n.d). https://papers.ssrn.com/sol3/papers.cfm?abstract_id=3259344

Artificial Intelligence Strategy, U. A. E. (2024). *Artificial Intelligence in the UAE*. Retrieved November 27, 2024, from https://ai.gov.ae/strategy/

Arya, M. (2024, February 22). Case Study: Unilever's Sustainable Living Plan. *Medium*. https://medium.com/@thisismayank/case-study-unilevers-sustainable-living-plan-2226b2e5d5f5

Askin, J. (2024). From data co-opting to data co-oping: Using new corporate structures, data corporate-hood, and data personhood to prioritize data privacy. *UC Law Business Journal*, *20*(2), 177.

Attwood, A. I. (2020). Changing social learning theory through reliance on the Internet of Things and artificial intelligence. *Journal of Social Change*, *12*(1), 103–111. DOI: 10.5590/JOSC.2020.12.1.08

Australian Government. (2024). National Framework for Assurance of Artificial Intelligence. *Department of Finance*. Retrieved March 15, 2025 from https://www.finance.gov.au/government/public-data/data-and-digital-ministers-meeting/national-framework-assurance-artificial-intelligence-government

Automating Immigration and Asylum. (2023, April 13). https://www.rsc.ox.ac.uk/files/files-1/automating-immigration-and-asylum_afar_9-1-23.pdf

Awad, E., Dsouza, S., Kim, R., Schulz, J., Henrich, J., Shariff, A., Bonnefon, J.-F., & Rahwan, I. (2018). The moral machine experiment. *Nature*, *563*(7729), 59–64. DOI: 10.1038/s41586-018-0637-6 PMID: 30356211

Babikian, S. (2023). The role of legal frameworks in protecting digital artifacts: Artists' and collectors' perspectives. *ISSLP*, *2*(4), 18–24. DOI: 10.61838/kman.isslp.2.4.4

Bacco M, Berton A, Ferro E, Gennaro C, Gotta A, Matteoli S, Paonessa F, Ruggeri M, Virone G, Zanella A (2018) Smart farming: opportunities, challenges and technology enablers. 2018 IoT Vert Top Summit Agric Tuscany (IOT Tuscany). 1–6. https://doi.org/DOI: 10.1109/IOT-TUSCANY.2018.8373043

Baduge, S. K., Thilakarathna, S., Perera, J. S., Arashpour, M., Sharafi, P., Teodosio, B., Shringi, A., & Mendis, P. (2022). Artificial intelligence and smart vision for building and construction 4.0: Machine and deep learning methods and applications. *Automation in Construction*, *141*, 104440. DOI: 10.1016/j.autcon.2022.104440

Bag, A., Chaturvedi, A., Sneha., & Tiwari, R. (2024, April 4). Navigating the Legal Landscape of AI-Induced Property Damage. IGI Global, undefined(undefined), 185-210. DOI: 10.4018/979-8-3693-1565-1.ch011

Bag, S., Wood, L. C., Xu, L., Dhamija, P., & Kayikci, Y. (2020). Big data analytics as an operational excellence approach to enhance sustainable supply chain performance. *Resources, Conservation and Recycling*, *153*, 104559. DOI: 10.1016/j.resconrec.2019.104559

Bai, C., & Sarkis, J. (2020). A supply chain transparency and sustainability technology appraisal model for blockchain technology. *International Journal of Production Research*, *58*(7), 2142–2162. DOI: 10.1080/00207543.2019.1708989

Bakiner, O. (2023). The promises and challenges of addressing artificial intelligence with human rights. *Big Data & Society*, *10*(2), 20539517231205476. Advance online publication. DOI: 10.1177/20539517231205476

Bakiner, O. (2023). What do academics say about artificial intelligence ethics? An overview of the scholarship. *AI and Ethics*, *3*(2), 513–525. DOI: 10.1007/s43681-022-00182-4

Bakshy, E., Messing, S., & Adamic, L. A. (2015). Exposure to ideologically diverse news and opinion on Facebook. *Science*, *348*(6239), 1130–1132. DOI: 10.1126/science.aaa1160 PMID: 25953820

Bansak, K., Ferwerda, J., Hainmueller, J., Dillon, A., Hangartner, D., Lawrence, D., & Weinstein, J. M. (2018, January 18). Improving refugee integration through data-driven algorithmic assignment. *Science*, *359*(6373), 325–329. DOI: 10.1126/science.aao4408 PMID: 29348237

Barocas, S., & Selbst, A. D. (2016). Big data's disparate impact. *SSRN*, *104*(3), 671–732. Retrieved November 27, 2024, from. DOI: 10.2139/ssrn.2477899

Baron-Cohen, S. (1995). *Mindblindness: An essay on autism and theory of mind*. MIT Press. Retrieved March 15, 2025 from https://doi.org/DOI: 10.7551/mit-press/4635.001.0001

Baron-Cohen, S. (2003). *The essential difference: The truth about the male and female brain*. Basic Books.

Baron-Cohen, S., Knickmeyer, R. C., & Belmonte, M. K. (2005). Sex differences in the brain: Implications for explaining autism. *Science*, *310*(5749), 819–823. Retrieved March 15, 2025 from. DOI: 10.1126/science.1115455 PMID: 16272115

Barrett, S., & al. (2023). *Canada's Pan-Canadian AI Strategy: Progress and Prospects*. CIFAR. Retrieved March 15, 2025 from https://cifar.ca

Bartlett, B. (2023). The possibility of ai-induced medical manslaughter: Unexplainable decisions, epistemic vices, and a new dimension of moral luck. *Medical Law International*, *23*(3), 241–270. DOI: 10.1177/09685332231193944

Batson, C. D. (2009). These things called empathy: Eight related but distinct phenomena. In J. Decety & W. Ickes (Eds.), *The social neuroscience of empathy* (pp. 3–15). Boston Review. Retrieved March 15, 2025 from https://doi.org/DOI: 10.7551/mitpress/9780262012973.003.0002

Bayisa, G., Dagnew, T., & Gebremeskel, T. (2023). An analysis of the presentation of human and democratic rights abuses in amelmal's yältäkoäche guzo (1974- 2018). *ERJSSH*, *10*(1), 173–184. DOI: 10.4314/erjssh.v10i1.10

Begishev, I., Khisamova, Z., & Gaifutdinov, R. (2019). On methods to legal regulation of artificial intelligence in the world. *International Journal of Innovative Technology and Exploring Engineering*, *9*(1), 5159–5162. DOI: 10.35940/ijitee.A9220.119119

Bell, A., Nov, O., & Stoyanovich, J. (2023). Think about the stakeholders first! toward an algorithmic transparency playbook for regulatory compliance. *Data & Policy*, *5*, e12. Advance online publication. DOI: 10.1017/dap.2023.8

Bell, E., Bryman, A., & Harley, B. (2018). *Business Research Methods*. Oxford University Press.

Benefo, E. O., Tingler, A., White, M., Cover, J., Torres, L., Broussard, C., Shirmohammadi, A., Pradhan, A. K., & Patra, D. (2022). Ethical, legal, social, and economic (ELSE) implications of artificial intelligence at a global level: A scientometrics approach. *AI and Ethics*, *2*(4), 667–682. DOI: 10.1007/s43681-021-00124-6

Benzinger, L., Ursin, F., Balke, W.-T., Kacprowski, T., & Salloch, S. (2023). Should Artificial Intelligence be used to support clinical ethical decision-making? A systematic review of reasons. *BMC Medical Ethics*, *24*(1), 48. DOI: 10.1186/s12910-023-00929-6 PMID: 37415172

Berman, M., & Dalvi, R. (2011). *Heroes, monsters and values: Science fiction films of the 1970s*. Cambridge Scholars Publishing.

Bhattacharjee, S. (2008, March 1). India Needs a Refugee Law. RELX Group (Netherlands). https://papers.ssrn.com/sol3/Delivery.cfm/SSRN_ID2066397_code396686.pdf?abstractid=1760460&mirid=1

Biller-Andorno, N., Ferrario, A., Joebges, S., Krones, T., Massini, F., Barth, P., Arampatzis, G., & Krauthammer, M. (2022). AI support for ethical decision-making around resuscitation: Proceed with care. *Journal of Medical Ethics*, *48*(3), 175–183. DOI: 10.1136/medethics-2020-106786 PMID: 33687916

Binns, R. (2018, January). Fairness in machine learning: Lessons from political philosophy. In Conference on fairness, accountability, and transparency (pp. 149-159). PMLR.

Binns, R. (2018). Fairness in Machine Learning: Lessons from Political Philosophy. In *Proceedings of the 2018 Conference on Fairness, Accountability, and Transparency* (pp. 149-159). ACM.

Boella, G., van der Torre, L., & Verhagen, H. (2001). An action-based ontology of legal relations. In *Proceedings of the 8th International Conference on Artificial Intelligence and Law* (pp. 227–228). ACM Press.

Bohr, A., & Memarzadeh, K. (Eds.). (2020). *Artificial Intelligence in healthcare [1 ed.] 0128184388, 9780128184387*. dokumen.pub. https://dokumen.pub/artificial-intelligence-in-healthcare-1nbsped-0128184388-9780128184387.html

Boncheva, A., & Serrano, R. (2022). Climate change, human rights and sustainability. *Revista Mexicana de Economía y Finanzas*, *17*(4), 1–20. DOI: 10.21919/remef.v17i4.802

Bostrom, N. (2014). *Superintelligence: Paths, dangers, strategies*. Oxford University Press.

Boswell, C. (2017, May 15). The Ethics of Refugee Policy. DOI: 10.4324/9781315239910

Bouhouita-Guermech, S., Gogognon, P., & Bélisle-Pipon, J.-C. (2023). Specific challenges posed by artificial intelligence in research ethics. *Frontiers in Artificial Intelligence*, *6*, 1149082. Advance online publication. DOI: 10.3389/frai.2023.1149082 PMID: 37483869

Bozdag, E. (2013). Bias in algorithmic filtering and personalization. *Ethics and Information Technology*, *15*(3), 209–227. DOI: 10.1007/s10676-013-9321-6

Bradley, A. S. (2019). Human rights racism. *Harvard Human Rights Journal*, *32*, 1–45.

Braubach, M., & Fairburn, J. (2010). Social inequities in environmental risks associated with housing and residential location—A review of evidence. *European Journal of Public Health*, *20*(1), 36–42. DOI: 10.1093/eurpub/ckp221 PMID: 20047933

Brauner, P., Hick, A., Philipsen, R., & Ziefle, M. (2023). What does the public think about artificial intelligence?—A criticality map to understand bias in the public perception of AI. *Frontiers of Computer Science*, *5*, 1113903. Advance online publication. DOI: 10.3389/fcomp.2023.1113903

Braun, V., & Clarke, V. (2006). Using Thematic Analysis in Psychology. *Qualitative Research in Psychology*, *3*(2), 77–101. DOI: 10.1191/1478088706qp063oa

Breczko, A., Filipkowski, W., & Kraśnicka, I. (2021). Ethics and the development of artificial intelligence: Challenges and dilemmas in the context of the 2030 United Nations agenda for sustainable development. In *Crime Prevention and Justice in 2030* (pp. 355–380). The UN and the Universal Declaration of Human Rights. DOI: 10.1007/978-3-030-56227-4_17

Bronzin, T., Milinovic, M., & Maricic, D. (2021). Artificial Intelligence (AI) brings enhanced personalized user experience. *2021 44th International Convention on Information, Communication and Electronic Technology (MIPRO)*. https://doi.org/ DOI: 10.23919/MIPRO52101.2021.9370446

Brown, R. D. (2020). Property ownership and the legal personhood of artificial intelligence. *Information & Communications Technology Law*, *30*(2), 208–234. DOI: 10.1080/13600834.2020.1861714

Brundage, M., Avin, S., Clark, J., Toner, H., Eckersley, P., Garfinkel, B., . . . Amodei, D. (2018). The malicious use of artificial intelligence: Forecasting, prevention, and mitigation. *arXiv preprint arXiv:1802.07228*.

Brundage, M., Avin, S., Clark, J., Toner, H., Eckersley, P., Garfinkel, B., Dafoe, A., Scharre, P., Zeitzoff, T., Filar, B., & Anderson, H. (2018). The malicious use of artificial intelligence: Forecasting, prevention, and mitigation. *arXiv preprint*, arXiv:1802.07228.

Brusseau, J. (2023). AI human impact: Toward a model for ethical investing in AI-intensive companies. *Journal of Sustainable Finance & Investment, 13*(2), 1030–1057. DOI: 10.1080/20430795.2021.1874212

Brusseau, J. (2023). From the ground truth up: Doing AI ethics from practice to principles. *AI & Society, 38*(4), 1651–1657. DOI: 10.1007/s00146-021-01336-4

Bryman, A. (2012). *Social Research Methods*. Oxford University Press.

Bryson, J., & Winfield, A. (2017). Standardizing Ethical Design for Artificial Intelligence and Autonomous Systems. *Computer, 50*(5), 116–119. *Computer*. Advance online publication. DOI: 10.1109/MC.2017.154

Bublitz, J. C. (2022). Might artificial intelligence become part of the person, and what are the key ethical and legal implications? *AI & Society*. Advance online publication. DOI: 10.1007/s00146-022-01432-7

Bullard, R. D., Johnson, G. S., & Torres, A. O. (Eds.). (2004). *Highway robbery: Transportation racism & new routes to equity*. South End Press.

Buolamwini, J., & Gebru, T. (2018). Gender Shades: Intersectional Accuracy Disparities in Commercial Gender Classification. *Proceedings of the 1st Conference on Fairness, Accountability and Transparency, Proceedings of Machine Learning Research,* 81, 77–91. Retrieved March 15, 2025 from https://proceedings.mlr.press/v81/buolamwini18a/buolamwini18a.pdf

Buolamwini, J., & Gebru, T. (2018, January). Gender shades: Intersectional accuracy disparities in commercial gender classification. In *Conference on fairness, accountability and transparency* (pp. 77-91). PMLR.

Burns, J. M. (1978). *Leadership*. HarperCollins.

Calo, R. (2017). "Artificial Intelligence Policy: A Primer and Roadmap." University of California, Davis, Law Review.

Calo, R. (2017). Artificial Intelligence Policy: A Primer and Roadmap. *UCLA Law Review. University of California, Los Angeles. School of Law*, *51*, 199–217.

Cameron, H E., Goldfarb, A., & Morris, L. (2021, March 30). Artificial Intelligence for a Reduction of False Denials in Refugee Claims. Oxford University Press, 35(1), 493-510. DOI: 10.1093/jrs/feab054

Campbell, M., Hoane, A. J.Jr, & Hsu, F. (2002). Deep Blue. *Artificial Intelligence, 134*(1–2), 57–83. DOI: 10.1016/S0004-3702(01)00129-1

Campolo, A., Sanfilippo, M., Whittaker, M., & Crawford, K. (2017). *AI Now Report 2017*. AI Now Institute.

Canadian Institute for Advanced Research. (2017). *Pan-Canadian Artificial Intelligence Strategy*.

Cardie, C. (1999). Integrating case-based learning and cognitive biases for machine learning of natural language. *Journal of Experimental & Theoretical Artificial Intelligence, 11*(3), 297–337. DOI: 10.1080/095281399146445

Carney, T. (2020). Artificial intelligence in welfare: Striking the vulnerability balance? *Monash University*, 25-26. DOI: 10.26180/13370369.v2

Carranza-García, M., García-Gutiérrez, J., & Riquelme, J. C. (2019). A framework for evaluating land use and land cover classification using convolutional neural networks. *Remote Sensing (Basel), 11*(3), 274. DOI: 10.3390/rs11030274

Cartabuke, M., Westerman, J. W., Bergman, J. Z., Whitaker, B. G., Westerman, J., & Beekun, R. I. (2019). Empathy as an antecedent of social justice attitudes and perceptions. *Journal of Business Ethics, 157*(3), 605–615. Retrieved March 15, 2025 from. DOI: 10.1007/s10551-017-3677-1

Cath, C. (2018). Governing artificial intelligence: Ethical, legal and technical opportunities and challenges. *Philosophical Transactions. Series A, Mathematical, Physical, and Engineering Sciences, 376*(2133), 20180080. DOI: 10.1098/rsta.2018.0080 PMID: 30322996

Cath, C. (2018). Governing Artificial Intelligence: Ethical, Legal, and Technical Opportunities and Challenges. *Philosophy & Technology, 31*(4), 655–680. PMID: 30873343

Cath, C., & Floridi, L. (2016). The design of the internet's architecture by the internet engineering task force (ietf) and human rights. *Science and Engineering Ethics, 23*(2), 449–468. DOI: 10.1007/s11948-016-9793-y PMID: 27255607

Cattaruzza, D., Absi, N., Feillet, D., & Gonzalez-Feliu, J. (2017). Vehicle routing problems for city logistics. *EURO Journal on Transportation and Logistics, 6*(1), 51–79. DOI: 10.1007/s13676-014-0074-0

Chae, B. (2015). Insights from hashtag #supplychain and Twitter Analytics: Considering Twitter and Twitter data for supply chain practice and research. *International Journal of Production Economics, 165*, 247–259. DOI: 10.1016/j.ijpe.2014.12.037

Chapman, S. (2020, May 17). *IBM launches its blockchain produce tracker, IBM Food Trust.* https://supplychaindigital.com/technology/ibm-launches-its-blockchain-produce-tracker-ibm-food-trust

Charmaz, K. (2006). Constructing Grounded Theory: A Practical Guide through Qualitative Analysis. *Sage (Atlanta, Ga.).*

Chatterjee, S. (2020, June 3). AI strategy of India: policy framework, adoption challenges and actions for government. Emerald Publishing Limited, 14(5), 757-775. DOI: 10.1108/TG-05-2019-0031

Chatterjee, I., & Chakraborty, P. (2021). Use of Information Communication Technology by Medical Educators Amid COVID-19 Pandemic and Beyond. *Journal of Educational Technology Systems, 49*(3), 310–324. DOI: 10.1177/0047239520966996

Chatterjee, S., Sreenivasulu, N., & Hussain, Z. (2021). Evolution of artificial intelligence and its impact on human rights: From sociolegal perspective. *International Journal of Law and Management, 64*(2), 184–205. DOI: 10.1108/IJLMA-06-2021-0156

Chauncey, S. A., & McKenna, H. P. (2023). A framework and exemplars for ethical and responsible use of AI Chatbot technology to support teaching and learning. *Computers and Education: Artificial Intelligence, 5*, 100182. Advance online publication. DOI: 10.1016/j.caeai.2023.100182

Chekovik, T., & Achkoski, J. (2019, January 1). Border Control and Using Analysis Tools due to the Humanitarian Aspect of the Immigrant Crisis. *SciPress Ltd., 85*, 1–13. . DOI: 10.18052/www.scipress.com/ILSHS.85.1

Chen, A., Hannon, O., Koegel, S., & Ciriello, R. (2024, March). Feels like empathy: How "emotional" AI challenges human essence. In *Australasian Conference on Information Systems.*

Cheng, L., Varshney, K. R., & Liu, H. (2021). Socially responsible AI algorithms: Issues, purposes, and challenges. *Journal of Artificial Intelligence Research, 1139-1140*, 1137–1181. Advance online publication. DOI: 10.1613/jair.1.12814

Cheng, X., Lin, X., Shen, X.-L., Zarifis, A., & Mou, J. (2022). The dark sides of AI. *Electronic Markets, 32*(1), 11–15. DOI: 10.1007/s12525-022-00531-5 PMID: 35600917

Chen, L., Msigwa, G., Yang, M., Osman, A. I., Fawzy, S., Rooney, D. W., & Yap, P.-S. (2022b). Strategies to achieve a carbon neutral society: A review. *Environmental Chemistry Letters, 20*(4), 2277–2310. DOI: 10.1007/s10311-022-01435-8 PMID: 35431715

Cheong, P. H. (2020). Rethinking piety, practices, and pedagogy in the era of artificial intelligence. In *Religion in the Age of Digitalization: From New Media to Spiritual Machines* (pp. 75-89). Routledge.

Chesterman, S. (2020). Artificial intelligence and the problem of autonomy. *Notre Dame Journal on Emerging Technologies*, *1*, 210.

Chiang, W. C., Li, Y., Shang, J., & Urban, T. L. (2019). Impact of drone delivery on sustainability and cost: Realizing the UAV potential through vehicle routing optimization. *Applied Energy*, *242*, 1164–1175. DOI: 10.1016/j.apenergy.2019.03.117

Chilton, A., & Posner, E. (2016). Respect for human rights: law and history. SSRN Electronic Journal. DOI: 10.2139/ssrn.2815272

China Briefing. (2021, August 24). *The PRC Personal Information Protection Law (Final): A Full Translation*. China Briefing News. https://www.china-briefing.com/news/the-prc-personal-information-protection-law-final-a-full-translation/

Chiu, T. K. F. (2021). A holistic approach to the design of Artificial Intelligence (AI) education for K-12 schools. *TechTrends*, *65*(5), 796–807. DOI: 10.1007/s11528-021-00637-1

Chohlas-Wood, A., Coots, M., Goel, S., & Nyarko, J. (2023, July 24). Designing equitable algorithms. *Nature Computational Science*, *3*(7), 601–610. DOI: 10.1038/s43588-023-00485-4 PMID: 38177749

Chowdhury, M. H., Mahmud, M., Islam, M. T., & Rahman, M. M. (2022). Unfair AI: It isn't just biased data. In *2022 IEEE International Conference on Data Mining (ICDM)*. DOI: 10.1109/ICDM54844.2022.00058

Chowdhury, P., Lau, C., & Pittayachawan, S. (2016, July 3). *Supply risk mitigation of small and medium enterprises: A social capital approach*. DOI: 10.13140/RG.2.2.23117.03045

CIFAR. Retrieved March 15, 2025 from https://cifar.ca/ai/ai-strategy/

Clapham, A. (2006). Human rights obligations of non-state actors.DOI: 10.1093/acprof:oso/9780199288465.001.0001

Cobbe, J., Lee, M S A., & Singh, J. (2021, March 1). Reviewable Automated Decision-Making. DOI: 10.1145/3442188.3445921

Coeckelbergh, M. (2020). *AI Ethics*. The MIT Press. DOI: 10.7551/mitpress/12549.001.0001

Cohen, L., Manion, L., & Morrison, K. (2018). *Research Methods in Education*. Routledge.

Cohrs, J., Maes, J., Moschner, B., & Kielmann, S. (2007). Determinants of human rights attitudes and behavior: A comparison and integration of psychological perspectives. *Political Psychology*, *28*(4), 441–469. DOI: 10.1111/j.1467-9221.2007.00581.x

Consejo Nacional de Política Económica y Social. República de Colombia, & Departamento Nacional de Planeación. (2019). *Política Nacional de Transformación Digital e Inteligencia Artificial* [National Council for Economic and Social Policy. Republic of Colombia, & National Planning Department. (2019). National Policy on Digital Transformation and Artificial Intelligence] (CONPES No. 3975). Retrieved March 15, 2025 from https://colaboracion.dnp.gov.co/CDT/Conpes/Econ%C3%B3micos/3975.pdf

Convention on the Rights of the Child. (1990). https://www.ohchr.org/en/instruments-mechanisms/instruments/convention-rights-child

Coplan, A., & Goldie, P. (Eds.). (2011). *Empathy: Philosophical and psychological perspectives*. Oxford University Press. Retrieved March 15, 2025 from https://doi.org/DOI: 10.1093/acprof:oso/9780199539956.001.0001

Corrêa, N K., Galvão, C., Santos, J W., Pino, C D., Pinto, E P., Barbosa, C., Massmann, D., Mambrini, R., Galvão, L., Terem, E., & Oliveira, N D. (2023, October 1). Worldwide AI ethics: A review of 200 guidelines and recommendations for AI governance. Elsevier BV, 4(10), 100857-100857. DOI: 10.1016/j.patter.2023.100857

Creswell, J. W., & Plano Clark, V. L. (2017). Designing and Conducting Mixed Methods Research. *Sage (Atlanta, Ga.)*.

Cuff, B. M. P., Brown, S. J., Taylor, L., & Howat, D. J. (2016). Empathy: A review of the concept. *Emotion Review*, *8*(2), 144–153. Retrieved March 15, 2025 from. DOI: 10.1177/1754073914558466

D'Antonoli, T. A. (2020). Ethical considerations for artificial intelligence: An overview of the current radiology landscape. *Diagnostic and Interventional Radiology (Ankara, Turkey)*, *26*(5), 504–511. DOI: 10.5152/dir.2020.19279 PMID: 32755879

Damoah, I. S., Ayakwah, A., & Tingbani, I. (2021). Artificial intelligence (AI)-enhanced medical drones in the healthcare supply chain (HSC) for sustainability development: A case study. *Journal of Cleaner Production*, *328*, 129598. DOI: 10.1016/j.jclepro.2021.129598

Das, U., Pathak, P., Meena, M., & Mallikarjun, N. (2018). Precision farming a promising technology in horticulture: A review. *Int J Pure Appl Biosci*, *6*(1), 1596–1606. DOI: 10.18782/2320-7051.3088

Daugherty, P. R., & Wilson, H. J. (2018). *Human+ machine: Reimagining work in the age of AI*. Harvard Business Press.

Davenport, T. H. (2018, January 30). Artificial Intelligence for the Real World. *Harvard Business Review*. https://hbr.org/webinar/2018/02/artificial-intelligence-for-the-real-world

Davenport, T. H., & Harris, J. G. (2007). *Competing on Analytics: The New Science of Winning*. Harvard Business Review Press.

Davis, M. H. (1983). Measuring individual differences in empathy: Evidence for a multidimensional approach. *Journal of Personality and Social Psychology*, *44*(1), 113–126. Retrieved March 15, 2025 from. DOI: 10.1037/0022-3514.44.1.113

Dawes, J. (2009). Human rights in literary studies. *Human Rights Quarterly*, *31*(2), 394–409. DOI: 10.1353/hrq.0.0071

Dawes, J. (2009). *That the world may know: Bearing witness to atrocities*. Harvard University Press.

de Oliveira Rodrigues, D. E. F. (2024). Personhood global whispers: Ethical echoes of decentralization? *Procedia Computer Science*, *237*, 767–774. DOI: 10.1016/j.procs.2024.05.164

Decety, J., & Ickes, W. (Eds.). (2009). *The social neuroscience of empathy*. MIT Press. Retrieved March 15, 2025 from https://doi.org/DOI: 10.7551/mitpress/9780262012973.001.0001

Decety, J., & Jackson, P. L. (2004). The functional architecture of human empathy. *Behavioral and Cognitive Neuroscience Reviews*, *3*(2), 71–100. Retrieved March 15, 2025 from. DOI: 10.1177/1534582304267187 PMID: 15537986

Decree of the President of the Russian Federation No. 490. (2019). *On the development of artificial intelligence in the Russian Federation*. Retrieved March 15, 2025 from http://www.kremlin.ru/acts/bank/44731

Denzin, N. K. (1978). *The Research Act: A Theoretical Introduction to Sociological Methods*. McGraw-Hill.

Dewey, J. (1922). *Democracy and education: An introduction to the philosophy of education*. Macmillan.

Dewey, J. (1938). *Experience and education*. Macmillan.

Dewey, J. (1994). The moral writings of John Dewey. In Gouinlock, J. (Ed.), *Prometheus Books* (Revised ed.).

Dhirani, L. L., Mukhtiar, N., Chowdhry, B. S., & Newe, T. (2023). Ethical Dilemmas and Privacy Issues in Emerging Technologies: A Review. *Sensors (Basel)*, *23*(3), 3. Advance online publication. DOI: 10.3390/s23031151 PMID: 36772190

Dignum, V. (2019). *Responsible Artificial Intelligence: How to Develop and Use AI in a Responsible Way*. DOI: 10.1007/978-3-030-30371-6

Dignum, V. (2018). *Responsible Artificial Intelligence: How to Develop and Use AI in a Responsible Way*. Springer.

Ding, Z., Chen, Z., Liu, J., Evrendilek, F., He, Y., & Xie, W. (2022). Co-combustion, life-cycle circularity, and artificial intelligence-based multi-objective optimization of two plastics and textile dyeing sludge. *Journal of Hazardous Materials*, *426*, 128069. DOI: 10.1016/j.jhazmat.2021.128069 PMID: 34959215

Donert, C. (2023). Women's rights as human rights after the end of history. *Gender & History*, *35*(3), 862–880. DOI: 10.1111/1468-0424.12729

Dulka, A. (2023). The use of artificial intelligence in international human rights law. *Stanford Technology Law Review : STLR : An Online High-Technology Law Journal from Stanford Law School*, *26*(2), 357.

Dwork, C., & Minow, M. (2022). Distrust of artificial intelligence: Sources & responses from computer science & law. *Daedalus*, *151*(2), 309–321. DOI: 10.1162/daed_a_01918

Ebrahimi, M., Sarikhani, M. R., Safari Sinegani, A. A., Ahmadi, A., & Keesstra, S. (2019). Estimating the soil respiration under different land uses using artificial neural network and linear regression models. *Catena*, *174*, 371–382. DOI: 10.1016/j.catena.2018.11.035

Elahi, E., Weijun, C., Zhang, H., & Abid, M. (2019a). Use of artificial neural networks to rescue agrochemical-based health hazards: A resource optimisation method for cleaner crop production. *Journal of Cleaner Production*, *238*, 117900. DOI: 10.1016/j.jclepro.2019.117900

Elendu, C., Amaechi, D. C., Elendu, T. C., Jingwa, K. A., Okoye, O. K., John Okah, M., Ladele, J. A., Farah, A. H., & Alimi, H. A. (2023). Ethical implications of ai and robotics in healthcare: A review. *Medicine*, *102*(50), e36671. DOI: 10.1097/MD.0000000000036671 PMID: 38115340

Elkins, D. (2024, June 12). *Impact of Technology on Nursing Education: Revolutionizing Learning*. Davis & Elkins College. https://www.dewv.edu/impact-of-technology-on-nursing-education-revolutionizing-learning/

Elliott, A. (2021). *Making sense of AI: Our algorithmic world*. John Wiley & Sons.

Erdélyi, O., & Goldsmith, J. (2020). Regulating artificial intelligence: proposal for a global solution. https://doi.org//arxiv.2005.11072DOI: 10.48550

Esteva, A., Robicquet, A., Ramsundar, B., Kuleshov, V., DePristo, M., Chou, K., Cui, C., Corrado, G., Thrun, S., & Dean, J. (2019). A guide to deep learning in healthcare. *Nature Medicine*, *25*(1), 24–29. DOI: 10.1038/s41591-018-0316-z PMID: 30617335

Estrada, D. (2020). Human supremacy as posthuman risk. *The Journal of Sociotechnical Critique*, *1*(1), 5.

Eubanks, V. (2018). *Automating Inequality: How High-tech Tools Profile, Police, and Punish the Poor* (Illustrated Edition). St Martin's Press.

Eubanks, V. (2018). *Automating Inequality: How High-Tech Tools Profile, Police, and Punish the Poor*. New York: Picador, St. Martin's Press. Retrieved March 15, 2025 from DOI: 10.5204/lthj.v1i0.1386

Eubanks, V. (2018). *Automating inequality: How high-tech tools profile, police, and punish the poor*. St. Martin's Press.

Eubanks, V. (2018). *Automating Inequality: How High-Tech Tools Profile, Police, and Punish the Poor*. St. Martin's Press.

European Commission. (2021). *Proposal for a Regulation of the European Parliament and of the Council Laying Down Harmonised Rules on Artificial Intelligence (Artificial Intelligence Act)*. European Commission.

European Parliament & Council of the European Union. (2016). *Regulation (EU) 2016/679 of the European Parliament and of the Council of 27 April 2016 on the protection of natural persons with regard to the processing of personal data and on the free movement of such data*. Official Journal of the European Union. Retrieved March 15, 2025 from https://eur-lex.europa.eu/legal-content/EN/TXT/?uri=CELEX%3A32016R0679

European Parliament & Council of the European Union. (2024). *Regulation (EU) 2024/1689 of the European Parliament and of the Council of 13 June 2024 laying down harmonised rules on artificial intelligence*. Official Journal of the European Union. Retrieved March 15, 2025 from https://artificialintelligenceact.eu/the-act

European Union Agency for Cybersecurity (ENISA). (2015, January 12). *Privacy and Data Protection by Design* [Report/Study]. ENISA. https://www.enisa.europa.eu/publications/privacy-and-data-protection-by-design

EU-Startups. (2024, October 1). *SuperAwesome*. EU-Startups. https://www.eu-startups.com/directory/superawesome/

Facchinetti, D., Santoro, S., Galli, L. E., Fontana, G., Fedeli, L., Parisi, S., Bonacchi, L. B., Šušnjar, S., Salvai, F., Coppola, G., Matteucci, M., & Pessina, D. (2021). Reduction of pesticide use in fresh-cut salad production through artificial intelligence. *Applied Sciences (Basel, Switzerland)*, *11*(5), 1992. DOI: 10.3390/app11051992

Fairly.ai. (2024). *Map of global AI regulations*. Retrieved November 27, 2024, from https://www.fairly.ai/blog/map-of-global-ai-regulations

Fang, B., Yu, J., Chen, Z., Osman, A. I., Farghali, M., Ihara, I., Hamza, E. H., Rooney, D. W., & Yap, P.-S. (2023). Artificial intelligence for waste management in smart cities: A review. *Environmental Chemistry Letters*, *21*(4), 1959–1989. Advance online publication. DOI: 10.1007/s10311-023-01604-3 PMID: 37362015

Farag, A., Greeley, L., & Swindell, A. (2022). Freire 2.0: Pedagogy of the digitally oppressed. *Educational Philosophy and Theory*, *54*(13), 2214–2227. DOI: 10.1080/00131857.2021.2010541

Feldstein, S. (2019). *The global expansion of AI surveillance* (Working Paper). Carnegie Endowment for International Peace. https://carnegieendowment.org/files/WP-Feldstein-AISurveillance_final1.pdf

Feldstein, S. (2019). The global expansion of AI surveillance. *Carnegie Endowment for International Peace* (Working paper). https://carnegieendowment.org/files/WP-Feldstein-AISurveillance_final1.pdf

Ferrara, E. (2023, December 26). Fairness and Bias in Artificial Intelligence: A Brief Survey of Sources, Impacts, and Mitigation Strategies. *Multidisciplinary Digital Publishing Institute*, *6*(1), 3–3. DOI: 10.3390/sci6010003

Ferrara, E. (2023, January 1). Fairness And Bias in Artificial Intelligence: A Brief Survey of Sources, Impacts, And Mitigation Strategies. Cornell University. https://doi.org/DOI: 10.48550/arXiv.2304

Ferrari, V. (2018, January 1). EU Blockchain Observatory and Forum Workshop on GDPR, Data Policy and Compliance. RELX Group (Netherlands). DOI: 10.2139/ssrn.3247494

Fesmire, S. (2003). *John Dewey and moral imagination: Pragmatism in ethics*. Indiana University Press.

Feuerriegel, S., Dolata, M., & Schwabe, G. (2020, May 11). Fair AI. *Business & Information Systems Engineering*, *62*(4), 379–384. DOI: 10.1007/s12599-020-00650-3

Field, A. (2013). Discovering Statistics Using IBM SPSS Statistics. *Sage (Atlanta, Ga.)*.

Fine, C. (2008). Will working mothers' brains explode? The popular new genre of neurosexism. *Neuroethics*, *1*(1), 69–72. Retrieved March 15, 2025 from. DOI: 10.1007/s12152-007-9004-2

Fiske, A., Henningsen, P., & Buyx, A. (2019). Your robot therapist will see you now: Ethical implications of embodied artificial intelligence in psychiatry, psychology, and psychotherapy. *Journal of Medical Internet Research*, *21*(5), e13216. Advance online publication. DOI: 10.2196/13216 PMID: 31094356

Flick, U. (2018). An Introduction to Qualitative Research. *Sage (Atlanta, Ga.)*.

Floridi, L., & Cowls, J. (2022). A unified framework of five principles for AI in society. *Machine learning and the city: Applications in architecture and urban design*, 535-545.

Floridi, L., Cowls, J., King, T. C., & Taddeo, M. (2021). How to design AI for social good: Seven essential factors. *Ethics, Governance, and Policies in Artificial Intelligence*, 125-151.

Floridi, L., & Cowls, J. (2019). A Unified Framework of Five Principles for AI in Society. *Harvard Data Science Review*. Advance online publication. DOI: 10.1162/99608f92.8cd550d1

Floridi, L., Cowls, J., Beltrametti, M., Chatila, R., Chazerand, P., Dignum, V., Luetge, C., Madelin, R., Pagallo, U., Rossi, F., Schafer, B., Valcke, P., & Vayena, E. (2018). AI4People—an ethical framework for a good AI society: Opportunities, risks, principles, and recommendations. *Minds and Machines*, *28*(4), 689–707. DOI: 10.1007/s11023-018-9482-5 PMID: 30930541

Floridi, L., & Taddeo, M. (2016). What is data ethics? *Philosophical Transactions. Series A, Mathematical, Physical, and Engineering Sciences*, *374*(2083), 20160360. DOI: 10.1098/rsta.2016.0360 PMID: 28336805

Floridi, L., & Taddeo, M. (2018). What is data ethics? *Philosophical Transactions of the Royal Society A*. PMID: 30322997

Fourneret, E., & Yvert, B. (2020). Digital normativity: A challenge for human subjectivation. *Frontiers in Artificial Intelligence*, *3*, 27. Advance online publication. DOI: 10.3389/frai.2020.00027 PMID: 33733146

Freeman, S. (1994, October 1). Utilitarianism, Deontology, and the Priority of Right. Wiley, 23(4), 313-349. DOI: 10.1111/j.1088-4963.1994.tb00017.x

Freire, P. (2014). *Pedagogy of the oppressed* (M. B. Ramos, Trans.; Thirtieth anniversary edition.). Bloomsbury.

Freund, D., Lykouris, T., Paulson, E., Sturt, B., & Weng, W. (2023, July 7). Group fairness in dynamic refugee assignment. DOI: 10.1145/3580507.3597758

Friedman, B., & Nissenbaum, H. (1996). Bias in computer systems. *ACM Transactions on Information Systems*, *14*(3), 330–347. DOI: 10.1145/230538.230561

Frith, C. D., & Frith, U. (2006). The neural basis of mentalizing. *Neuron*, *50*(4), 531–534. Retrieved March 15, 2025 from. DOI: 10.1016/j.neuron.2006.05.001 PMID: 16701204

Fritsch, T. (2024). Chatbots: An Overview of Current Issues and Challenges. In Arai, K. (Ed.), *Advances in Information and Communication* (pp. 84–104). Springer Nature Switzerland., DOI: 10.1007/978-3-031-53960-2_7

FTC. (2013, July 25). *Children's Online Privacy Protection Rule ("COPPA")*. Federal Trade Commission. https://www.ftc.gov/legal-library/browse/rules/childrens-online-privacy-protection-rule-coppa

FTC. (2020, July 20). *Complying with COPPA: Frequently Asked Questions*. Federal Trade Commission. https://www.ftc.gov/business-guidance/resources/complying-coppa-frequently-asked-questions

Fukuda-Parr, S., & Gibbons, E. (2021, June 19). Emerging Consensus on 'Ethical AI'. *Global Policy*, *12*(S6), 32–44. DOI: 10.1111/1758-5899.12965

Gaggioli, A., Villani, D., Serino, S., Banos, R., & Botella, C. (2019). Editorial: Positive Technology: Designing E-experiences for Positive Change. *Frontiers in Psychology*, *10*, 1571. Advance online publication. DOI: 10.3389/fpsyg.2019.01571 PMID: 31338053

Gasser, U., & Almeida, V. A. (2017). A layered model for AI governance. *IEEE Internet Computing*, *21*(6), 58–62. DOI: 10.1109/MIC.2017.4180835

Gaur, L., & Sahoo, B. M. (2022). Explainable AI in ITS: Ethical concerns. In *Explainable Artificial Intelligence for Intelligent Transportation Systems* (pp. 79–90). Springer. https://doi.org/DOI: 10.1007/978-3-030-95923-5_5

Gautam, K., Sharma, P., Dwivedi, S., Singh, A., Gaur, V. K., Varjani, S., Srivastava, J. K., Pandey, A., Chang, J.-S., & Ngo, H. H. (2023). A review on control and abatement of soil pollution by heavy metals: Emphasis on artificial intelligence in recovery of contaminated soil. *Environmental Research*, *225*, 115592. DOI: 10.1016/j.envres.2023.115592 PMID: 36863654

Gaviria, C. I. G. (2022). The role of artificial intelligence in pushing the boundaries of US regulation: A systematic review. *Santa Clara High-Technology Law Journal*, *38*, 123.

Gellert, R. (2015). Data protection: A risk regulation? Between the risk management of everything and the precautionary alternative. *International Data Privacy Law*, *5*(1), 3–19. DOI: 10.1093/idpl/ipu035

Gerver, M. (2020, June 23). Sufficiency, Priority, and Selecting Refugees. Wiley, 37(5), 713-730. DOI: 10.1111/japp.12445

Ghadami, N., Gheibi, M., Kian, Z., Faramarz, M. G., Naghedi, R., Eftekhari, M., Fathollahi-Fard, A. M., Dulebenets, M. A., & Tian, G. (2021). Implementation of solar energy in smart cities using an integration of artificial neural network, photovoltaic system and classical Delphi methods. *Sustainable Cities and Society*, *74*, 103149. DOI: 10.1016/j.scs.2021.103149

Ghassemi, M., Oakden-Rayner, L., & Beam, A. L. (2020). The False Hope of Current Approaches to Explainable Artificial Intelligence in Health Care. *The Lancet. Digital Health*. PMID: 34711379

Gibbons, E. (2021, March 1). Toward a More Equal World: The Human Rights Approach to Extending the Benefits of Artificial Intelligence. *IEEE Technology and Society Magazine*, *40*(1), 25–30. DOI: 10.1109/MTS.2021.3056295

Gibson, R. (1999). Paulo Freire and pedagogy for social justice. *Theory and Research in Social Education*, *27*(2), 129–159. DOI: 10.1080/00933104.1999.10505876

Gigerenzer, G., & Gaissmaier, W. (2011). Heuristic decision making. *Annual Review of Psychology*, *62*(1), 451–482. Retrieved March 15, 2025 from. DOI: 10.1146/annurev-psych-120709-145346 PMID: 21126183

Gillespie, N., Lockey, S., Curtis, C., Pool, J., & Akbari, A. (2023). *Trust in artificial intelligence: A global study*. The University of Queensland and KPMG Australia., DOI: 10.14264/00d3c94

Gillespie, T. (2020). Content moderation, AI, and the question of scale. *Big Data & Society*, *7*(2), 2053951720943234. DOI: 10.1177/2053951720943234

Gîrju, R. (2021, May 14). Adaptive Multimodal and Multisensory Empathic Technologies for Enhanced Human Communication. Cornell University. https://doi.org/DOI: 10.48550/arXiv.2110

Glikson, E., & Woolley, A. W. (2020). Human trust in artificial intelligence: Review of empirical research. *The Academy of Management Annals*, *14*(2), 627–660. DOI: 10.5465/annals.2018.0057

Global Education Monitoring Report Team. (2024). *Gender equality in STEM education: Breaking down the barriers*. UNESCO. Retrieved March 15, 2025 from https://doi.org/DOI: 10.54676/WVCF2762

Gold, S., Trautrims, A., & Trodd, Z. (2015). Modern slavery challenges to supply chain management. *Supply Chain Management*, *20*(5), 485–494. DOI: 10.1108/SCM-02-2015-0046

Golub, A., Marcantonio, R. A., & Sanchez, T. W. (2013). Race, space, and struggles for mobility: Transportation impacts on African Americans in Oakland and the East Bay. *Urban Geography*, *34*(5), 699–728. DOI: 10.1080/02723638.2013.778598

Goodfellow, I., Bengio, Y., & Courville, A. (2016). *Deep Learning*. MIT Press.

Google. (2024, October 1). *Be Internet Awesome*. Be Internet Awesome. https://beinternetawesome.withgoogle.com/en-us

Gorlick, B. (2001, October 15). Human Rights and Refugees: Enhancing Protection Through International Human Rights Law. RELX Group (Netherlands). https://papers.ssrn.com/sol3/papers.cfm?abstract_id=2331753

Government of Canada. (2023). *Artificial Intelligence and Data Act*. Retrieved March 15, 2025 from https://ised-isde.canada.ca/site/innovation-better-canada/en/artificial-intelligence-and-data-act-aida-companion-document

GPAI. (2024, October 1). *Global Partnership on Artificial Intelligence—GPAI*. https://gpai.ai/

Graef, I., Petročnik, T., & Tombal, T. (2023). Conceptualizing autonomy in an era of collective data processing: From theory to practice. *Digital Society : Ethics, Socio-Legal and Governance of Digital Technology*, *2*(2), 19. DOI: 10.1007/s44206-023-00045-3

Greenberg, D. M., Warrier, V., Allison, C., & Baron-Cohen, S. (2018). Testing the Empathizing-Systemizing theory of sex differences and the Extreme Male Brain theory of autism in half a million people. *Proceedings of the National Academy of Sciences of the United States of America, 115*(48), 12152–12157. Retrieved March 15, 2025 from. DOI: 10.1073/pnas.1811032115 PMID: 30420503

Greene, D., Hoffmann, A. L., & Stark, L. (2019). "Better, Nicer, Clearer, Fairer: A Critical Assessment of the Movement for Ethical Artificial Intelligence and Machine Learning." *Proceedings of the 52nd Hawaii International Conference on System Sciences*. DOI: 10.24251/HICSS.2019.258

Greenleaf, G. (2019). *Global Data Privacy Laws 2019: 132 National Laws & Many Bills* (SSRN Scholarly Paper 3381593). https://papers.ssrn.com/abstract=3381593

Greenleaf, R. K. (2002). *Servant Leadership: A Journey Into the Nature of Legitimate Power and Greatness*. Paulist Press.

Greer, C. (2010). *Crime and media: A reader* (1st ed.). Routledge.

Gunn, F., Haegele, M., Langellotti, A., & Zanolini, F. (2020). *An exploration of how artists use immersive technologies to promote inclusivity, diversity, and deep public engagement in ethical ways*. Electronic Workshops in Computing., DOI: 10.14236/ewic/EVA2020.36

Gurkaynak, G., Yilmaz, I., & Haksever, G. (2016). Stifling artificial intelligence: Human perils. *Computer Law & Security Report, 32*(5), 749–758. DOI: 10.1016/j.clsr.2016.05.003

Guy, M. E., & McCandless, S. A. (2012). Social equity: Its legacy, its promise. *Public Administration Review, 72*(s1, Suppl. S1), S5–S13. DOI: 10.1111/j.1540-6210.2012.02635.x

Haddad, G., & Chedrawi, C. (2019). Artificial intelligence and the inclusion-exclusion paradox: A sociomaterial perspective. *ICTO Conference*.

Haenssgen, M. J. (2018). The struggle for digital inclusion: Phones, healthcare, and marginalisation in rural India. *World Development, 104*, 358–374. DOI: 10.1016/j.worlddev.2017.12.023

Hagendorff, T. (2020). The ethics of AI ethics: An evaluation of guidelines. *Minds and Machines, 30*(1), 99–120. DOI: 10.1007/s11023-020-09517-8

Haleem, A., Javaid, M., Asim Qadri, M., Pratap Singh, R., & Suman, R. (2022). Artificial Intelligence (AI) applications for marketing: A literature-based study. *International Journal of Intelligent Networks, 3*, 119–132. DOI: 10.1016/j.ijin.2022.08.005

Hallamaa, J., & Kalliokoski, T. (2022). AI Ethics as Applied Ethics. *Frontiers of Computer Science*, *4*, 776837. Advance online publication. DOI: 10.3389/fcomp.2022.776837

Hamm, P., & Klesel, M. (2021). Success factors for the adoption of artificial intelligence in organizations: A literature review. *AMCIS 2021 Proceedings*, 10. https://aisel.aisnet.org/amcis2021/art_intel_sem_tech_intelligent_systems/art_intel_sem_tech_intelligent_systems/10

Hanifa M, Agarwal R, Sharma U, Thapliyal PC, Singh LP (2023) A review on CO2 capture and sequestration in the construction industry: emerging approaches and commercialised technologies. J Co2 Util 67:102292. DOI: 10.1016/j.jcou.2022.102292

Hannan, M. A., Al-Shetwi, A. Q., Ker, P. J., Begum, R. A., Mansor, M., Rahman, S. A., Dong, Z. Y., Tiong, S. K., Mahlia, T. M. I., & Muttaqi, K. M. (2021). Impact of renewable energy utilization and artificial intelligence in achieving sustainable development goals. *Energy Reports*, *7*, 5359–5373. DOI: 10.1016/j.egyr.2021.08.172

Hansen, A. S. (2022). *Relationships, religion, and robotics: The soul and the ethical implications of AI* (Doctoral dissertation).

Han, X., Wang, J., Zhang, M., & Wang, X. (2020). Using social media to mine and analyze public opinion related to COVID-19 in China. *International Journal of Environmental Research and Public Health*, *17*(8), 2788. DOI: 10.3390/ijerph17082788 PMID: 32316647

Harasimiuk, D., & Braun, T. (2021). *Regulating artificial intelligence: Binary ethics and the law*. Routledge. DOI: 10.4324/9781003134725

Hassan, A. (2022). The Influence of Artificial Intelligence on Smart Learning: An Overview. In Hamdan, A., Hassanien, A. E., Mescon, T., & Alareeni, B. (Eds.), *Technologies, Artificial Intelligence and the Future of Learning Post-COVID-19: The Crucial Role of International Accreditation* (pp. 591–615). Springer International Publishing., DOI: 10.1007/978-3-030-93921-2_31

Heeks, R. (2020). ICT4D 3.0? Part 1—The components of an emerging "digital-for-development" paradigm. *The Electronic Journal on Information Systems in Developing Countries*, *86*(3), e12124. DOI: 10.1002/isd2.12124

Henry, P. (2015). Analyzing the integration of human rights into social work education. *Issues in Social Science*, *3*(2), 100. DOI: 10.5296/iss.v3i2.8041

Hermann, I. (2021). Artificial intelligence in fiction: Between narratives and metaphors. *AI & Society*, *38*(2), 319–329. DOI: 10.1007/s00146-020-01119-y

Herzfeld, N. L. (2023). *The artifice of intelligence: Divine and human relationship in a robotic age*. Augsburg Fortress Publishers.

Hine, E., & Floridi, L. (2022). Artificial intelligence with American values and Chinese characteristics: A comparative analysis of American and Chinese governmental AI policies. *AI & Society*. Advance online publication. DOI: 10.1007/s00146-022-01425-9

Hoek, R. (2020). Research opportunities for a more resilient post-COVID-19 supply chain – closing the gap between research findings and industry practice. *International Journal of Operations & Production Management, ahead-of-print*. DOI: 10.1108/IJOPM-03-2020-0165

Hofmann, V., Kalluri, P. R., Jurafsky, D., & King, S. (2024). AI generates covertly racist decisions about people based on their dialect. *Nature*, *633*(8028), 147–154. Retrieved March 15, 2025 from. DOI: 10.1038/s41586-024-07856-5 PMID: 39198640

Hohfeld, W. N. (1917). Fundamental legal conceptions as applied in judicial reasoning. *The Yale Law Journal*, *26*(8), 710–770. DOI: 10.2307/786270

Hohma, E., Boch, A., Trauth, R., & Lütge, C. (2023). Investigating accountability for artificial intelligence through risk governance: A workshop-based exploratory study. *Frontiers in Psychology*, *14*, 1073686. Advance online publication. DOI: 10.3389/fpsyg.2023.1073686 PMID: 36760454

Holcomb, S. D., Porter, J., & Ballard, A. (2018). Overview on DeepMind and its AlphaGo Zero AI. In *Proceedings of the 2018 International Conference on Big Data and Education* (pp. 7–11). DOI: 10.1145/3206157.3206174

Holmes, W., Bialik, M., & Fadel, C. (2019). *Artificial Intelligence in Education*. Promise and Implications for Teaching and Learning.

Horváth, I. (2022). AI in interpreting: Ethical considerations. *Across Languages and Cultures*, *23*(1), 1–13. DOI: 10.1556/084.2022.00108

Huang, C., Samek, T., & Shiri, A. (2021). AI and Ethics: Ethical and Educational Perspectives for LIS. *Journal of Education for Library and Information Science*, *62*(4), 351–365. DOI: 10.3138/jelis-62-4-2020-0106

Huang, M.-H., & Rust, R. (2018). Artificial Intelligence in Service. *Journal of Service Research*, *21*(2), 109467051775245. DOI: 10.1177/1094670517752459

Human Rights Group. (2020). *Human rights in the digital age: Making digital technology work for human rights*. Glion Human Rights Dialogue., DOI: 10.1787/7ec52ca6-

Hunkenschroer, A., & Kriebitz, A. (2022). Is ai recruiting (un)ethical? a human rights perspective on the use of ai for hiring. *AI and Ethics*, *3*(1), 199–213. DOI: 10.1007/s43681-022-00166-4 PMID: 35909984

Huq, A. Z. (2021). Artificial intelligence and the rule of law. SSRN *Electronic Journal*. DOI: 10.2139/ssrn.3794777

Huriye, A. (2023). The ethics of artificial intelligence: Examining the ethical considerations surrounding the development and use of ai. *American Journal of Technology*, *2*(1), 37–45. DOI: 10.58425/ajt.v2i1.142

IA2030Mx. (2018). *Agenda Nacional Mexicana de IA 2030*. Retrieved March 15, 2025 from https://wp.oecd.ai/app/uploads/2022/01/Mexico_Agenda_Nacional_Mexicana_de_IA_2030.pdf

Ibhawoh, B. (2014). Testing the atlantic charter: Linking anticolonialism, self-determination and universal human rights. *International Journal of Human Rights*, *18*(7-8), 842–860. DOI: 10.1080/13642987.2014.951340

Ibiricu, B., & Made, M L V D. (2020, June 10). Ethics by design: a code of ethics for the digital age. Emerald Publishing Limited, 30(3), 395-414. DOI: 10.1108/RMJ-08-2019-0044

ICO. (2018). *Applications—Children and the GDPR*. https://ico.org.uk/media/for-organisations/guide-to-the-general-data-protection-regulation-gdpr/children-and-the-gdpr-1-0.pdf

ICO. (2023, September 14). *Children and the UK GDPR*. ICO. https://ico.org.uk/for-organisations/uk-gdpr-guidance-and-resources/childrens-information/children-and-the-uk-gdpr/

IEEE. (2019, April 4). *Now Available! Ethically Aligned Design, First Edition! - IEEE Robotics and Automation Society*. https://www.ieee-ras.org/about-ras/latest-news/now-available-ethically-aligned-design-first-edition

India, A. I. 2023: Expert Group Report. (2023, October 14). https://www.meity.gov.in/content/indiaai-2023-expert-group-report-%E2%80%93-first-editionthe-ministry-electronics-and-information

Ingrams, A., Manoharan, A., & Kang, D. (2021). In AI we trust? Citizen perceptions of AI in government decision making. *Policy and Internet*, *14*(3), 390–409. DOI: 10.1002/poi3.249

Institute, A. N. (2019, December 12). AI Now 2019 Report. *AI Now Institute*. https://ainowinstitute.org/publication/ai-now-2019-report-2

Institute, A. I. N. (2019). *Discriminating Systems: Gender, Race, and Power in Ai - Report*. Ai Now Institute.

Inverardi, P. (2021, November 23). The Challenge of Human Dignity in the Era of Autonomous Systems. *Springer Nature, 25-29*, 25–29. Advance online publication. DOI: 10.1007/978-3-030-86144-5_4

Ise, T., & Oba, Y. (2019). Forecasting climatic trends using neural networks: An experimental study using global historical data. *Frontiers in Robotics and AI, 6*, 32. DOI: 10.3389/frobt.2019.00032 PMID: 33501048

ITU. (2024, October 1). *Artificial Intelligence*. https://www.itu.int/en/action/ai/Pages/default.aspx

Izzidien, A. (2023). Using the interest theory of rights and Hohfeldian taxonomy to address a gap in machine learning methods for legal document analysis. *Humanities & Social Sciences Communications, 10*(1), 251. Advance online publication. DOI: 10.1057/s41599-023-01693-z

Jahanger, A., Ozturk, I., Chukwuma Onwe, J., Joseph, T. E., & Razib Hossain, M. (2023). Do technology and renewable energy contribute to energy efficiency and carbon neutrality? Evidence from top ten manufacturing countries. *Sustainable Energy Technologies and Assessments, 56*, 103084. DOI: 10.1016/j.seta.2023.103084

Jahani, H., Khosravi, Y., Kargar, B., Ong, K. L., & Arisian, S. (2024). Exploring the role of drones and UAVs in logistics and supply chain management: A novel text-based literature review. *International Journal of Production Research*, •••, 1–25.

Janssen, M., Hartog, M. H. D., Matheus, R., Ding, A. Y., & Kuk, G. (2020, December 28). Will Algorithms Blind People? The Effect of Explainable AI and Decision-Makers' Experience on AI-supported Decision-Making in Government. *Social Science Computer Review, 40*(2), 478–493. DOI: 10.1177/0894439320980118

Jarek, K., & Mazurek, G. (2019). Marketing and Artificial Intelligence. *Central European Business Review, 8*(2), 46–55. DOI: 10.18267/j.cebr.213

Jaynes, T. L. (2019). Legal personhood for artificial intelligence: Citizenship as the exception to the rule. *AI & Society, 35*(2), 343–354. DOI: 10.1007/s00146-019-00897-9

Jelodar, H., Wang, Y., Yuan, C., Feng, X., Jiang, X., Li, Y., & Zhao, L. (2019). Latent Dirichlet Allocation (LDA) and topic modeling: Models, applications, a survey. *Multimedia Tools and Applications, 78*(11), 15169–15211. DOI: 10.1007/s11042-018-6894-4

Jobin, A., Ienca, M., & Vayena, E. (2019). The global landscape of ai ethics guidelines. *Nature Machine Intelligence*, *1*(9), 389–399. DOI: 10.1038/s42256-019-0088-2

Johnson, T. L., Labbé, S., & Dunn, M. A. (2022). Facial recognition systems in policing and racial disparities in arrests. *Government Information Quarterly*, *39*(4), 101753. DOI: 10.1016/j.giq.2022.101753

Jordan, T. (2025). *Parental Control: A Guide to Raising Balanced Kids in the Digital Era*. Wiley.

Kaack, L. H., Donti, P. L., Strubell, E., Kamiya, G., Creutzig, F., & Rolnick, D. (2022). Aligning artificial intelligence with climate change mitigation. *Nature Climate Change*, *12*(6), 518–527. DOI: 10.1038/s41558-022-01377-7

Kadam, K., Godbole, S., Joijode, D., Karoshi, S., Jadhav, P., & Shilaskar, S. (2022). Multilingual Information Retrieval Chatbot. In Gunjan, V. K., & Zurada, J. M. (Eds.), *Modern Approaches in Machine Learning & Cognitive Science: A Walkthrough* (pp. 107–121). Springer International Publishing., DOI: 10.1007/978-3-030-96634-8_10

Kadow, C., Hall, D. M., & Ulbrich, U. (2020). Artificial intelligence reconstructs missing climate information. *Nature Geoscience*, *13*(6), 408–413. DOI: 10.1038/s41561-020-0582-5

Kaesling, K. (2023). A Rights-Based Approach to Children's Digital Participation in the Multi-Level System of the European Union. In Dethloff, N., Kaesling, K., & Specht-Riemenschneider, L. (Eds.), *Families and New Media: Comparative Perspectives on Digital Transformations in Law and Society* (pp. 73–110). Springer Fachmedien., DOI: 10.1007/978-3-658-39664-0_4

Kahneman, D. (2011). *Thinking, fast and slow*. Farrar, Straus and Giroux.

Kahneman, D., & Tversky, A. (1979). Prospect theory: An analysis of decision under risk. *Econometrica*, *47*(2), 263–291. Retrieved March 15, 2025 from. DOI: 10.2307/1914185

Kalluri, P. (2020). Don't ask if artificial intelligence is good or fair, ask how it shifts power. *Nature*, *583*(7815), 169. DOI: 10.1038/d41586-020-02003-2 PMID: 32636520

Kamble, S., Gunasekaran, A., & Arha, H. (2019). Understanding the Blockchain technology adoption in supply chains-Indian context. *International Journal of Production Research*, *57*(7), 2009–2033. DOI: 10.1080/00207543.2018.1518610

Karimian, G., Petelos, E., & Evers, S. M. A. A. (2022). The ethical issues of the application of artificial intelligence in healthcare: A systematic scoping review. *AI and Ethics*, *2*(4), 539–551. DOI: 10.1007/s43681-021-00131-7

Karpa, D., Müller, M., & Ernst, E. (2022). Artificial intelligence, surveillance, and big data. In *Advanced Studies in Diginomics and Digitalization* (pp. 145–172). Springer., DOI: 10.1007/978-3-030-97759-8_7

Katyal, S. K. (2022). Democracy & distrust in an era of artificial intelligence. *Daedalus, 151*(2), 322–334. DOI: 10.1162/daed_a_01919

Kaurin, D. (2019, May 15). Data Protection and Digital Agency for Refugees. https://www.cigionline.org/sites/default/files/documents/WRC%20Research%20Paper%20no.12.pdf

Kaur, N. (2013, January 1). Protection of Refugees in India: A Critical Analysis. RELX Group (Netherlands). DOI: 10.2139/ssrn.2214274

Kazim, E., & Koshiyama, A. S. (2021). A high-level overview of AI ethics. *Patterns (New York, N.Y.), 2*(9), 100314. Advance online publication. DOI: 10.1016/j.patter.2021.100314 PMID: 34553166

Kelly, B., King, M., Psy, L., Chapman Mnd, K., Boyland, E., Bauman, A. E., & Baur, L. A. (2015). A hierarchy of unhealthy food promotion effects: Identifying methodological approaches and knowledge gaps. *American Journal of Public Health, 105*(4), e86–e95. DOI: 10.2105/AJPH.2014.302476 PMID: 25713968

Khan, R., Tausif, S., & Javed Malik, A. (2019). Consumer acceptance of delivery drones in urban areas. *International Journal of Consumer Studies, 43*(1), 87–101. DOI: 10.1111/ijcs.12487

Kinchin, N. (2021, September 3). Technology, Displaced? The Risks and Potential of Artificial Intelligence for Fair, Effective, and Efficient Refugee Status Determination. Cambridge University Press, 37(3). DOI: 10.26826/law-in-context.v37i3.157

Kinchin, N., & Mougouei, D. (2022, October 1). What Can Artificial Intelligence Do for Refugee Status Determination? A Proposal for Removing Subjective Fear. Oxford University Press, 34(3), 373-397. DOI: 10.1093/ijrl/eeac040

Klimova, B., Pikhart, M., & Kacetl, J. (2023). Ethical issues of the use of AI-driven mobile apps for education. *Frontiers in Public Health, 10*, 1118116. Advance online publication. DOI: 10.3389/fpubh.2022.1118116 PMID: 36711343

Kollewe, J. (2022, June 1). *World's first raspberry picking robot cracks the toughest nut: Soft fruit.* The Guardian. https://www.theguardian.com/business/2022/jun/01/uk-raspberry-picking-robot-soft-fruit

Koniakou, V. (2023). From the "rush to ethics" to the "race for governance" in artificial intelligence. *Information Systems Frontiers*, *25*(1), 71–102. DOI: 10.1007/s10796-022-10300-6

Konidena, B. K., Malaiyappan, J. N. A., & Tadimarri, A. (2024). Ethical Considerations in the Development and Deployment of AI Systems. *European Journal of Technology*, *8*(2), 2. Advance online publication. DOI: 10.47672/ejt.1890

Kosta, E. (2020). Algorithmic state surveillance: Challenging the notion of agency in human rights. *Regulation & Governance*, *16*(2), 212–224. DOI: 10.1111/rego.12250

Kotzé, L. (2014). Human rights and the environment in the anthropocene. *The Anthropocene Review*, *1*(3), 252–275. DOI: 10.1177/2053019614547741

Kreps, S. (2021, November). *Democratizing harm: Artificial intelligence in the hands of nonstate actors*. Foreign Policy at Brookings. https://www.brookings.edu/wp-content/uploads/2021/11/FP_20211122_ai_nonstate_actors_kreps.pdf

Krivenko, E. Y. (2020). Gender and human rights: Expanding concepts. In *Gender and Human Rights* (pp. 21–43). Edward Elgar Publishing.

Krogh, C., & Herrestad, H. (1999). Artificial intelligence and legal reasoning: A survey of the field. *Artificial Intelligence and Law*, *7*(2), 81–96. DOI: 10.1023/A:1008367514393

Kroll, J. A., Huey, J., Barocas, S., & Felten, E. W. (2017). Accountable Algorithms. *University of Pennsylvania Law Review*, *165*(3), 633–705.

Krysa, J., & Impett, L. (2022). *Rethinking curating in the age of artificial intelligence*. Art Books Publishing Ltd.

Kuhlman, C., Jackson, L., & Chunara, R. (2020). No computation without representation: Avoiding data and algorithm biases through diversity. In *Proceedings of the 26th ACM SIGKDD International Conference on Knowledge Discovery & Data Mining* (pp. 1257–1265). DOI: 10.1145/3394486.3411074

Kuźniacki, B., Almada, M., Tyliński, K., & Górski, Ł. (2022). Requirements for tax xai under constitutional principles and human rights., 221-238. DOI: 10.1007/978-3-031-15565-9_14

Kvale, S., & Brinkmann, S. (2009). InterViews: Learning the Craft of Qualitative Research Interviewing. *Sage (Atlanta, Ga.)*.

Labelvisor. (2023). Ensuring label fairness and bias reduction in data labeling. *Labelvisor*. https://www.labelvisor.com/ensuring-label-fairness-and-bias-reduction-in-data-labeling/#:~:text=Label%20bias%20can%20result%20from,unfair%20outcomes%20for%20certain%20groups

Lam, K T., Lange, B., Blili-Hamelin, B., Davidovic, J., Brown, S., & Hasan, A. (2024, June 3). A Framework for Assurance Audits of Algorithmic Systems., 38, 1078-1092. DOI: 10.1145/3630106.3658957

Latonero, M. (2018). Governing artificial intelligence: Upholding human rights & dignity. *Data & Society, 38*.

Law, L. (2024). Application of generative artificial intelligence (GenAI) in language teaching and learning: A scoping literature review. *Computers and Education Open*, *6*, 100174. DOI: 10.1016/j.caeo.2024.100174

Lee, J., & Yoo, H. J. (2021). An overview of energy-efficient hardware accelerators for on-device deep-neural-network training. *IEEE Open Journal of the Solid-State Circuits Society*, *1*, 115–128. DOI: 10.1109/OJSSCS.2021.3119554

Lee, W. J., Wu, H., Yun, H., Kim, H., Jun, M. B. G., & Sutherland, J. W. (2019) Predictive maintenance of machine tool systems using artificial intelligence techniques applied to machine condition data. *Proc CIRP 80*:506–511. https://doi.org/ DOI: 10.1016/j.procir.2018.12.019

LEGO. (2024a, October 1). *Datenschutzrichtlinie*. https://kids.lego.com/de-de/legal/privacy-policy

LEGO. (2024b, October 1). *Privacy—Sustainability—LEGO.com*. https://www.lego.com/de-de/sustainability/digital-child-safety/privacy?locale=de-de

Leite, E., Leite, M., & Leite, A. (2023). Ai's impact on human rights: The need for legal evolution. *JER*, *1*(2), 81–86. DOI: 10.29073/jer.v1i2.16

Lemaignan, S., Newbutt, N., Rice, L., Daly, J., & Charisi, V. (2021). UNICEF guidance on AI for children: Application to the design of a social robot for and with autistic children. *arXiv preprint arXiv:2108.12166*.

Leslie, D. (2019). Understanding artificial intelligence ethics and safety. *arXiv preprint arXiv:1906.05684*.

Leslie, D., Mazumder, A., Peppin, A., Wolters, M. K., & Hagerty, A. (2021). Does "AI" stand for augmenting inequality in the era of COVID-19 healthcare? *BMJ (Clinical Research Ed.)*, *372*(304), n304. Advance online publication. DOI: 10.1136/bmj.n304 PMID: 33722847

Li, D., Tong, T. W., & Xiao, Y. (2021). Is China emerging as the global leader in AI? *Harvard Business Review*. https://hbr.org/2021/02/is-china-emerging-as-the-global-leader-in-ai

Lievens, E., & Verdoodt, V. (2018). Looking for needles in a haystack: Key issues affecting children's rights in the General Data Protection Regulation. *Computer Law & Security Report*, *34*(2), 269–278. DOI: 10.1016/j.clsr.2017.09.007

Li, J., Xu, Q., Cuomo, R., Purushothaman, V., & Mackey, T. (2020). Data mining and content analysis of the Chinese social media platform Weibo during the early COVID-19 outbreak: Retrospective observational infoveillance study. *JMIR Public Health and Surveillance*, *6*(2), e18700. DOI: 10.2196/18700 PMID: 32293582

Lima, J. A. de O., de Oliveira, L. C., & Nascimento, F. L. (2021). Casting the light of the theory of opposition onto Hohfeld's fundamental legal concepts. *Legal Theory*, *27*(1), 2–35. DOI: 10.1017/S1352325221000070

Lin, P., Selinger, E., & Hadley, P. (2020). "Ethics and AI: The Role of Interdisciplinary Education." Journal of Ethical AI.

Lin, C.-C., Huang, A. Y., & Lu, O. H. (2023). Artificial intelligence in intelligent tutoring systems toward sustainable education: A systematic review. *Smart Learning Environments*, *10*(1), 41. Advance online publication. DOI: 10.1186/s40561-023-00260-y

Ling, X., Gao, M., & Wang, D. (2020, November 6). Intelligent document processing based on RPA and machine learning., 33, 1349-1353. DOI: 10.1109/CAC51589.2020.9326579

Lippert-Rasmussen, K., & Vitikainen, A. (2020, January 1). The ethics of refugee prioritization: reframing the debate. Taylor & Francis, 13(1), 1-5. DOI: 10.1080/16544951.2020.1735019

Litman, T. (2017). *Evaluating transportation equity*. Victoria Transport Policy Institute.

Livingstone, S., Third, A., Burton, P., Witting, S., Lievens, E., Steinberg, S., Gilutz, S., & Souter, D. (2022). The UN committee on the rights of the child's general comment on the digital environment. *AoIR Selected Papers of Internet Research*. DOI: 10.5210/spir.v2022i0.12960

Li, X. (2021). Analysis of racial discrimination in artificial intelligence from the perspective of social media, search engines, and future crime prediction systems. *Advances in Social Science, Education and Humanities Research*, *575*, 172–177. DOI: 10.2991/assehr.k.210902.029

Li, X., Tupayachi, J., Sharmin, A., & Martinez Ferguson, M. (2023). Drone-aided delivery methods, challenges, and the future: A methodological review. *Drones (Basel)*, *7*(3), 191. DOI: 10.3390/drones7030191

Lockey, S., Gillespie, N., Holm, D., & Someh, I. A. (2021). A review of trust in artificial intelligence: Challenges, vulnerabilities and future directions. In *Proceedings of the 54th Hawaii International Conference on System Sciences* (pp. 5468–5469). DOI: 10.24251/HICSS.2021.664

Long, D., & Magerko, B. (2020). What is AI Literacy? Competencies and Design Considerations. *Proceedings of the 2020 CHI Conference on Human Factors in Computing Systems*, 1–16. DOI: 10.1145/3313831.3376727

Lovell, J. (2023). *Legal aspects of artificial intelligence personhood: Exploring the possibility of granting legal personhood to advanced AI systems and the implications for liability, rights, and responsibilities*. Rights and Responsibilities.

Lovell, J. J. (2024).. . *International Journal of Artificial Intelligence and Machine Learning*.

Lukyanenko, R., Maass, W., & Storey, V. C. (2022). Trust in artificial intelligence: From a foundational trust framework to emerging research opportunities. *Electronic Markets*, *32*(4), 1993–2020. DOI: 10.1007/s12525-022-00605-4

Luminita, D. (2013). Evolution of the human rights issue. *International Journal of Academic Research in Economics and Management Sciences*, *2*(1), 45–58. DOI: 10.6007/IJAREMS/v2-i1/738

Lum, K., & Isaac, W. (2016). To predict and serve? *Significance*, *13*(5), 14–19. DOI: 10.1111/j.1740-9713.2016.00960.x

Lund, B. D., Wang, T., Mannuru, N. R., Nie, B., Shimray, S., & Wang, Z. (2023). ChatGPT and a new academic reality: Artificial Intelligence-written research papers and the ethics of the large language models in scholarly publishing. *Journal of the Association for Information Science and Technology*, *74*(5), 570–581. DOI: 10.1002/asi.24750

Lu, S. (2022). Data privacy, human rights, and algorithmic opacity. *California Law Review*, *110*(4), 2081–2103. DOI: 10.15779/Z38804XM07

Lyreskog, D. M., Zohny, H., Savulescu, J., & Singh, I. (2023). Merging minds: The conceptual and ethical impacts of emerging technologies for collective minds. *Neuroethics*, *16*(1), 12. DOI: 10.1007/s12152-023-09516-3 PMID: 37009261

Madariaga Zeballos, J. (2023). *Perú: Proyecto de ley para la regulación de la inteligencia artificial*. Congreso de la República del Perú. Retrieved March 15, 2025 from https://wb2server.congreso.gob.pe/spley-portal-service/archivo/MTYzNzgw/pdf

Maier, D., Waldherr, A., Miltner, P., Wiedemann, G., Niekler, A., Keinert, A., Pfetsch, B., Heyer, G., Reber, U., Häussler, T., Schmid-Petri, H., & Adam, S. (2018). Applying LDA topic modeling in communication research: Toward a valid and reliable methodology. *Communication Methods and Measures*, *12*(2-3), 93–118. DOI: 10.1080/19312458.2018.1430754

Manap, N., & Abdullah, A. (2020). Regulating artificial intelligence in malaysia: The two-tier approach. *UUM Journal of Legal Studies*, *11*. Advance online publication. DOI: 10.32890/uumjls.11.2.2020.8779

Manheim, K., & Kaplan, L. (2019). Artificial intelligence: Risks to privacy and democracy. *The Yale Journal of Law & Technology*, *21*, 118–119.

Manoogian, J., & Benson, B. (2017). *Cognitive Bias Codex*. Retrieved November 27, 2024 from https://commons.wikimedia.org/wiki/File:Cognitive_Bias_Codex_-_180%2B_biases,_designed_by_John_Manoogian_III(jm3).jpg

Mantelero, A. (2018). AI and Big Data: A blueprint for a human rights, social and ethical impact assessment. *Computer Law & Security Report*, *34*(4), 754–772. DOI: 10.1016/j.clsr.2018.05.017

Marginson, S. (2016). The worldwide trend to high participation higher education: Dynamics of social stratification in inclusive systems. *Higher Education*, *72*(4), 413–434. DOI: 10.1007/s10734-016-0016-x

Marino, D., Cananzi, D., & Aragona, F. (2024). Artificial intelligence and ethics challenges of technologies. In *Ethics and Artificial Intelligence: Towards a Moral Technology* (pp. 41–55). Springer Nature Switzerland. DOI: 10.1007/978-3-031-50902-5_3

Markovich, R. (2019). Understanding Hohfeld and formalizing legal rights: The Hohfeldian conceptions and their conditional consequences. *Studia Logica*, *108*(1), 129–158. DOI: 10.1007/s11225-019-09870-5

Marr, B. (2019, March 11). *The Amazing Ways Coca Cola Uses Artificial Intelligence And Big Data To Drive Success*. Forbes. https://www.forbes.com/sites/bernardmarr/2017/09/18/the-amazing-ways-coca-cola-uses-artificial-intelligence-ai-and-big-data-to-drive-success/

Mars. (2020, September 8). *Mars Builds on Cocoa Supply Chain Transparency | Mars Global*. https://www.mars.com/news-and-stories/articles/cocoa-supply-chain-transparency

Marsch, N. (2019). Artificial intelligence and the fundamental right to data protection: Opening the door for technological innovation and innovative protection. In *Regulating Artificial Intelligence* (pp. 33–52).

Marshall, C., & Rossman, G. B. (2016). Designing Qualitative Research. *Sage (Atlanta, Ga.)*.

Martin, K. E., & Freeman, R. E. (2004). The Separation of Technology and Ethics in Business Ethics. *Journal of Business Ethics*, *53*(4), 353–364. DOI: 10.1023/B:BUSI.0000043492.42150.b6

Maslej, N., Fattorini, L., Perrault, R., Parli, V., Reuel, A., Brynjolfsson, E., Etchemendy, J., Ligett, K., Lyons, T., Manyika, J., Niebles, J. C., Shoham, Y., Wald, R., & Clark, J. (2024). *The AI Index 2024 Annual Report*. AI Index Steering Committee, Institute for Human-Centered AI, Stanford University. Retrieved March 15, 2025 from https://aiindex.stanford.edu

Ma, W., Qiu, Z., Song, J., Li, J., Cheng, Q., Zhai, J., & Ma, C. (2018). A deep convolutional neural network approach for predicting phenotypes from genotypes. *Planta*, *248*(5), 1307–1318. DOI: 10.1007/s00425-018-2976-9 PMID: 30101399

Mayer-Schönberger, V., & Cukier, K. (2013). *Big data: A revolution that will transform how we live, work, and think*. Houghton Mifflin Harcourt.

McCauley, L. (2007). The Frankenstein complex and Asimov's three laws. *AI & Society*. Advance online publication. DOI: 10.1007/s00146-006-0047-2

McDaniel, J., & Pease, K. (2021). *Predictive policing and artificial intelligence* (1st ed.). Routledge., DOI: 10.4324/9780429265365

McDowall, R. (2015). Periodic evaluation: Independent review to ensure continued validation of computerized systems. In *EU Annex 11 Guide to Computer Validation Compliance for the Worldwide Health Agency GMP* (pp. 91–110). Springer., DOI: 10.1201/b18315-15

McGregor, L., Murray, D., & Ng, V. (2019, April 1). International Human Rights Law as a Framework for Algorithmic Accountability. Cambridge University Press, 68(2), 309-343. DOI: 10.1017/S0020589319000046

Mehrabi, N., Morstatter, F., Saxena, N., Lerman, K., & Galstyan, A. (2021). A Survey on Bias and Fairness in Machine Learning. *ACM Comput. Surv., 54*(6), 115:1-115:35. DOI: 10.1145/3457607

Mehrabi, N., Morstatter, F., Saxena, N., Lerman, K., & Galstyan, A. (2021). A Survey on Bias and Fairness in Machine Learning. *ACM Computing Surveys.*

Memon, A., Given-Wilson, Z., Ozkul, D., Richmond, K. M., Muraszkiewicz, J., Weldon, E., & Katona, C. (2024, January 30). Artificial Intelligence (AI) in the asylum system. *Medicine, Science, and the Law*, *64*(2), 87–90. DOI: 10.1177/00258024241227721 PMID: 38291676

Mendonca, N., Murphy, J. R., Patterson, M. D., Alexander, R., Juárez, G., & Harper, C. (2022). *Advanced air mobility vertiport considerations: A list and overview.* American Institute of Aeronautics and Astronautics, Inc.

Microsoft. (2024, October 1). AI for Health. *Microsoft Research.* https://www.microsoft.com/en-us/research/project/ai-for-health/

Milic-Frayling, N. (2023). On the cusp: Computing thrills and perils and professional awakening. *Proceedings of the VLDB Endowment International Conference on Very Large Data Bases*, *16*(12), 4152–4159. DOI: 10.14778/3611540.3611640

Milkaite, I., & Lievens, E. (2019). Children's Rights to Privacy and Data Protection Around the World: Challenges in the Digital Realm. *European Journal of Law and Technology*, *10*(1). https://ejlt.org/index.php/ejlt/article/view/674

Milkaite, I., & Lievens, E. (2020). Child-friendly transparency of data processing in the EU: From legal requirements to platform policies. *Journal of Children and Media, 14*(1), 5–21. DOI: 10.1080/17482798.2019.1701055

Min, H. (2010). Artificial intelligence in supply chain management: Theory and applications. *International Journal of Logistics*, *13*(1), 13–39. DOI: 10.1080/13675560902736537

Ministry of Science. Technology, Knowledge and Innovation. (2021). *Política Nacional de Inteligencia Artificial.* Retrieved March 15, 2025 from https://www.minciencia.gob.cl/uploads/filer_public/bc/38/bc389daf-4514-4306-867c-760ae7686e2c/documento_politica_ia_digital_.pdf

MIT. (2017, September 7). *IBM and MIT to pursue joint research in artificial intelligence, establish new MIT-IBM Watson AI Lab.* MIT News | Massachusetts Institute of Technology. https://news.mit.edu/2017/ibm-mit-joint-research-watson-artificial-intelligence-lab-0907

MIT. (2024.). *AI risk domain taxonomy*. Retrieved March 15, 2025 from https://airisk.mit.edu/#Domain-Taxonomy-of-AI-Risks

Mittelstadt, B. D., Allo, P., Taddeo, M., Wachter, S., & Floridi, L. (2016). The ethics of algorithms: Mapping the debate. *Big Data & Society*, *3*(2), 2053951716679679. DOI: 10.1177/2053951716679679

Mocanu, D. M. (2022). Gradient legal personhood for AI systems—Painting continental legal shapes made to fit analytical molds. *Frontiers in Robotics and AI*, *8*, 740500. Advance online publication. DOI: 10.3389/frobt.2021.788179 PMID: 35127836

Mökander, J., & Axente, M. (2023). Ethics-based auditing of automated decision-making systems: Intervention points and policy implications. *AI & Society*, *38*(1), 153–171. DOI: 10.1007/s00146-021-01286-x

Molnar, P. (2019, December 1). Technology on the margins: AI and global migration management from a human rights perspective. Edward Elgar Publishing, 8(2), 305-330. DOI: 10.4337/cilj.2019.02.07

Monshipouri, M. (2016). Human rights in the digital age: Opportunities and constraints. *Public Integrity*, *19*(2), 123–135. DOI: 10.1080/10999922.2016.1230690

Monteith, S., Glenn, T., Geddes, J., Whybrow, P., Achtyes, E., & Bauer, M. (2022). Expectations for artificial intelligence (ai) in psychiatry. *Current Psychiatry Reports*, *24*(11), 709–721. DOI: 10.1007/s11920-022-01378-5 PMID: 36214931

Montgomery, K. C., Chester, J., & Milosevic, T. (2017). Children's Privacy in the Big Data Era: Research Opportunities. *Pediatrics*, *140*(Suppl 2), S117–S121. DOI: 10.1542/peds.2016-1758O PMID: 29093045

Morales, V. (2023). *Proyecto de ley: Régimen de regulación de la inteligencia artificial* (Exp. 2505-D-2023). Cámara de Diputados de la Nación Argentina. Retrieved March 15, 2025 from https://www4.hcdn.gob.ar/dependencias/dsecretaria/Periodo2023/PDF2023/TP2023/2505-D-2023.pdfhttps://www.diputados.gov.ar/diputados/vmoralesg/proyecto.html?exp=2505-D-2023

Morley, J., Machado, C., Burr, C., Cowls, J., Taddeo, M., & Floridi, L. (2019). The debate on the ethics of ai in health care: a reconstruction and critical review. SSRN Electronic Journal. DOI: 10.2139/ssrn.3486518

Morley, J., Floridi, L., Kinsey, L., & Elhalal, A. (2020). From What to How: An Initial Review of Publicly Available AI Ethics Tools, Methods and Research to Translate Principles into Practices. *Science and Engineering Ethics*, *26*(4), 2141–2168. DOI: 10.1007/s11948-019-00165-5 PMID: 31828533

Moshref-Javadi, M., & Winkenbach, M. (2021). Applications and research avenues for drone-based models in logistics: A classification and review. *Expert Systems with Applications*, *177*, 114854. DOI: 10.1016/j.eswa.2021.114854

Mou, X. (2019). Artificial-intelligence-investment-trends-and-selected- ... https://documents1.worldbank.org/curated/en/617511573040599056/pdf/Artificial-Intelligence-Investment-Trends-and-Selected-Industry-Uses.pdf

Mühlhoff, R. (2023, January 1). Predictive privacy: Collective data protection in the context of artificial intelligence and big data. *Big Data & Society*, *10*(1), 2053 95172311668–205395172311668. DOI: 10.1177/20539517231166886

Multilingual AI Chatbot Frameworks | Restackio. (n.d.). Retrieved October 17, 2024, from https://www.restack.io/p/ai-for-language-translation-answer-multilingual-ai-chatbot-frameworks-cat-ai

Murgia, M. (2024). *Code dependent: Living in the shadow of AI*. Henry Holt and Company.

Musa, A. Elsadig. (2024). *Developing Digital Inclusion Through Globalization and Digitalization*. IGI Global.

Nairn, A., & Fine, C. (2008). Who's messing with my mind?: The implications of dual-process models for the ethics of advertising to children. *International Journal of Advertising*, *27*(3), 447–470. DOI: 10.2501/S0265048708080062

Nalbandian, L. (2022, August 3). An eye for an 'I:' a critical assessment of artificial intelligence tools in migration and asylum management. *Comparative Migration Studies*, *10*(1), 32. Advance online publication. DOI: 10.1186/s40878-022-00305-0 PMID: 35967601

National Conference of State Legislatures. (2024). Social media and children 2024 legislation. https://www.ncsl.org/technology-and-communication/social-media-and-children-2024-legislation

Nawaz, R., Akhtar, R., Shahid, M. A., Qureshi, I. M., & Mahmood, M. H. (2021). Machine learning based false data injection in smart grid. *International Journal of Electrical Power & Energy Systems*, *130*, 106819. DOI: 10.1016/j.ijepes.2021.106819

Nazer, L., Zatarah, R., Waldrip, S., Ke, J. X. C., Moukheiber, M., Khanna, A. K., Hicklen, R. S., Moukheiber, L., Moukheiber, D., Ma, H., & Mathur, P. (2023, June 22). Bias in artificial intelligence algorithms and recommendations for mitigation. *PLOS Digital Health*, *2*(6), e0000278–e0000278. DOI: 10.1371/journal.pdig.0000278 PMID: 37347721

Nersessian, D., & Mancha, R. (2020). From automation to autonomy: Legal and ethical responsibility gaps in artificial intelligence innovation. *Michigan Technology Law Review*, *27*, 55.

New Zealand Government. (2023). *AI Strategy Blueprint: Trust and Innovation*. Ministry of Innovation and Employment. Retrieved March 15, 2025 from https://www.govt.nz

Newman, M. E., McLaren, K. P., & Wilson, B. S. (2014). Assessing deforestation and fragmentation in a tropical moist forest over 68 years; the impact of roads and legal protection in the Cockpit Country, Jamaica. *Forest Ecology and Management*, *315*, 138–152. DOI: 10.1016/j.foreco.2013.12.033

Ng, D. T. K., Leung, J. K. L., Chu, K. W. S., & Qiao, M. S. (2021). AI Literacy: Definition, Teaching, Evaluation and Ethical Issues. *Proceedings of the Association for Information Science and Technology*, *58*(1), 504–509. DOI: 10.1002/pra2.487

Niemiec, E. (2022). Will the eu medical device regulation help to improve the safety and performance of medical ai devices? *Digital Health*, *8*, 205520762210890. DOI: 10.1177/20552076221089079 PMID: 35386955

Nikolinakos, N. Th. (2023). Ethical Principles for Trustworthy AI. In Nikolinakos, N. Th. (Ed.), *EU Policy and Legal Framework for Artificial Intelligence, Robotics and Related Technologies—the AI Act* (pp. 101–166). Springer International Publishing., DOI: 10.1007/978-3-031-27953-9_3

Nissenbaum, H. (2019). Contextual integrity up and down the data food chain. *Theoretical Inquiries in Law*, *20*(1), 221–256. DOI: 10.1515/til-2019-0008

Noble, S. U. (2018). *Algorithms of oppression: How search engines reinforce racism*. NYU Press. DOI: 10.18574/nyu/9781479833641.001.0001

Noorbakhsh-Sabet, N., Zand, R., Zhang, Y., & Abedi, V. (2019). Artificial intelligence transforms the future of health care. *The American Journal of Medicine*, *132*(7), 795–801. DOI: 10.1016/j.amjmed.2019.01.017 PMID: 30710543

Nourbakhsh, I. R., & Keating, J. (2020). *AI and humanity*. MIT Press.

Novelli, C., Di Marco, A., & Mazzoccoli, M. (2021). A conceptual framework for legal personality and its application to AI. *Jurisprudence*, *13*(2), 194–219. DOI: 10.1080/20403313.2021.2010936

Nuenen, T. V., Ferrer, X., Such, J. M., & Coté, M. (2020, October 22). Transparency for Whom? Assessing Discriminatory Artificial Intelligence. *Computer*, *53*(11), 36–44. DOI: 10.1109/MC.2020.3002181

Nweke, O., & Nweke, G. I. (2024). Legal and ethical conundrums in the ai era: A multidisciplinary analysis. *International Law Research, 13*(1), 1. DOI: 10.5539/ilr.v13n1p1

Nyst, C., & Falchetta, T. (2017). The right to privacy in the digital age. *Journal of Human Rights Practice, 9*(1), 104–118. DOI: 10.1093/jhuman/huw026

O'Brolcháin, F., Jacquemard, T., Monaghan, D., O'Connor, N., Novitzky, P., & Gordijn, B. (2016). The Convergence of Virtual Reality and Social Networks: Threats to Privacy and Autonomy. *Science and Engineering Ethics, 22*(1), 1–29. DOI: 10.1007/s11948-014-9621-1 PMID: 25552240

O'Neil, C. (2016). *Weapons of math destruction: How big data increases inequality and threatens democracy*. Crown Publishing Group.

O'Neil, C. (2016). *Weapons of Math Destruction: How Big Data Increases Inequality and Threatens Democracy*. Crown Publishing Group.

Obermeyer, Z., & Emanuel, E. J. (2016). Predicting the Future—Big Data, Machine Learning, and Clinical Medicine. *The New England Journal of Medicine, 375*(13), 1216–1219. DOI: 10.1056/NEJMp1606181 PMID: 27682033

Oduntan, O., & Ruthven, I. (2020, May 10). People and places: Bridging the information gaps in refugee integration. Wiley, 72(1), 83-96. DOI: 10.1002/asi.24366

OECD (2024b). *Assessing potential future artificial intelligence risks, benefits, and policy imperatives, OECD Artificial Intelligence Papers*, No. 27, OECD Publishing, Paris. Retrieved March 15, 2025 from DOI: 10.1787/dee339a8-en

OECD. (2019). *Child labour risks in global supply chains*. Ending Child Labour, Forced Labour and Human Trafficking in Global Supply Chains. https://mneguidelines.oecd.org/child-labour-risks-in-global-supply-chain.htm

OECD. (2020). Artificial-intelligence-machine-learning-big-data-in-finance. ... https://www.oecd.org/finance/financial-markets/Artificial-intelligence-machine-learning-big-data-in-finance.pdf

OECD. (2021, January 7). *Children in the digital environment*. OECD. https://www.oecd.org/en/publications/children-in-the-digital-environment_9b8f222e-en.html

OECD. (2022). Rights in the digital age: Challenges and ways forward. *OECD Digital Economy Papers*, (347), 12–13. DOI: 10.1787/7ec52ca6-en

OECD. (2024). *MNE Guidelines—Organisation for Economic Co-operation and Development*. OECD Guidelines for Multinational Enterprises on Responsible Business Conduct. https://mneguidelines.oecd.org/mneguidelines/

OECD. (2024a). *OECD AI Policy Observatory: Overview*. Retrieved November 27, 2024, from https://oecd.ai/en/dashboards/overview

Olabi, A. G., Abdelghafar, A. A., Maghrabie, H. M., Sayed, E. T., Rezk, H., Radi, M. A., Obaideen, K., & Abdelkareem, M. A. (2023). Application of artificial intelligence for prediction, optimization, and control of thermal energy storage systems. *Thermal Science and Engineering Progress*, *39*, 101730. DOI: 10.1016/j.tsep.2023.101730

Organization for Economic Cooperation and Development. (2019). *OECD Principles on Artificial Intelligence*. OECD Publishing.

Osasona, F., Amoo, O. O., Atadoga, A., Abrahams, T. O., Farayola, O. A., & Ayinla, B. S. (2024). Reviewing the Ethical Implications of AI in Decision Making Processes. *International Journal of Management & Entrepreneurship Research*, *6*(2), 2. Advance online publication. DOI: 10.51594/ijmer.v6i2.773

Ouchchy, L., Coin, A., & Dubljević, V. (2020). AI in the headlines: The portrayal of the ethical issues of artificial intelligence in the media. *AI & Society*, *35*(4), 927–936. DOI: 10.1007/s00146-020-00965-5

Oxford Commission on AI & Good Governance. (2021, December). *AI in the public service: From principles to practice* (p. 13). https://oxcaigg.oii.ox.ac.uk/wp-content/uploads/sites/11/2021/12/AI-in-the-Public-Service-Final.pdf

Oxford Insights. (2024.). *AI readiness index*. Retrieved March 15, 2025 from https://oxfordinsights.com/ai-readiness/ai-readiness-index/

Pagallo, U., Durante, M., & Monteleone, S. (2017). What Is New with the Internet of Things in Privacy and Data Protection? Four Legal Challenges on Sharing and Control in IoT. In Leenes, R., van Brakel, R., Gutwirth, S., & De Hert, P. (Eds.), *Data Protection and Privacy: (In)visibilities and Infrastructures* (pp. 59–78). Springer International Publishing., DOI: 10.1007/978-3-319-50796-5_3

Palumbo, D. (2008). The monomyth in James Cameron's The Terminator: Sarah as monomythic heroine. *Journal of Popular Culture*, *41*(3), 413–427. DOI: 10.1111/j.1540-5931.2008.00528.x

Papadopoulos, T., Gunasekaran, A., Dubey, R., & Fosso Wamba, S. (2017). Big data and analytics in operations and supply chain management: Managerial aspects and practical challenges. *Production Planning and Control*, *28*(11-12), 873–876. DOI: 10.1080/09537287.2017.1336795

Paraman, P., & Anamalah, S. (2023). Ethical artificial intelligence framework for a good AI society: Principles, opportunities and perils. *AI & Society*, *38*(2), 595–611. DOI: 10.1007/s00146-022-01458-3

Park, J., Kim, S., & Suh, K. (2018). A comparative analysis of the environmental benefits of drone-based delivery services in urban and rural areas. *Sustainability (Basel), 10*(3), 888. DOI: 10.3390/su10030888

Partel, V., Charan Kakarla, S., & Ampatzidis, Y. (2019). Development and evaluation of a low-cost and smart technology for precision weed management utilizing artificial intelligence. *Computers and Electronics in Agriculture, 157*, 339–350. DOI: 10.1016/j.compag.2018.12.048

Pasquale, F. (2015). *The Black Box Society: The Secret Algorithms That Control Money and Information*. Harvard University Press. https://www.jstor.org/stable/j.ctt13x0hch

Pasquale, F. (2015). *The black box society: The secret algorithms that control money and information*. Harvard University Press. DOI: 10.4159/harvard.9780674736061

Patterson, D., Laing, P., & Johnson, T. (2017, September 27). How today's smart tech inspired *Star Trek: Discovery*. *TechRepublic*. https://www.techrepublic.com/article/review-how-todays-smart-tech-inspired-star-trek-discovery/

Patton, M. Q. (2015). Qualitative Research and Evaluation Methods. *Sage (Atlanta, Ga.)*.

Paz, M. (2014). The tower of Babel: Human rights and the paradox of language. *European Journal of International Law, 25*(2), 473–496. DOI: 10.1093/ejil/chu037

Perset, K., Russo, L., & Oder, N. (2024, June 11). *OECD AI Review of Germany: A new phase for Germany's AI strategy*. OECD.AI Policy Observatory. https://oecd.ai/en/wonk/review-germany-2024

Peters, D., Vold, K., Robinson, D., & Calvo, R. A. (2020). Responsible AI—Two Frameworks for Ethical Design Practice. *IEEE Transactions on Technology and Society, 1*(1), 34–47. *IEEE Transactions on Technology and Society*. Advance online publication. DOI: 10.1109/TTS.2020.2974991

Petković, D. (2023, January 30). It is Not "Accuracy vs. Explainability"—We Need Both for Trustworthy AI Systems. *IEEE Transactions on Technology and Society, 4*(1), 46–53. DOI: 10.1109/TTS.2023.3239921

Pizzi, M., Romanoff, M., & Engelhardt, T. (2020). Ai for humanitarian action: Human rights and ethics. *International Review of the Red Cross, 102*(913), 145–180. DOI: 10.1017/S1816383121000011

Policy on the Protection of Personal Data of Persons of Concern to UNHCR. (2023, February 6). https://www.refworld.org/docid/55643c1d4.html

Pomaza-Ponomarenko, A., Leonenko, N., Cherniahivska, V., Lehan, I., & Puzanova, G. (2023). Legal reform and change: Research on legal reform processes and their impact on society. analysis of factors that facilitate or hinder legal change, including political, social, and economic factors. *Journal of Law and Sustainable Development, 11*(10), e1854. DOI: 10.55908/sdgs.v11i10.1854

Porter, M. E., & Kramer, M. R. (2019). Creating Shared Value. In Lenssen, G. G., & Smith, N. C. (Eds.), *Managing Sustainable Business: An Executive Education Case and Textbook* (pp. 323–346). Springer Netherlands., DOI: 10.1007/978-94-024-1144-7_16

Porter, S. R. (2006). Institutional Structures and Student Engagement. *Research in Higher Education, 47*(5), 521–558. DOI: 10.1007/s11162-005-9006-z

Premack, D., & Woodruff, G. (1978). Does the chimpanzee have a theory of mind? *Behavioral and Brain Sciences, 1*(4), 515–526. Retrieved March 15, 2025 from. DOI: 10.1017/S0140525X00076512

Prentzas, N., Kakas, A. C., & Pattichis, C. S. (2023, January 1). Explainable AI applications in the Medical Domain: a systematic review. Cornell University. https://doi.org/DOI: 10.48550/arXiv.2308

Primer: Defending the rights of refugees and migrants in the digital age. (2024, February 5). https://www.amnesty.org/en/documents/pol40/7654/2024/en/

Proposal for a Regulation of the European Parliament and of the Council Laying down Harmonised Rules on Artificial Intelligence (Artificial Intelligence Act) and Amending Certain Union Legislative Acts (2021). https://eur-lex.europa.eu/legal-content/EN/TXT/?uri=celex%3A52021PC0206

Pusca, C. A. (2020). Should we share rights and obligations with artificial intelligence robots? In *Lecture Notes of the Institute for Computer Sciences, Social Informatics and Telecommunications Engineering* (pp. 412–419). Springer. DOI: 10.1007/978-3-030-51005-3_33

Qadhi, S. M., Alduais, A., Chaaban, Y., & Khraisheh, M. (2024). Generative AI, Research Ethics, and Higher Education Research: Insights from a Scientometric Analysis. *Information (Basel), 15*(6), 6. Advance online publication. DOI: 10.3390/info15060325

Qorbani, M. (2020). *Humanity in the age of AI: How to thrive in a post-human world*. Bloomsbury.

Quincey, J. (2022). *The Coca-Cola Company 2022 Business & Sustainability Report.* https://www.coca-colacompany.com/content/dam/company/us/en/reports/coca-cola-business-sustainability-report-2022.pdf

Raghavan, M. (2023). Mitigating bias in algorithmic decision-making: Evaluating claims and practices. In *The Societal Impacts of Algorithmic Decision-Making.* Springer., DOI: 10.1145/3603195.3603212

Rahwan, I., Cebrian, M., Obradovich, N., Bongard, J., Bonnefon, J. F., Breazeal, C., Crandall, J. W., Christakis, N. A., Couzin, I. D., Jackson, M. O., Jennings, N. R., Kamar, E., Kloumann, I. M., Larochelle, H., Lazer, D., McElreath, R., Mislove, A., Parkes, D. C., Pentland, A. S., & Wellman, M. (2019). Machine behavior. *Nature, 568*(7753), 477–486. DOI: 10.1038/s41586-019-1138-y PMID: 31019318

Raja, S., Jebadurai, D. J., Ivan, L., Mykola, R. V., Ruslan, K., & Nadiia, P. R. (2024). Impact of Artificial Intelligence in Students' Learning Life. In Khamis, R., & Buallay, A. (Eds.), *AI in Business: Opportunities and Limitations* (Vol. 2, pp. 3–17). Springer Nature Switzerland., DOI: 10.1007/978-3-031-49544-1_1

Raj, E. F. I., Appadurai, M., & Athiappan, K. (2021). Precision farming in modern agriculture. *Smart Agriculture Automation Using Advanced Technologies, 65,* 294–329. DOI: 10.1007/978-981-16-6124-2_4

Raji, I., Bender, E., Smith, T., & Boyd, K. (2020). "Out of One, Many: The Rise of AI Ethics." Journal of AI Research.

Raji, I. D., Smart, A., White, R. N., Mitchell, M., Gebru, T., Hutchinson, B., Smith-Loud, J., Theron, D., & Barnes, P. (2020). Closing the AI accountability gap: Defining an end-to-end framework for internal algorithmic auditing. *Proceedings of the 2020 Conference on Fairness, Accountability, and Transparency*, 33–44. DOI: 10.1145/3351095.3372873

Ramakrishnan, T., Mohan Gift, M. D., Chitradevi, S., Jegan, R., & Subha Hency Jose, P. Nagaraja, Rajneesh Sharma H.N., Selvakumar P., Sintayehu Mekuria Hailegiorgis, (2022). "Study of Numerous Resins Used in Polymer Matrix Composite Materials", Advances in Materials Science and Engineering, vol. 2022, Article ID 1088926, 8 pages, 2022. https://doi.org/DOI: 10.1155/2022/1088926

Ransbotham, S., Gerbert, P., Reeves, M., Kiron, D., & Spira, M. (2018). Artificial Intelligence in Business Gets Real. *MIT Sloan Management Review.* https://sloanreview.mit.edu/projects/artificial-intelligence-in-business-gets-real/

Rashid, A. B., & Kausik, M. A. K. (2024). AI revolutionizing industries worldwide: A comprehensive overview of its diverse applications. *Hybrid Advances, 7*, 100277. Regulation - 2016/679 - EN - GDPR - EUR-Lex (2016). https://eur-lex.europa.eu/eli/reg/2016/679/ojDOI: 10.1016/j.hybadv.2024.100277

Raveh, A. R., & Tamir, B. (2018). From homo sapiens to Robo sapiens: The evolution of intelligence. *Information (Basel)*, *10*(2), 2. DOI: 10.3390/info10010002

Razmetaeva, Y., Barabash, Y., & Lukianov, D. (2022). The concept of human rights in the digital era: Changes and consequences for judicial practice. *Access to Justice in Eastern Europe*, *5*(3), 41–56. DOI: 10.33327/AJEE-18-5.3-a000327

Reddy, K. S. P., & Roopa, Y. M. L.N KR, Nandan NS (2020) IoT based smart agriculture using machine learning. In: 2020 Second international conference on inventive research in computing applications (ICIRCA), pp 130–134. https://doi.org/DOI: 10.1109/ICIRCA48905.2020.918337

Reed, S., Campbell, A. M., & Thomas, B. W. (2022). The value of autonomous vehicles for last-mile deliveries in urban environments. *Management Science*, *68*(1), 280–299. DOI: 10.1287/mnsc.2020.3917

Refugee protection in the artificial intelligence era. (2022, September 7). https://www.chathamhouse.org/2022/09/refugee-protection-artificial-intelligence-era

Refugee Status Determination. (2023, February 6). https://www.refworld.org/rsdproceduralstandards.html

Rejeb, A., Rejeb, K., Simske, S. J., & Treiblmaier, H. (2023). Drones for supply chain management and logistics: A review and research agenda. *International Journal of Logistics*, *26*(6), 708–731. DOI: 10.1080/13675567.2021.1981273

Resnik, D. B., & Hosseini, M. (2024). The ethics of using artificial intelligence in scientific research: New guidance needed for a new tool. *AI and Ethics*. Advance online publication. DOI: 10.1007/s43681-024-00493-8

Rest, J. R., Narvaez, D., Bebeau, M. J., & Thoma, S. J. (1999). *Postconventional Moral Thinking: A Neo-Kohlbergian Approach*. Psychology Press. DOI: 10.4324/9781410603913

Richards, N. M., & Smart, W. D. (2017). *How Should the Law Think About Robots?* The Law and AI Journal.

Richardson, R., Schultz, J., & Crawford, K. (2019). *Dirty Data, Bad Predictions: How Civil Rights Violations Impact Police Data, Predictive Policing Systems, and Justice* (SSRN Scholarly Paper 3333423). https://papers.ssrn.com/abstract=3333423

Risse, M. (2021). The fourth generation of human rights: Epistemic rights in digital lifeworlds. *Moral Philosophy and Politics*, *8*(3), 351–378. DOI: 10.1515/mopp-2020-0039

Rizzolatti, G., & Craighero, L. (2004). The mirror-neuron system. *Annual Review of Neuroscience*, *27*(1), 169–192. Retrieved March 15, 2025 from. DOI: 10.1146/annurev.neuro.27.070203.144230 PMID: 15217330

Rizzolatti, G., & Luppino, G. (2001). The cortical motor system. *Neuron*, *31*(6), 889–901. Retrieved November 27, 2024, from. DOI: 10.1016/S0896-6273(01)00423-8 PMID: 11580891

Roberts, C. (2014). The contentious history of the international bill of human rights.. DOI: 10.1017/CBO9781139035675

Roberts, H., Cowls, J., Morley, J., Taddeo, M., Wang, V., & Floridi, L. (2020). The Chinese approach to artificial intelligence: An analysis of policy, ethics, and regulation. *AI & Society*, *36*(1), 59–77. DOI: 10.1007/s00146-020-00992-2

Robertson, M., Morris, K., & Walter, G. (2007, September 7). Overview of Psychiatric Ethics V: Utilitarianism and the Ethics of Duty. *Australasian Psychiatry*, *15*(5), 402–410. DOI: 10.1080/10398560701439640 PMID: 17828638

Rolf, S. (2023). *China's regulations on algorithms: Context, impact, and comparisons with the EU*. Friedrich-Ebert-Stiftung. Retrieved March 15, 2025 from https://library.fes.de/pdf-files/bueros/bruessel/19904.pdf

Ronanki, K. (2023). Re-centric recommendations for the development of trustworthy(er) autonomous systems., 1-8. DOI: 10.1145/3597512.3599697

Rud, A. G., & Attwood, A. I. (2016). Dewey Lab School at the University of Chicago. In Peters, M. A. (Ed.), *Encyclopedia of educational philosophy and theory* (pp. 480–484). Springer., DOI: 10.1007/978-981-287-532-7_46-1

Rudin, C. (2019). Stop Explaining Black Box Machine Learning Models for High Stakes Decisions and Use Interpretable Models Instead. *Nature Machine Intelligence*, *1*(5), 206–215. DOI: 10.1038/s42256-019-0048-x PMID: 35603010

Rupp, G. (2016, September 1). Rethinking the Assumptions of Refugee Policy: Beyond Individualism to the Challenge of Inclusive Communities. *Journal on Migration and Human Security*, *4*(3), 76–82. DOI: 10.1177/233150241600400303

Russell, S. J., & Norvig, P. (2016). *Artificial intelligence: a modern approach*. Pearson.

Russell, S., Dewey, D., & Tegmark, M. (2015). Research Priorities for Robust and Beneficial Artificial Intelligence. *AI Magazine*, *36*(4), 105–114. DOI: 10.1609/aimag.v36i4.2577

Russell, S., & Norvig, P. (2020). *Artificial Intelligence: A Modern Approach* (4th ed.). Pearson.

Ryan, M. (2020). In AI we trust: Ethics, artificial intelligence, and reliability. *Science and Engineering Ethics*, *26*(5), 2749–2767. DOI: 10.1007/s11948-020-00228-y PMID: 32524425

Ryan, M., Antoniou, J., Brooks, L., Jiya, T., Macnish, K., & Stahl, B. (2021). Research and Practice of AI Ethics: A Case Study Approach Juxtaposing Academic Discourse with Organisational Reality. *Science and Engineering Ethics*, *27*(2), 16. DOI: 10.1007/s11948-021-00293-x PMID: 33686527

Saberi, S., Kouhizadeh, M., Sarkis, J., & Shen, L. (2019). Blockchain technology and its relationships to sustainable supply chain management. *International Journal of Production Research*, *57*(7), 2117–2135. DOI: 10.1080/00207543.2018.1533261

Saenz, A., Harned, Z., Banerjee, O., Abràmoff, M. D., & Rajpurkar, P. (2023, October 6). Autonomous AI systems in the face of liability, regulations and costs. *NPJ Digital Medicine*, *6*(1), 185. Advance online publication. DOI: 10.1038/s41746-023-00929-1 PMID: 37803209

Saheb, T., Dehghani, M., & Saheb, T. (2022). Artificial intelligence for sustainable energy: A contextual topic modeling and content analysis. *Sustainable Computing : Informatics and Systems*, *35*, 100699. DOI: 10.1016/j.suscom.2022.100699

Sainani, D. (2024, September 24). *Innovative Partnership for Transparent Logistics and Transportation Missions: Mars Launches Pioneering Project With Rigterink, Shippeo and Elain*. AFP.Com. https://www.afp.com/en/news/1313/innovative-partnership-transparent-logistics-and-transportation-missions-mars-launches-pioneering-project-rigterink-shippeo-and-elain-202409243139011

Samek, W., Montavon, G., Vedaldi, A., Hansen, L. K., & Müller, K.-R. (Eds.). (2019). *Explainable AI: Interpreting, Explaining and Visualizing Deep Learning* (1st ed. 2019 Edition). Springer.

Samuel, G., Chubb, J., & Derrick, G. (2021). Boundaries between Research Ethics and Ethical Research Use in Artificial Intelligence Health Research. *Journal of Empirical Research on Human Research Ethics; JERHRE*, *16*(3), 325–337. DOI: 10.1177/15562646211002744 PMID: 33733915

Sánchez-Monedero, J., & Dencik, L. (2020, August 3). The politics of deceptive borders: 'biomarkers of deceit' and the case of iBorderCtrl. Routledge, 25(3), 413-430. DOI: 10.1080/1369118X.2020.1792530

Sandvik, K B. (2021, September 30). Digital Refugee Lawyering: Risk, Legal Knowledge, and Accountability. Oxford University Press, 40(4), 414-432. DOI: 10.1093/rsq/hdab013

Sarker, I. H. (2021). Machine learning: Algorithms, real-world applications and Research Directions. *SN Computer Science*, *2*(3), 160. Advance online publication. DOI: 10.1007/s42979-021-00592-x PMID: 33778771

Save the Children Deutschland e. V. (2023, March 28). *Child Protection in the Digital World: Why is it needed?* Save the Children International. https://www.savethechildren.net/blog/child-protection-digital-world-why-it-needed

Scherer, M. U. (2015). Regulating artificial intelligence systems: Risks, Challenges, competencies, and strategies. SSRN *Electronic Journal*. DOI: 10.2139/ssrn.2609777

Schoemaker, E., Baslan, D., Pon, B., & Dell, N. (2020, July 1). Identity at the margins: data justice and refugee experiences with digital identity systems in Lebanon, Jordan, and Uganda. Taylor & Francis, 27(1), 13-36. DOI: 10.1080/02681102.2020.1785826

Schwartz, R., Vassilev, A., White, C., Grady, N., & Johnson, J. (2022). *Towards a standard for identifying and managing bias in artificial intelligence* (NIST Special Publication 1270). National Institute of Standards and Technology. DOI: 10.6028/NIST.SP.1270

Schweitzer, L., & Valenzuela, A.Jr. (2004). Environmental injustice and transportation: The claims and the evidence. *Journal of Planning Literature*, *18*(4), 383–398. DOI: 10.1177/0885412204262958

Selvakumar, P. (2023). Biomedical applications of polymers in biosensors, cancer vaccines and drug delivery systems. *Biopolimery i Kletka*, *39*(4), 245–256. DOI: 10.7124/bc.000AA0

Selvakumar, P., Mariappan, P., Shunmugam, S., Paul, R., & Jabez, S. (2024). Removal of chemicals from effluent using photobioreactor technology to improve environmental and health impacts. *Zeitschrift für Physikalische Chemie*, *2024*(7), 1295–1302. Advance online publication. DOI: 10.1515/zpch-2023-0450

Senado de la República. (2023). *Por medio de la cual se establecen los lineamientos de política pública para el desarrollo, uso e implementación de inteligencia artificial y se dictan otras disposiciones* (Proyecto de Ley 059/23). Retrieved March 15, 2025 from https://leyes.senado.gov.co/proyectos/index.php/proyectos-ley/cuatrenio-2022-2026/2023-2024/article/59-por-medio-de-la-cual-se-establecen-los-lineamientos-de-politica-publica-para-el-desarrollo-uso-e-implementacion-de-inteligencia-artificial-y-se-dictan-otras-disposiciones

Shahriari, K., & Shahriari, M. (2017). IEEE standard review — Ethically aligned design: A vision for prioritizing human wellbeing with artificial intelligence and autonomous systems. *2017 IEEE Canada International Humanitarian Technology Conference (IHTC)*, 197–201. DOI: 10.1109/IHTC.2017.8058187

Shams, R. A., Zowghi, D., & Bano, M. (2023, January 1). Challenges and Solutions in AI for All. Cornell University. https://doi.org/DOI: 10.48550/arXiv.2307

Sharma, K., Yang, Y., & Wu, Q. (2017). A complex network analysis of ethnic conflicts and human rights violations. *Scientific Reports*, *7*(1), 1–12. DOI: 10.1038/s41598-017-09101-8 PMID: 28811628

Shelley, M. (1818). *Frankenstein; or, The Modern Prometheus*. Lackington, Hughes, Harding, Mavor & Jones.

Shin, D. D. (2022). Algorithmic bias. In *Algorithms, Humans, and Interactions* (pp. 103–125). Springer. DOI: 10.1201/b23083-6

Shneiderman, B. (2020). Bridging the Gap between Ethics and Practice: Guidelines for Reliable, Safe, and Trustworthy Human-centered AI Systems. *ACM Trans. Interact. Intell. Syst.*, *10*(4), 26:1-26:31. DOI: 10.1145/3419764

Sichelman, T. M. (2017). Very tight 'bundles of sticks': Hohfeld's complex jural relations. In S. Balganesh, T. Sichelman, & H. Smith (Eds.), *The legacy of Wesley Hohfeld: Edited major works, select personal papers, and original commentaries* (Forthcoming). Cambridge University Press. *San Diego Legal Studies Paper No. 17*. https://doi.org/DOI: 10.2139/ssrn.2950912

Silverman, D. (2013). Doing Qualitative Research: A Practical Handbook. *Sage (Atlanta, Ga.).*

Singapore Government. (2019). *National Artificial Intelligence Strategy (NAIS) 2019*. Retrieved March 15, 2025 from https://file.go.gov.sg/nais2019.pdf

Singapore Government. (2023). *National Artificial Intelligence Strategy (NAIS) 2023*. Retrieved March 15, 2025 from https://file.go.gov.sg/nais2023.pdf

Singer, T., & Lamm, C. (2009). The social neuroscience of empathy. *Annals of the New York Academy of Sciences*, *1156*(1), 81–96. Retrieved March 15, 2025 from. DOI: 10.1111/j.1749-6632.2009.04418.x PMID: 19338504

Sjoberg, L., Bondesson, M., & Olsson, K. (2020). "Interdisciplinary Education for Ethical AI: A Study on Morality in AI Curriculums." Journal of Ethical AI.

Skutnabb-Kangas, T., & Phillipson, R. (2023). *Handbook of linguistic human rights*. John Wiley & Sons.

Sloane, M., & Moss, E. (2019). "AI Ethics and Education: A Framework for a Responsible Future." *Proceedings of the AAAI Conference on AI Ethics and Society*.

Sloane, M., Moss, E., Awomolo, O., & Forlano, L. (2022). Participation Is not a Design Fix for Machine Learning. *Proceedings of the 2nd ACM Conference on Equity and Access in Algorithms, Mechanisms, and Optimization*, 1–6. DOI: 10.1145/3551624.3555285

Smrkolj, M. (2009, October 14). International Institutions and Individualized Decision-Making: An Example of UNHCR's Refugee Status Determination. Springer Nature, 165-193. DOI: 10.1007/978-3-642-04531-8_7

Sodhi, M. S., & Tang, C. S. (2018). *Research Opportunities in Supply Chain Transparency* (SSRN Scholarly Paper 3189990). https://papers.ssrn.com/abstract=3189990

Srinivasan, R., & González, B. (2022). The role of empathy for artificial intelligence accountability. *Journal of Responsible Technology*, *9*, 100021. Retrieved March 15, 2025 from. DOI: 10.1016/j.jrt.2021.100021

Stahl, B. C. (2021). *Artificial intelligence for a better future: An ecosystem perspective on the ethics of AI and emerging digital technologies*. Springer Nature. DOI: 10.1007/978-3-030-69978-9

Stahl, B. C. (2022). From computer ethics and the ethics of AI towards an ethics of digital ecosystems. *AI and Ethics*, *2*(1), 65–77. DOI: 10.1007/s43681-021-00080-1

Stahl, B. C., Antoniou, J., Ryan, M., Macnish, K., & Jiya, T. (2022). Organisational responses to the ethical issues of artificial intelligence. *AI & Society*, *37*(1), 23–37. DOI: 10.1007/s00146-021-01148-6

Stahl, B. C., Brooks, L., Hatzakis, T., Santiago, N., & Wright, D. (2023). Exploring ethics and human rights in artificial intelligence: A Delphi study. *Technological Forecasting and Social Change*, *191*, 123439. DOI: 10.1016/j.techfore.2023.122502

Stahl, B. C., & Eke, D. (2024). The ethics of ChatGPT – Exploring the ethical issues of an emerging technology. *International Journal of Information Management*, *74*, 102700. DOI: 10.1016/j.ijinfomgt.2023.102700

Stahl, B., & Wright, D. (2018). Ethics and Privacy in AI and Big Data: Implementing Responsible Research and Innovation. *IEEE Security and Privacy*, *16*(3), 26–33. DOI: 10.1109/MSP.2018.2701164

Stanford University. (2024). *AI policy tracker*. Center for Research on Foundation Models. Retrieved March 15, 2025 from https://aipolicy.stanford.edu/

Stanford, S., & Wilson, C. (2022). Bias in schools. In *Systems of Thought and Behaviors within Bias and Inequity*. Springer., DOI: 10.4135/9781071900611

State Council of the People's Republic of China. (2017). *New generation artificial intelligence development plan* (No. 2017-35). State Council of the People's Republic of China. Retrieved March 15, 2025 from http://fi.china-embassy.gov.cn/eng/kxjs/201710/P020210628714286134479.pdf

State of California. (2018). *California Consumer Privacy Act (CCPA) | State of California—Department of Justice—Office of the Attorney General*. https://oag.ca.gov/privacy/ccpa

Stoilova, M., Livingstone, S., & Khazbak, R. (2021). *Investigating Risks and Opportunities for Children in a Digital World: A Rapid Review of the Evidence on Children's Internet Use and Outcomes*. Innocenti Discussion Papers. [Innocenti Discussion Papers], DOI: 10.18356/52f3f4b1-en

Stotland, E., Matthews, K. E., Sherman, S., Hansson, R. O., & Richardson, B. Z. (1978). Empathy, fantasy, and helping. *Sage (Atlanta, Ga.)*.

Sugumaran, K. M. D., Pandiyarajan V., Y. B S, Selvakumar P. and M. S, (2024). "Analysis of Electromyography Signals for Control of Mechanical Prosthesis using Machine Learning Techniques," 2024 2nd International Conference on Intelligent Data Communication Technologies and Internet of Things (IDCIoT), Bengaluru, India, 2024, pp. 1059-1067, DOI: 10.1109/IDCIoT59759.2024.10467707

Su, J., Guo, K., Chen, X., & Chu, S. K. W. (2024). Teaching artificial intelligence in K–12 classrooms: A scoping review. *Interactive Learning Environments*. *32*(9), 5207-5226. DOI: 10.1080/10494820.2023.2212706

SuperAwesome. (2024, October 1). *Youth Marketing Solutions*. SuperAwesome. https://www.superawesome.com/

Suresh, H., & Guttag, J. (2021). A framework for understanding sources of harm throughout the machine learning life cycle. In *Equity and Access in Algorithms, Mechanisms, and Optimization (EAAMO '21)* (Article 17, pp. 1–9). Association for Computing Machinery. Retrieved March 15, 2025 from DOI: 10.1145/3465416.3483305

Taddeo, M., & Floridi, L. (2018). How AI can be a force for good. *Science, 361*(6404), 751–752. DOI: 10.1126/science.aat5991 PMID: 30139858

Talhouk, R., Montague, K., Garbett, A., Araújo-Soares, V., Akik, C., Ghattas, H., Ahmad, B., & Balaam, M. (2019, June 3). A Call For Embedding Dignity In Humanitarian Technologies., undefined(undefined). DOI: 10.1145/3328320.3328373

Tangen, J. (2020). Timescapes in public policy – constructing the "victim of trafficking". *Journal of Borderlands Studies, 37*(3), 475–492. DOI: 10.1080/08865655.2020.1787189

Tarasova, Y., Bobkova, T., & Bobkov, O. (2020). The contradictions of human rights definition in the russian civil law.DOI: 10.15405/epsbs.2020.04.80

Tashakkori, A., & Teddlie, C. (2003). *Handbook of Mixed Methods in Social & Behavioral Research.* Sage.

Technology and innovation report 2021: Catching technological waves innovation with equity. (2021). United Nations Publications. https://unctad.org/webflyer/technology-and-innovation-report-2021

Telkamp, J. B., & Anderson, M. H. (2022). The Implications of Diverse Human Moral Foundations for Assessing the Ethicality of Artificial Intelligence. *Journal of Business Ethics, 178*(4), 961–976. DOI: 10.1007/s10551-022-05057-6

Tepylo, N., Straubinger, A., & Laliberte, J. (2023). Public perception of advanced aviation technologies: A review and roadmap to acceptance. *Progress in Aerospace Sciences, 138*, 100899. DOI: 10.1016/j.paerosci.2023.100899

The Role of Data in AI. (2020). *Report for the Data Governance Working Group of the Global Partnership on AI.* Digital Curation Centre, Trilateral Research, School of Informatics, The University of Edinburgh.

The UK is dropping an immigration algorithm that critics say is racist. (2020, August 5)., undefined(undefined). Retrieved October 29, 2024 from https://www.technologyreview.com/2020/08/05/1006034/the-uk-is-dropping-an-immigration-algorithm-that-critics-say-is-racist/

Tlili, A., Shehata, B., Adarkwah, M. A., Bozkurt, A., Hickey, D. T., Huang, R., & Agyemang, B. (2023). What if the devil is my guardian angel: ChatGPT as a case study of using chatbots in education. *Smart Learning Environments*, *10*(1), 15–24. DOI: 10.1186/s40561-023-00237-x

Tomazzoli, C., Scannapieco, S., & Cristani, M. (2020). Internet of Things and artificial intelligence enable energy efficiency. *Journal of Ambient Intelligence and Humanized Computing*, *14*(5), 4933–4954. DOI: 10.1007/s12652-020-02151-3

Topol, E. J. (2019). High-performance medicine: The convergence of human and artificial intelligence. *Nature Medicine*, *25*(1), 44–56. DOI: 10.1038/s41591-018-0300-7 PMID: 30617339

Torija, A. J., Li, Z., & Self, R. H. (2020). Effects of a hovering unmanned aerial vehicle on urban soundscapes perception. *Transportation Research Part D, Transport and Environment*, *78*, 102195. DOI: 10.1016/j.trd.2019.11.024

Translate, C. L. (2022). Provisions on the management of algorithmic recommendations in internet information service. *China Law Translate*. Retrieved March 15, 2025 from https://www.chinalawtranslate.com/en/algorithms/(open in a new window)

Trocin, C., Mikalef, P., Papamitsiou, Z., & Conboy, K. (2023). Responsible AI for Digital Health: A Synthesis and a Research Agenda. *Information Systems Frontiers*, *25*(6), 2139–2157. DOI: 10.1007/s10796-021-10146-4

Trotta, A., Ziosi, M., & Lomonaco, V. (2023). The future of ethics in AI: Challenges and opportunities. *AI & Society*, *38*(2), 439–441. DOI: 10.1007/s00146-023-01644-x

Tsai, K. (2017, June 13). Transparency-One and Mars Food launch revolutionary supply chain solution. *Transparency-One*. https://www.transparency-one.com/de/mars-food-revolutionary-supply-chain-solution/

Tsakalogianni, I. (2023). *The twin transition: Mapping the use of artificial intelligence in sustainability—the role of law and ethics* (Master's thesis, Πανεπιστήμιο Πειραιώς).

Twiss, S. (2011). Global ethics and human rights: A reflection. *The Journal of Religious Ethics*, *39*(2), 204–222. DOI: 10.1111/j.1467-9795.2011.00473.x

U.S. Congress. (2023). *Algorithmic Accountability Act of 2023*, S. 2892, 118th Congress. Retrieved March 15, 2025 from https://www.congress.gov/bill/118th-congress/senate-bill/2892

U.S. Congress. (2024). *H.R.10092 - Eliminating Bias in Algorithmic Systems Act.* Congress.gov. Retrieved March 15, 2025 from https://www.congress.gov/bill/118th-congress/house-bill/10092/text

Ubaydullayeva, A. (2023). Artificial intelligence and intellectual property: Navigating the complexities of cyber law. *Irshad J. Law and Policy*, *1*(4). Advance online publication. DOI: 10.59022/ijlp.57

UK Government. (2021). *National AI Strategy: Leading the future of artificial intelligence.* Retrieved March 15, 2025 from https://www.gov.uk/government/publications/national-ai-strategy

UN Global Compact Netzwerk Deutschland e. V. (2015). *Guide to Corporate Sustainability | UN Global Compact*. https://unglobalcompact.org/library/1151

UN. U. N. S. D. D. D. and O. B. (2020). *The Sustainable Development Goals Report 2020*. https://unstats.un.org/sdgs/report/2020/

UNESCO. (2020). *Artificial intelligence and gender equality: Key findings of UNESCO's Global Dialogue*. (GEN/2020/AI/2 REV). https://unesdoc.unesco.org/ark:/48223/pf0000374174

UNESCO. (2021). *AI and education: Guidance for policy-makers*. UNESCO., DOI: 10.54675/PCSP7350

UNESCO. (2021). *Recommendation on the Ethics of Artificial Intelligence*. United Nations Educational, Scientific and Cultural Organization.

UNESCO. (2024). *Global AI ethics and governance observatory*. Retrieved March 15, 2025 from https://www.unesco.org/ethics-ai/en/global-hub

UNICEF. (2011). *Children's Rights and Business Principles | UNICEF*. https://www.unicef.org/documents/childrens-rights-and-business-principles

UNICEF. (2014a, April). *Children's Rights in Sustainability Reporting—A guide for integrating children's rights into the GRI reporting framework*. UNICEF. https://www.business-humanrights.org/en/latest-news/pdf-childrens-rights-in-sustainability-reporting-a-guide-for-integrating-childrens-rights-into-the-gri-reporting-framework/

UNICEF. (2014b, August). *Children Are Everyone's Business Workbook 2.0—A guide for integrating children's rights into policies, impact assessments and sustainability reporting*. UNICEF. https://www.business-humanrights.org/en/latest-news/pdf-children-are-everyones-business-workbook-20-a-guide-for-integrating-childrens-rights-into-policies-impact-assessments-and-sustainability-reporting/

UNICEF. (2020, December 10). *Magic Box Inspires Child Innovators | UNICEF USA*. https://www.unicefusa.org/stories/unicef-magic-box-commended-inspiring-kids-be-social-innovators

UNICEF. (2021, November 19). *Policy guidance on AI for children | Innocenti Global Office of Research and Foresight*. https://www.unicef.org/innocenti/reports/policy-guidance-ai-children

UNICEF. (2022, October 27). *Charting the course | UNICEF*. https://www.unicef.org/reports/charting-course

Unilever. (2017). *Unilever Responsible Sourcing Policy*. https://www.unilever.com/files/92ui5egz/production/f51492642f57b314b05466b6194792e02d075d76.pdf

Unilever. (2021, March). *Unilever Sustainable Living Plan 2010 to 2020*. https://www.unilever.com/files/92ui5egz/production/16cb778e4d31b81509dc5937001559f1f5c863ab.pdf

Unilever. (2024a). *Our History | Unilever global company website—2010—2020*. Unilever. https://www.unilever.com/our-company/our-history-and-archives/2010-2020/

Unilever. (2024b). *The importance of sustainable farming & sourcing*. Unilever. https://www.unilever.com/sustainability/nature/sustainable-sourcing/

United Nations Office of Counter-Terrorism (UNOCT) & United Nations Interregional Crime and Justice Research Institute (UNICRI). (2021). *Algorithms and terrorism: The malicious use of artificial intelligence for terrorist purposes: A joint report*.

United Nations. (1966). *International Covenant on Civil and Political Rights*. ICCPR.

United Nations. (1989). *Convention on the Rights of the Child*. CRC.

United Nations. (2003). *Human rights-based approach*. United Nations Sustainable Development Group. https://unsdg.un.org/2030-agenda/universal-values/human-rights-based-approach

Vainio-Pekka, H., Agbese, M. O.-O., Jantunen, M., Vakkuri, V., Mikkonen, T., Rousi, R., & Abrahamsson, P. (2023). The Role of Explainable AI in the Research Field of AI Ethics. *ACM Trans. Interact. Intell. Syst., 13*(4), 26:1-26:39. DOI: 10.1145/3599974

van der Hof, S. (2017). I Agree... Or Do I?: A Rights-Based Analysis of the Law on Children's Consent in the Digital World. *Wisconsin International Law Journal, 34*(2), 409–445.

van Dijk, J. (2020). *The Digital Divide*. John Wiley & Sons.

Van Giffen, B., Sternberg, Z., Meijer, S., & Batchelor, D. (2022). Overcoming the pitfalls and perils of algorithms: A classification of machine learning biases and mitigation methods. *Journal of Business Research*, *144*, 93–106. DOI: 10.1016/j.jbusres.2022.01.076

Van Kolfschooten, H. (2023). The AI cycle of health inequity and digital ageism: Mitigating biases through the EU regulatory framework on medical devices. *Journal of Law and the Biosciences*, *10*(2), lsad031. Advance online publication. Retrieved March 15, 2025 from. DOI: 10.1093/jlb/lsad031 PMID: 38075950

Vanlehn, K. (2011). The Relative Effectiveness of Human Tutoring, Intelligent Tutoring Systems, and Other Tutoring Systems. *Educational Psychologist*, *46*(4), 197–221. DOI: 10.1080/00461520.2011.611369

Veale, M., & Binns, R. (2017). Fairer machine learning in the real world: Mitigating discrimination without collecting sensitive data. *Big Data & Society*, *4*(2), 2053951717743530. DOI: 10.1177/2053951717743530

Vemuri, V. K. (2014). *the second machine age: Work, progress, and prosperity in a time of brilliant technologies*, by Erik Brynjolfsson and Andrew McAfee. *Journal of Information Technology Case and Application Research*, *16*(2), 112–115. DOI: 10.1080/15228053.2014.943094

Verma, S. (2019). Weapons of math destruction: How big data increases inequality and threatens democracy. *Vikalpa*, *44*(2), 97–98. DOI: 10.1177/0256090919853933

Vicente, L., & Matute, H. (2023). Humans inherit artificial intelligence biases. *Scientific Reports*, *13*(1), 15737. Retrieved March 15, 2025 from. DOI: 10.1038/s41598-023-42384-8 PMID: 37789032

Vijayakumar, G., Rajkumar, M., Rajiv Chandar, N., Selvakumar, P., & Duraisamy, R. (2024). Muniyandi Rajkumar, Rajiv Chandar N, Selvakumar P and Ramesh Duraisamy, (2024). Environmentally friendly TDS removal from waste water by electrochemical ion exchange batch-type recirculation (EIR) technique. *Environmental Science. Water Research & Technology*, *10*(4), 826–835. DOI: 10.1039/D3EW00793F

Villata, S., Araszkiewicz, M., Ashley, K., Bench-Capon, T., Branting, L. K., Conrad, J. G., & Wyner, A. (2022). Thirty years of artificial intelligence and law: The third decade. *Artificial Intelligence and Law*, *30*(4), 561–591. DOI: 10.1007/s10506-022-09327-6

Vinuesa, R., Azizpour, H., Leite, I., Balaam, M., Dignum, V., Domisch, S., Felländer, A., Langhans, S. D., Tegmark, M., & Fuso Nerini, F. (2020). The role of artificial intelligence in achieving the Sustainable Development Goals. *Nature Communications*, *11*(1), 1–10. DOI: 10.1038/s41467-019-14108-y PMID: 31932590

Voigt, P., & Von dem Bussche, A. (2017). The eu general data protection regulation (gdpr). *A Practical Guide, 1st Ed., Cham: Springer International Publishing*, *10*(3152676), 10-5555.

Voigt, P., & Von Dem Bussche, A. (2017). *The EU General Data Protection Regulation (GDPR)*. Springer International Publishing., DOI: 10.1007/978-3-319-57959-7

von Struensee, S. (2021). *Artificial intelligence and corporate social responsibility*. Global Research Initiative.

Vorster, J. (2009, July 26). The contribution of deontological Christian ethics to the contemporary human rights discourse. *In die Skriflig*, *43*(3). Advance online publication. DOI: 10.4102/ids.v43i3.234

Wang, G., Gunasekaran, A., Ngai, E. W. T., & Papadopoulos, T. (2016). Big data analytics in logistics and supply chain management: Certain investigations for research and applications. *International Journal of Production Economics*, *176*, 98–110. DOI: 10.1016/j.ijpe.2016.03.014

Wang, J., Mao, W., & Wenjie, W. (2023). The ethics of artificial intelligence: Sociopolitical and legal dimensions. *ISSLP*, *2*(2), 27–32. DOI: 10.61838/kman.isslp.2.2.6

Wang, N., & Lester, J. (2023). K-12 education in the age of AI: A call to action for K-12 AI literacy. *International Journal of Artificial Intelligence in Education*, *33*(2), 228–232. DOI: 10.1007/s40593-023-00358-x

Wang, T., Lund, B. D., Marengo, A., Pagano, A., Mannuru, N. R., Teel, Z. A., & Pange, J. (2023). Exploring the Potential Impact of Artificial Intelligence (AI) on International Students in Higher Education: Generative AI, Chatbots, Analytics, and International Student Success. *Applied Sciences (Basel, Switzerland)*, *13*(11), 11. Advance online publication. DOI: 10.3390/app13116716

Wang, Z., & Srinivasan, R. S. (2017). A review of artificial intelligence based building energy use prediction: Contrasting the capabilities of single and ensemble prediction models. *Renewable & Sustainable Energy Reviews*, *75*, 796–808. DOI: 10.1016/j.rser.2016.10.079

Ward, H. (2024). *Technologies of the self: Diachronic privacy and self-presentational autonomy* (Doctoral dissertation, UCL).

Weber, L., & Gerard, A. (2024, April 10). Robodeport or surveillance fantasy?: how automated is automatic visa cancellation in Australia?. Frontiers Media, 9(undefined). DOI: 10.3389/fsoc.2024.1336160

Weber, R H. (2019, November 15). Socio-ethical values and legal rules on automated platforms: The quest for a symbiotic relationship. Elsevier BV, 36, 105380-105380. DOI: 10.1016/j.clsr.2019.105380

Weidinger, L., Uesato, J., Rauh, M., Griffin, C., Huang, P.-S., Mellor, J., Glaese, A., Cheng, M., Balle, B., Kasirzadeh, A., Biles, C., Brown, S., Kenton, Z., Hawkins, W., Stepleton, T., Birhane, A., Hendricks, L. A., Rimell, L., Isaac, W., & Gabriel, I. (2022). Taxonomy of Risks posed by Language Models. *2022 ACM Conference on Fairness, Accountability, and Transparency*, 214–229. DOI: 10.1145/3531146.3533088

Wei, M. C. F., Maldaner, L. F., Ottoni, P. M. N., & Molin, J. P. (2020). Carrot yield mapping: A precision agriculture approach based on machine learning. *AI*, *1*(2), 229–241. DOI: 10.3390/ai1020015

Whittaker, M., Crawford, K., & Dobbe, R.. (2018). *AI Now 2018 Report*. AI Now Institute.

Whittaker, M., Crawford, K., Dobbe, R., Fried, G., Kaziunas, E., Mathur, V., & Schwartz, O. (2018). *AI now report 2018*. AI Now Institute at New York University.

Widder, D G., Zhen, D., Dabbish, L., & Herbsleb, J D. (2023, June 12). It's about power: What ethical concerns do software engineers have, and what do they (feel they can) do about them?. DOI: 10.1145/3593013.3594012

Wikipedia. (2024). Article 29 Data Protection Working Party. In *Wikipedia*. https://en.wikipedia.org/w/index.php?title=Article_29_Data_Protection_Working_Party

Williamson, B., Eynon, R., Knox, J., & Davies, H. (2023). *Critical perspectives on AI in education: Political economy, discrimination, commercialization, governance and ethics* (pp. 553–570). DOI: 10.4337/9781800375413.00037

Williamson, B., & Eynon, R. (2020). Historical threads, missing links, and future directions in AI in education. *Learning, Media and Technology*, *45*(3), 1–13. DOI: 10.1080/17439884.2020.1798995

Winfield, A., Michael, K., & Pitt, J. (2019). AI and Ethics: A Framework for Responsible AI. *AI & Society*.

Wolff, J. (2006, May 1). Making the World Safe for Utilitarianism. Cambridge University Press, 58(undefined), 1-22. DOI: 10.1017/S1358246106058012

Wolff, S. (2021, May 13). The Security Sector Governance–Migration Nexus. DOI: 10.2307/j.ctv1v3gqt4

Wolfram, M., & Kienesberger, M. (2023). Gender in sustainability transition studies: Concepts, blind spots, and future orientations. *Environmental Innovation and Societal Transitions, 46*, 100686. DOI: 10.1016/j.eist.2022.100686

Wong, A. (2021). Ethics and regulation of artificial intelligence. In *Artificial Intelligence for Knowledge Management: 8th IFIP WG 12.6 International Workshop, AI4KM 2021* (pp. 1-18). Springer International Publishing. DOI: 10.1007/978-3-030-80847-1_1

Wong, C. (2023). AI 'fairness' research held back by lack of diversity. *Nature*. Advance online publication. DOI: 10.1038/d41586-023-00935-z PMID: 36997714

Wong, L.-W., Cheng, T. C. E., & Tong, Y. (2023). The role of institutional and self in the formation of trust in artificial intelligence technologies. *Internet Research*. Advance online publication. DOI: 10.1108/INTR-09-2022-0455

World Bank. (2018). *World Development Report 2018: Learning to Realize Education's Promise* [Text/HTML]. World Bank. https://www.worldbank.org/en/publication/wdr2018

World Economic Forum. (2022). *A blueprint for equity and inclusion in artificial intelligence* (White Paper). https://www.weforum.org/reports/a-blueprint-for-equity-and-inclusion-in-artificial-intelligence

World Economic Forum. (2022). *Earning digital trust: Decision-making for trustworthy technologies.* In collaboration with Accenture, KPMG, and PwC. https://www.weforum.org/reports/earning-digital-trust-decision-making-for-trustworthy-technologies

Xiang, X., Li, Q., Khan, S., & Khalaf, O. I. (2021). Urban water resource management for sustainable environment planning using artificial intelligence techniques. *Environmental Impact Assessment Review, 86*, 106515. DOI: 10.1016/j.eiar.2020.106515

Xudaybergenov, A. (2023). Toward legal recognition of artificial intelligence proposals for limited subject of law status. *Irshad J. Law and Policy, 1*(4). Advance online publication. DOI: 10.59022/ijlp.55

Yamin, M. M., Ahmed, F., Shaikh, M. A., & Majeed, A. (2021). Weaponized AI for cyber attacks. *Journal of Information Security and Applications, 57*, 102722. DOI: 10.1016/j.jisa.2020.102722

Yan, B., Hao, F., & Meng, X. (2021). When artificial intelligence meets building energy efficiency, a review focusing on zero energy building. *Artificial Intelligence Review*, *54*(3), 2193–2220. DOI: 10.1007/s10462-020-09902-w

Yazici, E. (2018). Nationalism and human rights. *Political Research Quarterly*, *72*(1), 147–161. DOI: 10.1177/1065912918781187

Yin, R. K. (2014). Case Study Research: Design and Methods. *Sage (Atlanta, Ga.)*.

Young, M., Himmelreich, J., Honcharov, D., & Soundarajan, S. (2021, June 9). The Right Tool for The Job? Assessing the Use of Artificial Intelligence for Identifying Administrative Errors. DOI: 10.1145/3463677.3463714

Yousafzai, M., & Lamb, C. (2013). *I Am Malala: The girl who stood up for education and was shot by the Taliban*. Little, Brown and Company. • Orwell, G. (1949). *1984*. Secker & Warburg.

Yue, M., Jong, M. S.-Y., & Dai, Y. (2022). Pedagogical design of K-12 Artificial Intelligence education: A systematic review. *Sustainability (Basel)*, *14*(23), 15620. DOI: 10.3390/su142315620

Zajko, M. (2022). Artificial intelligence, algorithms, and social inequality: Sociological contributions to contemporary debates. *Sociology Compass*, *16*(2), e12966. DOI: 10.1111/soc4.12962

Zawacki-Richter, O., Marín, V. I., Bond, M., & Gouverneur, F. (2019). Systematic review of research on artificial intelligence applications in higher education – where are the educators? *International Journal of Educational Technology in Higher Education*, *16*(1), 39. DOI: 10.1186/s41239-019-0171-0

Zemel, R., Wu, Y., Swersky, K., Pitassi, T., & Dwork, C. (2013). Learning Fair Representations. *Proceedings of the 30th International Conference on Machine Learning*, 325–333. https://proceedings.mlr.press/v28/zemel13.html

Zenz, A., & Powles, J. (2024). Resisting technological inevitability: Google Wing's delivery drones and the fight for our skies. *Philosophical Transactions. Series A, Mathematical, Physical, and Engineering Sciences*, *382*(2285), 20240107. DOI: 10.1098/rsta.2024.0107 PMID: 39533913

Zerilli, J., Knott, A., Maclaurin, J., & Gavaghan, C. (2022). How transparency modulates trust in artificial intelligence. *Patterns (New York, N.Y.)*, *3*(7), 100455. DOI: 10.1016/j.patter.2022.100455 PMID: 35465233

Zeyer, A. (2018). Gender, complexity, and science for all: Systemizing and its impact on motivation to learn science for different science subjects. *Journal of Research in Science Teaching*, *55*(2), 147–171. Retrieved March 15, 2025 from. DOI: 10.1002/tea.21413

Zhai, X. (2022). Artificial AI bias. In *American Educational Research Association (AERA)2022 Annual Meeting*. https://doi.org/DOI: 10.3102/9781234567890

Zhang, D., Maslej, N., Brynjolfsson, E., Etchemendy, J., Lyons, T., Manyika, J., Ngo, H., Niebles, J. C., Sellitto, M., Sakhaee, E., Shoham, Y., Clark, J., & Perrault, R. (2022). *The AI index 2022 annual report*. AI Index Steering Committee, Stanford Institute for Human-Centered AI, Stanford University. https://hai.stanford.edu/ai-index-2022

Zhang, D., Mishra, S., Brynjolfsson, E., Etchemendy, J., Ganguli, D., Grosz, B., Lyons, T., Manyika, J., Niebles, J. C., Sellitto, M., Shoham, Y., Clark, J., & Perrault, R. (2021). *The AI index 2021 annual report*. AI Index Steering Committee, Stanford Human-Centered AI Institute, Stanford University. https://hai.stanford.edu/ai-index-2021

Zhang, M. (2015, July 1). Google Photos tags two African-Americans as gorillas through facial recognition software. *Forbes*. https://www.forbes.com/sites/mzhang/2015/07/01/google-photos-tags-two-african-americans-as-gorillas-through-facial-recognition-software/?sh=40c3811c713d

Zhang, R., Chen, Z.-Y., Xu, L.-J., & Ou, C.-Q. (2019a). Meteorological drought forecasting based on a statistical model with machine learning techniques in Shaanxi province, China. *The Science of the Total Environment*, *665*, 338–346. DOI: 10.1016/j.scitotenv.2019.01.431 PMID: 30772564

Zheng, Y. (2023). Redefining rights in the age of ai: Philosophical perspectives in shaping the future of international human rights law. *Lecture Notes in Education Psychology and Public Media*, *16*(1), 124–132. DOI: 10.54254/2753-7048/16/20231124

Zhu, X. (2019). Segmenting the public's risk beliefs about drone delivery: A belief system approach. *Telematics and Informatics*, *40*, 27–40. DOI: 10.1016/j.tele.2019.05.007

Zuboff, S. (2019). *The Age of Surveillance Capitalism: The Fight for a Human Future at the New Frontier of Power, edn*. Public Affairs.

Zuboff, S. (2019). *The Age of Surveillance Capitalism: The Fight for a Human Future at the New Frontier of Power*. PublicAffairs.

About the Contributors

Swati Chakraborty is a Visiting faculty (Assistant Professor), at the Institute of Legal Studies and Research, GLA University. Dr. Chakraborty is associated as an Adjunct Faculty with Concordia University, Canada. Dr. Chakraborty is a Writing Fellow, at COMPOSE, Universitas Islam International Indonesia. She is an International Fellow of KAICIID in Lisbon, Portugal. She has done her Ph.D. in Human Rights from the University of Calcutta with the thesis on "RIGHT TO EDUCATION: A Study on tribal women of Rural West Bengal". She was attached to National Human Rights Commission, India as a Research Consultant. Dr. Chakraborty had good teaching expertise as Lecturer at The English College (School of Management), Calicut, Kerala her special lectures are on Human Values, Peace dialogue and ethics to Management studies, and Dignity and Rights. She is also a member of the executive council at the Colibri Learning Foundation, Surrey, BC. Also, she is the editor of books like "Dynamics of Dialogue, Cultural Development, and Peace in the Metaverse", "Handbook of Dialogue and Development: India-China-EU", Handbook of Human Rights and Tribal Studies", "Handbook of Social Media for Digital and Social Inclusion", "TRIBAL DEVELOPMENT", "Gender Identity and Roles in India: Issues and Challenges", "Multidisciplinary Handbook of Social Exclusion and Human Rights" and "Handbook of International Relations: Issues of Human Rights and Foreign Policy Vol. I and Vol II". Dr. Chakraborty has a wide range of international connections with her network through different seminars, conferences, and workshops.

***

lSoundariya is an esteemed Assistant Professor at B.S Abdur Rahman Crescent institute of science and technology . With a robust background in these dynamic areas, she is dedicated to guiding and mentoring students, helping them navigate the complexities of modern data science. Her scholarly interest in Cloud computing and cyber security.Her ongoing commitment to excellence continues to make her a valuable asset to the institution and the broader academic landscape.

Daniel Opotamutale Ashipala (MN Sc) is a senior lecturer and Head of Department for the School of nursing and Public Health (SoNPH), General Nursing Science department in the Faculty of Health Sciences and Veterinary Medicine (FHSVM) at the University of Namibia, Namibia. His research interests include Primary health Care, Health care services and Nursing education, and Mental health. He has published a quite number of peer reviewed articles on the delivery of teaching and learning in higher education studies. He is a doctoral candidate at Stellenbosch University in Cape Town, South Africa. His doctoral work focuses on developing a framework to facilitate the implementation of a task shifting approach for nurses in the Namibia PHC services.

Adam I. Attwood, PhD, is a faculty member in the Eriksson College of Education at Austin Peay State University in Clarksville, Tennessee. Attwood was previously a social studies/humanities teacher and social studies department chair (grades 6-12) and has experience as an educator in Washington, Missouri, and Tennessee.

Niyati Kumari Behera, presently working as Asst. Professor in B.S. Abdur Rahman Crescent Institute of Science and Technology. She did her B.E from University College of Engineering, Burla in Electrical Engineering. She completed her Masters in CSE from N.I.T., Rourkela. She received the PhD in the field of Text Mining and Machine Learning from Anna University, Chennai. Her research interest includes Text Mining, NLP and Machine Learning.

Christopher Bohlens is a junior researcher at Leuphana University of Lüneburg and serves as the head of the Science Working Group at Transparency International Germany. With a strong background in student advocacy, Christopher has represented student interests on a national scale during his years in student self-administration. He is a dedicated expert in promoting student participation and ensuring their voices are heard in university quality assurance processes. He is also active in the field of quality assurance and quality management of Higher Education Institutions (HEIs), where he is involved in many functions and areas. Additionally, he is an active advocate for Freedom of Information, open data, transparency legislation, and the application of Administrative Law. Christopher's academic journey includes studies in business administration, economics, political science, and law at Hamburg University of Applied Sciences, Leuphana University, and FernUni Hagen.

Prithwijit Chakraborty is a Guest Faculty of English at University of Engineering and Management, Kolkata. His research and publications address twentieth century American literature, professional writing, and pedagogy. He was formerly engaged with Brainware Institution, Kingston Educational Institution and Heritage Institute, Kolkata.

Pankaj Singh Chandel is an Assistant Professor in the Department of Tourism Management at Dev Sanskriti Vishwavidyalaya, Haridwar, where he has been teaching since 2014. He holds a PhD in Tourism and a PGDM in Tourism and Travel Services Management from the Indian Institute of Tourism and Travel Management. His expertise lies in homestay tourism, and he is recognized as a skilled orator on travel knowledge. Dr. Chandel has published research articles in Scopus-indexed journals and has contributed to publications with IGI Global, demonstrating his commitment to academic scholarship in the field of tourism. His research focuses on sustainable tourism practices and the development of homestay tourism in India. As an active member of several professional organizations, including the Travel Agents Association of Uttarakhand and ATOAI, Dr. Chandel actively participates in initiatives promoting sustainability in the tourism industry. His involvement in these organizations allows him to bridge the gap between academic research and industry practices. Dr. Chandel's ORCID (https://orcid.org/0000-0002-4135-8194) showcases his research outputs and professional activities.

Anindita Das is an esteemed Associate Professor and the Head of the Department of Physical Education Pedagogy at Lakshmibai National Institute of Physical Education (LNIPE), Gwalior. With a prolific career, Dr. Das has authored numerous research papers, patents, and books, contributing significantly to the field of physical education. Her work emphasizes innovative teaching methodologies and educational strategies, aimed at fostering excellence in physical education pedagogy. Renowned for her dedication to advancing knowledge and practice in sports education, Dr. Das continues to inspire students and professionals through her leadership, research, and unwavering commitment to academic excellence.

Shivani Dhand is an Associate Professor (14+ years of experience),a Certified SHRM faculty in the department of Mittal School of Business of Lovely Professional University.She is NLP(Neuro Linguistics Programming) Practioner She is also a member of the Indian Society for Training and Development (ISTD) and the Indian Society of Labour Economics. She is also a core committee member and Mentor of an NGO(Shalini Fellowship) Shivani Dhand has worked in an IT company as an HR. She has published 15 research papers in various journals which include Scopus-indexed, UGC care and peer-reviewed journals. She is a certified POSH trainer.

She has worked with CPSC (Columbo Plan Staff College) Manila Philippines, and trained the faculty and Principals of various Institutions in Bhutan (2015). In 2016, She worked on one of the projects," Austria as an Innovation leader in EU in Austria and was awarded as best presentation and Excellent paper. Her research is situated in the field of Startups, HR Practices Human Resource Management, Entrepreneurship, and social entrepreneurship with a special focus on academic entrepreneurship. Shivani teaches several courses on Human Resource Management and Organisational Behaviour; Leadership. She teaches and developed a full E-content for Online education in the course "HR Analytics and was a guest lecturer in the "Entrepreneurship and Management" training. Shivani has delivered Lectures on Social Entrepreneurship. Completed two International Projects. Presented and published 13 papers at national and International Conferences. She has also served as a Training placement coordinator in Lovely Professional University. She has also invited as a Resource Person in the various faculty development programs sponsored by the Ministry of Human Resource Development, India. She has participated in 40+ conferences and seminars on data. Her experience includes lecturing in Bhutan, Austria, India, US. Research Publication: International: 13 National: 06 Authored Books:1 Patents: 3 Copyright: 2 Research Propsal:1 submitted Qualification: Ph.D. in Human Resource Management MBA from Punjab Technical University, Jalandhar Pist Graduation Diploma In Human Rights Qualified UGC-NET in Management Certified POSH Trainer Life Time member of Indian Society of Training and Development

Anwesha Ghosh is currently an Assistant Professor in the Amity Law School, Amity University, Kolkata, India. She was formerly working at Rajiv Gandhi National University of Law as an Assistant Professor. Before joining RGNUL Patiala she was a UGC-sponsored Senior Research Fellow (SRF) at the Central University of Punjab, India. Her topic of research is "Legislative Governance Towards Gender Diversity in Indian Corporate Boards: Review of Institutional Factors and Best Practices". She has completed the pre-submission of her Ph.D. thesis to the Department of Law, Central University of Punjab, and is now heading towards the final submission of her Ph.D. thesis. Her current Ph.D. research majorly focuses on the pivotal role of women directors in Indian companies, advocating for gender diversity and inclusive corporate governance. Ms. Ghosh achieved a University Gold Medal during her LL.B. degree at Haldia Law College affiliated with Vidyasagar University, West Bengal, India. Her academic excellence continued with a University Gold Medal for her outstanding performance in LL.M. and an Endowment Medal for securing the highest marks in Corporate and Commercial Laws from the National University of Study and Research in Law, Ranchi, India. Ms. Ghosh had previously worked

Sandeep Gupta is presently working as an Associate Professor at Graphic Era (Deemed to be University), Dehradun (India). He has a total of 15 years of experience. His areas of interest in research are the application of artificial intelligence to power system control design, FACTS devices, renewable energy, power electronics, and stability of power systems with machine learning. He is a member of numerous professional organizations, including IE, IRED, IDES, IAENG, IAER, and the Society of Professional Engineers. He has been the author or co-author of more than 90 papers published in international journals and presented at different national and international conferences. He is currently guiding the research scholars under Machine Learning, Control & Power Systems, Wireless Technology, and the Internet of Things.

Sakshi is a Research Scholar in the department of Mittal School of Business of Lovely Professional University. She earned her master's degree in management from Maharshi Dayanand University, Rohtak, Haryana. She is pursuing her PhD in the field of human resource management.

Debajit Karmakar is a dedicated Ph.D. scholar in Physical Education Pedagogy at Lakshmibai National Institute of Physical Education (LNIPE), Gwalior. An accomplished researcher, Mr. Karmakar has authored numerous research papers, patents, and books, contributing significantly to sports education and physical fitness. His work focuses on advancing training methodologies and innovative approaches in physical education, aligning both theory and practice. He brings his passion for sports into his academic pursuits, making impactful contributions to sports science and education. Mr. Karmakar's achievements mark him as a promising figure in physical education and sports science.

Awesh Khati is a dedicated Ph.D. scholar specializing in Physical Education Pedagogy at the Lakshmibai National Institute of Physical Education (LNIPE), Gwalior. With a strong academic foundation and passion for advancing physical education, he has contributed several research papers in the field. In addition to his scholarly pursuits, Mr. Khati is an accomplished football player, blending theoretical knowledge with practical expertise. His work reflects a commitment to enhancing sports education and promoting holistic physical development, making him a promising figure in the domain of physical education and sports science.

T.C. Manjunath is a distinguished Professor in the Department of Computer Science and Engineering (CSE) at Rajarajeswari College of Engineering, Bangalore. He specializes in areas such as Internet of Things (IoT) Cybersecurity and Blockchain Technology. Dr. Manjunath currently serves as the Dean of Research and Development (R&D) at the institution. His academic expertise and research focus on emerging technologies that aim to enhance digital security and enable innovative applications. Located on Mysore Road at Kumbalgodu Post, Bangalore, Dr. Manjunath is a leading figure in advancing cutting-edge technological solutions.

Abhinash Mohanta pursued a three-year LLB program at Ganjam Law College, Berhampur, under Berhampur University. He further advanced his academic journey by enrolling in the Master of Laws (LLM) program in Business Law at the P.G. Department of Law, Berhampur University, graduating in 2017. Currently, he is pursuing a Ph.D. in Law at KIIT Deemed to be University, with a research focus on critical areas such as refugee and human rights. In addition to his academic endeavors, Abhinash has authored several Scopus-indexed papers covering diverse topics, including AI, environment, migration, crime, intellectual property, and refugee studies.

Fransisco Chindongo Ntjamba is a registered nurse holding a Bachelor of Nursing Science Clinical, (Honours) from the University of Namibia, School of Nursing and Public Health, General Nursing Department in the Faculty of Health Sciences and Veterinary Medicine (FHSVM). He is also a NIMART-trained nurse and a Provider Initiated Testing and Counseling (PITC NURSE), specializing in sexual reproductive health and focusing on HIV testing, prevention, treatment initiation, and ongoing care. In addition to his qualifications, Fransisco is an academic researcher who has published research studies in nursing education, mental health, and interdisciplinary innovations, including Artificial Intelligence on human healthcare. His career is defined by a commitment to advancing healthcare through both direct patient care and research, with key interests in Primary Health Care, healthcare services, nursing education, and mental health. He is dedicated to improving healthcare through evidence-based practices and community outreach initiatives, particularly in HIV prevention and treatment. He can be reached at 05sisco@gmail.com.

P. Selvakumar is currently an Associate Professor in the Department of Science and Humanities at Nehru Institute of Technology, Coimbatore, Tamil Nadu, India. Prior to joining Nehru Institute of Technology, he held positions as an Assistant Professor in the Department of Science and Humanities various engineering colleges in Coimbatore, Tamil Nadu, India, spanning from 2011 to 2023. Dr.P. Selvakumar's academic and research interests lie primarily in the field of phytochemistry. He has made significant contributions to his field, with an impressive record that includes 30 journal publications, 12 patents, and 20 conference papers. Additionally, he has organized 5 seminars and conferences, further demonstrating his commitment to academic and scientific exchange. Throughout his career, Dr. P.Selvakumar has been recognized for his dedication and excellence in teaching, having been honored with several awards, including the Best Teacher Awards. He holds a prominent position in the academic community, contributing actively to research.

Sachin Singh Parihar, a UGC JRF recipient (2020) and Ph.D. Scholar at the Central University of Punjab's Department of Law is an accomplished legal scholar with a focus on renewable energy frameworks and sustainability dimensions. He eamed his LL.M. in Criminal Law from Jamia Millia Islamia University, New Delhi, India. Currently pursuing his Ph.D. at the Central University of Punjab under the guidance of Prof. (Dr.) Deepak Kumar Chauhan, Mr. Parihar's research is centered on the "Implementation of Renewable Energy Framework with Reference to Sustainability Dimensions: A Comparative Analysis of Laws in India, China, and Germany." His commitment to environmental and legal studies is evident in this ambitious and globally relevant project. Mr. Parihar has actively contributed to the academic discourse, having participated in numerous national and intemational conferences. Furthermore, his scholarly works have found a place in various reputed peer-reviewed journals, reflecting the depth and impact of his research contributions. With a strong foundation in criminal law and a keen interest in the intersection of law and sustainable energy, Mr. Parihar continues to be a noteworthy presence in the field of legal research and environmental law.

Hemlata Parmar, a luminary in the field of computer Science and Engineering. Her Journey is defined by a constellation of achievements from prolific publications and patents to enlightening guest lectures. Yet, her brilliance is matched only by her unwavering commitment to human ethics. With a heart as vast as her intellect, she champions women's empowerment, bridging the gap with empathy and insight. In every endeavor, she embodies the perfect fusion of academic process and compassionate advocacy, inspiring generations to come.

Ria Roy Choudhury is currently an Assistant Professor of English at the Department of Basic Science and Humanities, University of Engineering and Management, Kolkata. She had formerly taught at the Department of English, Amity University, Kolkata and was a Guest faculty at the Department of English, Deshbandhu College for Girls and Naba Ballygunge Mahavidyalaya, Kolkata. She is presently pursuing her PhD from University of Engineering and Management, Kolkata. Her M.Phil from Jadavpur University was on the politics involved in translating Indian writing in English. Her fictional and non-fictional writings have featured in several reputed journals and books. She has been felicitated with the Sarvepalli Radhakrishnan Teacher Excellence Award by Shikshak Kalyan

Foundation. Her areas of research include Translation Studies and Environmental Humanities.

Muhammad Usman Tariq has more than 16+ year's experience in industry and academia. He has authored more than 200+ research articles, 110+ case studies, 105+ book chapters and several books other than 4 patents. He is founder and CEO of The Case HQ, a unique repository for courses, narrative and video case studies. He has been working as a consultant and trainer for industries representing six sigma, quality, health and safety, environmental systems, project management, and information security standards. His work has encompassed sectors in aviation, manufacturing, food, hospitality, education, finance, research, software and transportation. He has diverse and significant experience working with accreditation agencies of ABET, ACBSP, AACSB, WASC, CAA, EFQM and NCEAC. Additionally, Dr. Tariq has operational experience in incubators, research labs, government research projects, private sector startups, program creation and management at various industrial and academic levels. He is Certified Higher Education Teacher from Harvard University, USA, Certified Online Educator from HMBSU, Certified Six Sigma Master Black Belt and has been awarded PFHEA, SFSEDA, SMIEEE, and CMBE.

Rita Tegon is a PhD candidate in Digital Humanities at the University of Salamanca. Classical receptions in videogames, digital cultures, neurosciences, and the cognitive and social impact of AI are among her primary research interests.

Narem Usha Amulya, Assistant Professor, School of Law, GITAM (Deemed-to-be University) and PhD Scholar at NALSAR University has been an avid writer and has published two academic books & two poetry anthologies available on Amazon and Flipkart. She has participated in various National and International Conferences and her papers are published in reputed journals. Having a keen interest in multi-disciplinary and interdisciplinary knowledge in her academic pursuits, she graduated with BA.LLB, (Hons.), LLM, MA in public administration PGDHRM & MBA in Finance. Apart from her academic pursuits, she is a prolific poet and has published & won several accolades nationally and internationally. Her areas of interest include Environmental Law, Constitutional law, AI, IPR and space Law. She can be contacted at ushaamulya.work@gmail.com

Index

A

B

C

D

E

F

G

H

I

J

K

M

N

P

R

S

T

www.ingramcontent.com/pod-product-compliance
Lightning Source LLC
LaVergne TN
LVHW080344011026
849161LV00009B/2183
9798369390351